Recent Advances in Chemical Information II

Recent Advances in Chemical Information II

Edited by

H. Collier
Infonortics Ltd., Calne, Wiltshire

The Proceedings of the Montreux 1992 International Chemical Information Conference, held at Annecy, France, 19–21 October 1992

Special Publication No. 120

ISBN 0-85186-235-7

A catalogue record for this book is available from the British Library

Published by The Royal Society of Chemistry,
Thomas Graham House, Science Park, Cambridge
CB4 4WF

Printed by Hartnolls Ltd., Bodmin

Preface

The 1992 International Chemical Information Meeting and Exhibition took place this year in Annecy in France 19–21 October 1992. This was the fourth annual meeting and, as previously, some 200 attendees were present.

The 'Montreux' meetings began in 1989; their subject area has always been the latest developments in chemical information in electronic form. Chemical Information — in a broad sense — has always had something of a pioneer role in the information world, due partly to its acute needs for highly specific searching and highly flexible retrieval, and partly to the fact that the large chemical, petroleum and pharmaceutical companies of the world have always placed a high value on good, accurate, high-quality information and have been prepared to support innovative new products and services in this area.

Building on the success of the previous three meetings, the 1992 conference continued to explore chemical information and patent information available via online systems, optical media such as CD-ROM and magnetic diskettes, as well as a diverse range of search and retrieval software, and chemical database management software. There featured a comparatively large number of papers on graphics, particularly molecular graphics.

The papers that make up the conference proceedings are assembled during the six weeks before the conference; this gives authors the maximum possible time to write their papers, and also ensures that the information given is as up-to-date as possible. Those who have ever been concerned with publishing will know that, in the real world, there is a definite trade-off between maximum currency and maximum scholarship. The papers are therefore presented in this present volume in the same basic form as they were written by the authors during the summer of 1992. This means, alas, the occasional textual error. But it also means that the information is as current as time and technology permit.

This current volume provides an interesting picture of just how modern information technology is affecting the complex areas of chemical and patent information; and thus, how technology will increasingly affect information in many other subject areas.

Harry Collier

Infonortics Ltd., Calne, England

Contents

Authors

Information systems development in a cost constrained environment

Aldona K. Valicenti

Manager, Systems Development, Amoco Chemical Company, Chicago, IL 60601, USA

My comments and examples will be directed primarily at experiences in the Amoco Chemical Company, but when appropriate, examples from other companies will also be cited. This presentation is intended to accomplish the following:

- Provide general information on the Amoco Chemical Company
- Indicate the computing environment in the Amoco Chemical Company
- Provide a general understanding of the systems development philosophy, funding and cost containment
- Review a general overview of systems development strategies and general practices in the industry
- Relate the systems development environment to this audience.

Amoco Chemical Company is a wholly-owned operating company of Amoco Corporation. It is approximately a $4 billion dollar petrochemical company with commodity chemicals as its main emphasis. The strategy for Amoco Chemical Company over the years has been to add value to the feedstocks available in the corporation. Over the years, additional businesses were bought that continued to add value to the commodity chemicals produced. Amoco Chemical has purchased and established fabrics, fibres, insulation, plastic food containers, oil processing chemical, petroleum additives and high performance plastics product lines during the last 20 years.

Included in the formation and acquisition of these various businesses were a variety of information systems. For the most part, these systems have continued to function independently and had little or no relationship to each other except when the financial performance of the Amoco Chemical Company was required to be collected or aggregated on a monthly, quarterly or annual basis. Over the years, Amoco Chemical has used systems that have been deployed on various hardware which have included Datapoint, IBM, Data General, DEC and Apple platforms. By mentioning the various vendors, you can infer that there has been little done over the years to arrive at a common computing architecture.

What is different at the present time? Why do we think that the conditions are significantly different to make companies such as Amoco Chemical act and respond differently? The following conditions make this a receptive environment:

- Globalisation of the chemical industry
- Downturn in the chemical business
- Cost containment by the chemical business
- Alignment of information systems plans to business plans
- Customers requiring more control over their applications and data
- Availability of many systems choices to run the business
- Changing technology cycles occurring rapidly
- Proliferation of personal computers, workstations, LANs and distributed computing

I believe that these conditions are leading information system managers to respond in a different manner than they would have done in past years.

Four of the eight reasons I have mentioned are business-related and represent an overall short list; more could be cited. The indication is obvious that information systems are at the core of many business functions. The financial performance, marketing, sales, distribution, inventory, scheduling and operations management employ various individual or integrated information systems. Publications such as *Information Week, Datamation, CIO, ComputerWorld* and many others are and continue to write about companies who are implementing and planning upgrades of information systems which are used directly to leverage the business. These implementations include customer service, distribution and logistics systems. Dow Chemical, DuPont, Mobil Chemical and Amoco are examples of companies who have been cited for having such projects underway.

The assessment is that strategic projects which deal with facilitating business practices will continue to be funded. These projects tend to be large, multi-million dollar investments in core systems which are intended to serve the implementing companies for many years. Most of the systems will be implemented in central or regional computing environments, both nationally and globally. In many cases, the basic software is purchased and then integrated to fit the companies' business practices. That trend will continue and companies, such as Amoco, will continue to search for software to help the business. Should companies develop their own software rather than purchase? Our position has been that we will purchase software in those areas where we find applicability to our business. The best example for the chemical industry may be the financial systems area. All businesses are required to show total revenue, taxes and ultimately a profit or loss statement. We felt that to produce custom software is not a good use of our information systems dollars. But, for a financial business, that may be exactly what is required to keep its competitive advantage.

The following three factors have played a key role in assessing information systems spending:

- Identification of business leverage
- Availability of key technologies or systems opportunities
- Recognition of a continued cost constrained environment.

The business factor is the key to any project initiation. Is there an advantage to the business to do the project? What does that really mean? It means that information systems projects will not be approved for purely operational or technical reasons. In order to have a clearer understanding of what is required by the Amoco Chemical Company, we are focusing on conducting planning projects with key departments and business units. The planning process identifies key business functions performed, the importance of those functions and then assesses to what extend the business function is supported by information systems. If we identify a critical business function, and determine that it has no or inadequate systems support, it is identified as a systems opportunity. Our longer term vision is that as systems opportunities are identified across the company, we will be able to prioritise which ones may have the greatest potential to help the business and it will also assist us to better allocate limited systems funding.

Up to this point, I have not mentioned computing technology, but I do not feel that technology in itself is the key driver. As practitioners, we are beginning to identify a series of important computing and technology axioms which are developing in our individual companies and are becoming key factors in evaluating systems plans:

- Personal computers have been deployed aggressively for five years and should be used for more than spreadsheets and word processing.
- Various standards for computing and telecommunications are beginning to be implemented.
- Global telecommunications capabilities are available and should be used productively, therefore, our traditional support structure may change.
- Commercial software is available to meet business needs.
- Application maintenance and support should be decided as a business issue, how much, when or not done at all.
- All systems do not need to run on a central host computer.
- Hardware costs are continuing to drive downward and have become a commodity item.
- System developers and support staff do not need to be permanent employees.

Systems plans need to take advantage of these kinds of events to be successfully executed in a company. We recognise that software which is strategic can be purchased and integrated into existing systems. The integration work can be successfully done by contractors or consultants. We consider that type of strategy attractive and in the best interest of being able to deliver functionality to our clients in a relatively quick time frame. The term quick may be relative itself, six months to two years. We are beginning to develop applications for the personal computer and for personal computers connected to local area networks. We have also learned that we can provide systems support, answer questions, gather new requirements, even though we are remotely located from our client. For one client, our support is located in downtown Chicago and the actual computing environment is in Hong Kong. If we were given a choice, I am not sure that we would have been that aggressive, but during the past two years we have stretched the limits of technology and continue to have a satisfied client.

In order to continue to look for cost reduction opportunities, we are beginning to evaluate the possibilities of downsizing applications. What applications could and should run on small departmental systems? I am referring to 'downsizing' as a term applied to moving applications from a large centralised host system to smaller and, in many cases, distributed platforms. Some of our chemical plants have been pursuing that strategy and identifying applications which could be successfully executed on local area networks.

The distributed platforms could be smaller, departmental systems or networked personal computers. Applications that may be appropriate to be considered for downsizing tend to have several characteristics. The application is only used by specified departments, the data need to be in a client's environment, the application may have a shorter life span and will need to be changed. A greater number of software applications exist for the smaller environment and there may be other reasons.

Another area that is ripe to be considered as a business issue is the area of application maintenance and support. I specifically exclude software maintenance which is part of a contractual agreement with a vendor and needs to be paid for or executed. Application maintenance and support have usually been carried on an as-needed basis. In many cases the need of one client was treated equivalently to the need of many. If business principles are applied to maintenance and support, different considerations emerge because different questions are asked. I exclude application failures; when an application may be unavailable due to systems or application crashes, immediate action is usually taken. The examples I am referring to are the continuous fixes, upgrades and enhancements that are applied to information systems. The maintenance may come as a new release from the vendor or continuous, small, incremental patches produced by the support organisation. In many cases, little is done in differentiating whether the maintenance is non-discretionary or discretionary. Are there government, legal or financial requirements which will be violated if the maintenance is not done? Discretionary maintenance usually take the form of doing wanted rather than needed items.

Until recently, issues and spending associated with maintenance and support were left to be decided by the technical support organisation. I propose that the decision of what, when, and how much support is provided should be a business issue. The level of spending and, consequently, the level of support should be negotiated with the business owner of the system or a client group who prioritises the work to be done. This level of involvement from the owner of the system is important to cultivate. Ultimately, the owner of the system will be a key player in making a decision when a system no longer meets the business needs and should be abandoned, rewritten or replaced. Most systems have been implemented in the past with the belief that somehow, through technical magic, they will live forever. Over the last few years the information community has experienced the same, rapid, seemingly perpetual change that has affected so many businesses worldwide. As an employee of a company which has divested itself from several businesses and added a few new ones, the effect on the information systems arena is dramatic. In some cases, a sale may depend on how quickly the information systems can be disengaged from a seller to the purchaser.

The legacy-systems that we inherit as part of our workload continue to take a greater and greater bite from our budgets in support. The effect is cumulative. As we deploy new systems, the maintenance expenditures grow. In the late 1970s and early 1980s that growth was probably not as steep and could be overcome by continued growth of information technology budgets. That is not the case today and, I believe, will not be the case again. Most information systems budgets will see little growth. Consequently, continued escalation of the maintenance and support expenditures leaves little for future systems investment. Some companies will continue to find ways of getting selected and strategic systems funded by seeking management approval as special events. But, the smaller, less visible, new opportunities for systems will have to be funded from existing budgets. The only way to do that is to link the maintenance and support objectives to business value.

Let us focus on our clients. Client independence of support may be another way of limiting the maintenance spending. If there are software and hardware solutions that offer a client independence, they should be explored and implemented. I refer to the availability of personal computers with easier to use interfaces, graphical and icon capabilities, and data extraction software which can be purchased, custom tailored to specific requirements, and then easily used by the client. It is one of the ways to give old systems a new look or to provide the client the query and report regeneration capability.

The challenge will be to continue to do more work with less resources. I do not believe that the information technology groups will be staffed to meet clients' needs in all areas. One solution will be to continue to make our clients as self sufficient as they are willing to become. That will differ with the comfort level of each client. In addition, the role of the information technology specialist is also changing from traditional programmer/analyst to facilitator and systems integrator. Superb technical skills and customer service skills are required. Not everyone will be comfortable in the role. Our industry will continue to struggle with the issue of balancing technical and customer focused skills.

Let us focus on another movement which has captured the attention of management. Michael Hammer, a former computer science professor at MIT, has been credited with inventing the phase 're-engineering'. Re-engineering advocates the radical redesign of work processes. Ford and General Motors are companies who have embraced the concept. Ford, as an example, totally restructured its order and invoice processing system before it automated. General Motors has published a set of principles which included a statement that simplification of business practices through elimination and integration will take precedence over automation. I am sure that all of us could identify an example from our own experience, where we took an inefficient work process, automated it, and now we can execute that inefficient process much faster. I propose that the re-engineering movement may offer us the greatest opportunities of applying information systems solutions to the business.

How do these examples apply to this audience? I feel that there are several opportunities that have relevance. First, let me commend you for your ability to work together. One area where this group has excelled is in the area of cooperating with each other through consortiums, industry and user groups to benefit each of our companies. Technology has been leveraged and deployed better because the people involved have insured that it would take place by facilitating the way for that implementation.

The downsizing of applications has been already accomplished to some degree by moving data from centralised facilities to serve individual companies or specific groups. That trend will continue to the degree that individual groups or business units will want to have control over data which serve their individual needs. Customers want some business, technical, literature and public information and they will want access to it when they want it, not when others can provide it. Low-cost personal computers, with easy to use capabilities, connected to central hosts, departmental systems or local area networks are a reality. Implementation of such systems is not trivial but when appropriate planning takes place, is feasible and cost effective.

The area of maintenance and support offers an opportunity for identifying some savings. I am least familiar with individual practices but I assume that support provided is done at several levels. The manner and timing of what is done could offer opportunities. I believe two specific areas that I talked about offer the greatest opportunities:

- Customer self sufficiency and independence in accessing information
- Re-engineering the work process at the information producer, supplier and intermediary levels.

I mention these two specific areas because technology has provided a number of choices which are in place. A variety of platforms with software choices, over a range of low to high cost solutions, is available. But, your creativity in finding the right opportunity is needed. We have the capability to do that now.

Re-engineering has the potential to help us make a gigantic leap because it offers the challenge of looking at our work as if we were beginning it today. If we started now, knowing what we know, with the foundation that we have, how should we proceed? Solutions that were implemented in the past have continued to limit the future. Disconnecting the past from the future may be the opportunity to take advantage of re-engineering efforts. I believe that is your challenge.

Supporting end users: one size does not fit all

Patricia L. Dedert and Patricia A. Lorenz

Exxon Research and Engineering Co. Annandale, New Jersey, USA

We will discuss our efforts to support end-user searching by the scientists at Exxon's basic research facility in Clinton, New Jersey. Our client base consists of about 300 principal and associate scientists. They require information in chemistry, biology, materials science, petroleum processing, energy resources and engineering, with occasional need for business and news information.

These scientists have access to the services of the Information Research and Analysis Unit, a group of technical searchers with advanced degrees in chemistry and extensive experience in information retrieval. Most of these chemical information specialists have been with the company at least 10 years, and are familiar with Exxon's business and the continuing needs of our clients. Two members of this group are dedicated to the basic research facility and are located within the Clinton Information Centre.

With such extensive services at their disposal, why would these scientists want to do their own searching? Not all of them do. However, there are a number of good reasons for bench scientists to try their hand at online searching.

One reason has to do with preferred research styles. Some people always prefer to try to do things for themselves before asking for help. You yourself may prefer to try the do-it-yourself tools at the public library before seeking the help of the reference staff. Is this due to a feeling that you can do it better yourself? A need to be self-reliant at all times? Or an unwillingness to bother the professional with your 'small' problem?

Some researchers prefer to do their own searching because of the serendipity associated with browsing through the references themselves. They fear that the professional searcher may weed out the 'false drop' that would stimulate a whole new line of creative thought.

Another reason is searcher accessibility. While the searching staff tries to provide some instant service for the quick question, with only two people on staff it is not always possible. When the need for a piece of information strikes suddenly, the wait of even a couple of hours can be too long; the urge to jump up and find the answer immediately (or turn to the computer and try a search) can be irresistible.

Finally, the impact of budget crunches cannot be ignored. The trend in our organisation is to decrease overhead services, trying to bill back as many costs as possible. The search requester is billed for both online time and the time spent by the searcher. Some researchers feel more comfortable keeping their hands directly on the purse strings by doing the work themselves; and they may tend to discount the value of their own time spent in searching.

These reasons and others have been mentioned in the literature as factors in the growth of end-user searching. Obviously, there are a number of reasons and their relative importance varies from company to company and researcher to researcher. Whatever the reason, interest in end-user searching has not died despite some early setbacks. While it has not grown as rapidly as the online services first expected and hoped, end users are an important part of our client base. This report updates an earlier description of our programmes to support end users [Dedert and Johnson, 1990].

History of support for end-user searching

We have tried various approaches to training and/or supporting end users over the years and have derived useful lessons from each approach. Our first experience was in 1978 with Drexel University's computer-mediated search system, called IIDA, for searching Engineering Index on Dialog [Landsberg, *et al.*, 1980]. Participation in this research project taught us that a surprising number of our clients wanted to try online searching, even for a short-term experimental programme and even if it meant dealing with a slow and cumbersome system.

In the next few years we tried developing detailed, customised courses to teach selected small groups the nuances of searching Chemical Abstracts, API Literature, and Predicasts PROMT on the SDC Orbit system [Walton and Dedert, 1983]. This was our first experience in trying to teach end users to love and appreciate database structure and controlled vocabulary indexing as much as we do. We were not successful. We tried to teach too much and did not allow enough time for online practice. We learned how frustrating telecommunications problems can be for those who are not accustomed to dealing with them. A year after these courses, none of the trainees were still searching.

For a few years after that we did little to facilitate end-user searching. A few determined candidates were given passwords and sent off to vendor training classes, but we provided no formal support beyond handling the monthly bills. We felt that the majority of our clients would be best served by a menu-driven, friendly front-end system and we waited for it to arrive. We read about Knowledge Index and BRS After Dark, but these services were intended for the after hours population at an academic library rather than our corporate scientists.

End-user searching in the information centre

SearchMaster — the early days

In the early 80s SDC's SearchMaster software became available and was installed on a publicly accessible PC in the Information Centre. One of our colleagues wrote several scripts to allow our clients to do author searches, reference searches and simple subject searches involving one or two concepts. These scripts seemed wordy and cumbersome to the professional searchers, but they were a big success with the research staff [Walton, 1986]. Some researchers bought SearchMaster to use in their offices. SearchMaster was used and appreciated by some of our research staff until 1988 when we had to remove it from our public PC due to lack of ORBIT support and continuing compatibility problems between software, hardware and vendor. However, the success of SearchMaster demonstrated to us that a program allowing only very simple searches, on just a few databases, could be very much appreciated by our clients.

The basic learning we derived from all of these experiences was that most of our end users wanted to do fairly simple searches. Most did not want to think about database structure and controlled vocabulary. They were not interested in taking long courses detailing database structure. They did not want to hear about the dangers lurking for those who dared to search without first consulting a thesaurus. We found that a number of users needed a menu-driven program that automated the process of connecting to, and communicating with, the host system. They did not mind trading flexibility for ease of use.

STN Express

In 1988 the STN Express software became available. The first version of Express had a number of features that made it attractive as an end-user tool, including an automated telecommunication and login process. Communications to Dialog could also be automated through Express. While the telecommunications choices were few and rather inflexible, this very inflexibility made choices simple and the whole logging-in process easy.

Express has a module called Guided Search to aid the inexperienced searcher in creation of search strategies in a non-threatening, offline mode. The user enters all synonyms for one concept on to an on-screen 'index card,' thereby transparently performing a Boolean OR. For searching *Chemical Abstracts*, Express provides three different types of searching index cards: one for subject terms, another for document aspects (i.e., authors, document type, language and date range), and one for chemical substances (i.e., chemical names, molecular formulae, Registry Numbers, and/or drawn structures). Terms on the chemical substance cards will be searched in File REGISTRY to find the appropriate Registry Numbers, which are then transferred to File CA for subject searching. The switch between File REGISTRY and File CA will be done transparently to the end-user.

Terms for each concept in [illegible] search are placed on separate cards. The cards are ANDed together — again, transparent to the user. In Guided Search, the user can construct a multiple concept strategy containing date ranges, multiword terms, and molecular formulae without ever using a Boolean or proximity operator — and with no need to know of the existence of database field qualifiers. Using the terms entered into the cards, Express creates a strategy in STN Messenger language, logs on to the system, and performs the search in the desired databases. Control returns to the searcher to decide whether to download answers or abort the search.

We placed Express on our public search terminal in June of 1988. In order to allow searching by those without a personal STN password, we put an STN ID and password belonging to the Information Centre on the public PC. To facilitate billing to the appropriate cost centre, our computing support group developed a front-end access menu, based on Paradox, requiring the user to enter a name and a valid cost centre code before using Express.

We advertised the arrival of STN Express searching with desktop fliers and newsletter blurbs, and we offered free, lunch-time demonstrations for several months. The new set-up was immediately popular with our clients. It offered far more flexibility in strategy construction than the SearchMaster scripts and it was also more visually appealing with its pull down menus and graphical display of concept groupings. Comments from users were very favourable.

Despite the nice features of Express and Guided Search, however, we did have a number of problems in supporting it. After the end user entered search terms into the Guided Search front end, the system would often 'hang' and lock up during the login process. A cold boot might be required. A second attempt to do the search (first necessitating a pretended modification to the strategy) might be successful — or it might not.

It took us a while to discover that these system hangs were more frequent with Guided Searches involving more than three 'cards' (ANDed concepts). The problem seemed to be caused by insufficient RAM memory, even though we had more free kilobytes than Express supposedly required. The Express experts at STN thought it probable that our Paradox front-end was stealing enough memory to cause the RAM-hogging Express to lock up at random. We needed the Paradox front-end to force people to give us their charge code, but we were tearing our hair out dealing with the frequent system problems and increasingly angry users. After several frustrated attempts to login, we would often resort to doing the search for the client on our own PCs, not billing for our time because the client had intended to do the search without us. It was an unacceptable situation.

The Windows/Macintosh version of Express promised to eliminate some of these problems, and in 1991 we managed to acquire a Mac as our second public workstation. This time our computer support people wrote us a Hypercard stack to allow the user to enter name and charge code before running Express. This Mac version has presented us with fewer support problems. While the PC/DOS version of Express is still accessible to users, we have tried to discourage its use.

Apart from software and equipment problems, there are some limitations to Guided Search that make it an imperfect solution to the question of menus for end users. In the early version of Express, a mistake in the Guided Search strategy — a misspelled word or too many restrictive concepts — required the end user to log off the system, revise the strategy, and log in all over again. Any search term charges would have to be paid again in the re-execution of the search. The inherent restrictions of a non-interactive, 'batch' search system were often evident. In the newer Express version, control is returned to the user at the end of the Guided Search. However, in order to modify the search without first logging off, the end user must know Messenger commands.

We also cannot forget that STN Express Guided Search can only give access to STN databases. While *Chemical Abstracts* is our most heavily used source by far, our clients do have need for other online databases. The most frequently requested of these was *Science Citation Index* online, not available on STN. For this combination of reasons, we were enthusiastic about the advent of Dialog Menus, which became available to us in late 1990.

Dialog Menus

When we added access to Dialog Menus to our public PC, we thought its chief appeal would be the fact that *Science Citation Index* was now, finally, available to our clients in a menu-searchable mode. Our desktop and newsletter advertising emphasised the new availability of *Scisearch*, which clients had been requesting for years. To our surprise, however, response to these ads on *Scisearch* availability was very slight.

Nevertheless, we found ourselves steering new users of the public PC toward Dialog Menus rather than STN Express Guided Search, for several reasons. First, since Menus software is resident on Dialog's mainframe computer, we have no problems with hardware-software compatibility or insufficient RAM memory.

Even more important,with Dialog Menus the end user has the benefits of a truly interactive search. Questions can be modified online based on the answers retrieved. Online help is accessible if the end user gets stuck. In contrast, with STN Express Guided Search the user is doing a non-interactive, 'batch' search.

Searching for authors via Dialog Menus is easy because the researcher can make use of the online Expand function to choose among various spellings, first names, and middle initials. In contrast, when using Guided Search the searcher must enter the author's last name with only a first initial. Alternatively, the searcher can try to OR together all possible variations. When you know that you are searching for papers by D. Jean Jones, you really do not want to use 'D Jones' as your search term. With Dialog Menus, you do not have to guess what the name variations might be; you can choose just the ones you want from the online expand, and if there are too many hits you can immediately narrow the search with added subject terms.

We have found Dialog Menus to be easy to teach and support. Few users require more than one, brief introduction to the system. The menus are well written, customised for each database. An early version of Dialog's Medical Connection gave access to *Science Citation Index*, but did not offer an option for searching cited references. This omission was corrected in Menus.

Problems that do arise with using Dialog Menus usually involve printing. When it is time to look at the search results, Menus gives the option of Prints or Displays, but Prints shows up as an earlier choice. Some users have requested Prints, expecting them to show up on the attached printer. They may repeat the print request several times before giving up. We may not know about the problem until we receive the offline prints — multiple copies of them — a few days later.

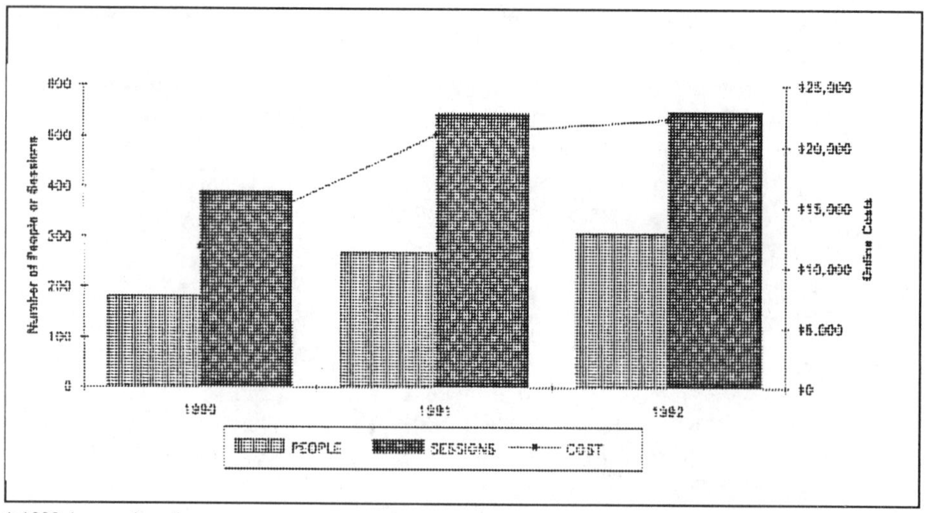

* 1992 data projected

Figure 1: Use of public searching PC Jan. 1990 – June 1992

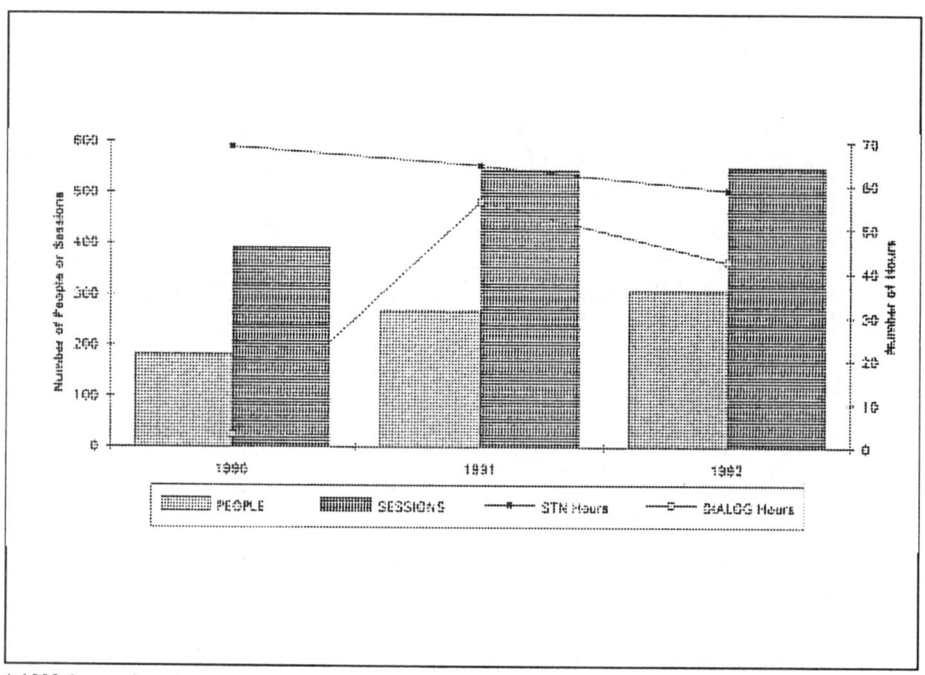

* 1992 data projected

Figure 2: Use of public searching PC Jan 1990 – June 1992

A related problem has to do with hard copy of screen displays. The user may display the answers on the screen, realise the desired answer is there, and then press the function key to turn on the printer. Too late! This problem is compounded by the fact that the attached laser printer does not print 'simultaneously;' instead, it waits until it has a full page to begin printing.

Figure 1 shows the use of the public searching workstations from 1990 until June of this year. Access to Dialog Menus was added to the public PC in December of 1990. The number of online users and search sessions at the public workstation increased by 49 and 40 percent, respectively, from 1990 to 1991. The number of connect hours jumped by 68%, while the dollars spent by the end users leaped by 79%. In contrast, there has been little change in the use of the public PC between 1991 and 1992. It would seem that the addition of Menus led to the increase.

Figure 2 again shows use of the public PCs this time superimposing the number of hours spent on STN and Dialog. This diagram makes it clear that the availability of Dialog Menus accounted for the great increase in usage between 1990 and 1991. STN use decreased over this period, but only slightly. It appears from these data that the introduction of Menus fulfilled a need previously unmet by the STN Express Guided Search package, reaching a new group of users — without materially affecting the use of STN.

End user 'guided search' strategies at the public PC

When using the Guided Search Module of STN Express, the end user develops the strategy offline and stores it on the hard disk before it is uploaded to the STN system. Most end users do not delete the strategies after their searches. The searching staff can look at the stored strategies to find out what types of searches our end users are doing.

We carried out an analysis of the Guided Search strategies employed by end users for the period January 1990 to June 1992. Over 300 sessions were evaluated. Table 1 shows a breakdown of the types of searches done during this period. [See Note 2].

Author searches represented 38% of the total search sessions conducted. Of these, 61% were simple searches for author(s) only. The other 39% were author searches qualified by date, language, document type, journal coden, or subject terms.

Searches for chemical substance terms (Registry Number, chemical substance name, or molecular formula) constituted another 15% of the search sessions evaluated. One third of these sessions used the chemical term alone; the rest qualified the Registry Number or substance name with subject search terms, year, molecular formula, or the flag for preparation.

Searches conducted using only subject terms represented 47% of the sessions conducted. Of these, 64% employed more than one 'card' in their strategies, which is equivalent to the use of the Boolean AND operator. These multi-card strategies constituted 30% of the total sessions in the study.

Search Type	Number of Searches		%
AUTHOR SEARCHES		**125**	**38%**
Author only	76		
Author + Subject	25		
Author + YR/LA/DT/CO	23		
Corporate Source only	1		
CHEMISTRY TERMS		**48**	**15%**
Registry Numbers (RNs)		24	
RNs only	12		
RN + Preparation	10		
RN + Subject Terms	2		
Chemical Substance Name (Name)		22	
Name only	4		
Name + Preparation	1		
Name + Subject Terms	14		
Name + Molecular Formula	1		
Name + Year	2		
Molecular Formula + Subject Terms		2	
SUBJECT TERMS ONLY		**152**	**47%**
One Card	42		
Multiple Subject Cards	97		
Subject Terms + LA/YR/DT	13		
	TOTAL	**325**	**100%**

Table 1: STN 'Guided Searches' at public PC. Sample of strategies (Jan. 1990 – June 1992)

The data in Table 1 demonstrate that 70% of the search sessions at the public PC were 'simple' searches involving authors, chemical substance terms, or 'one card' subject terms. Most of these searches would have retrieved some useful references. These are probably what the end user was looking for.

Table 2 analyses the sessions that used subject terms, including those in which the subject terms were used as qualifiers for author or chemical substance terms. Subject term searching gives the greatest possibility for comprehensive retrieval and also the greatest possibility for serious strategy errors. The most common error in strategy noted in Table 2 is the lack of proper truncation. Less than 50% of subject searches used any truncation. At the same time, 41% of the errors noted in the subject strategies were due to improper truncation. Many of the end users who used truncation did not apply it to all relevant terms.

	Subject only		Subject Plus ...					
Comments	Multiple cards	One card	LA/YR /DT	Author	RN/MF	Substance name	Total	%
USED TRUNCATION	66	7	4	6	1	7	91	47%
USED 'OR' LOGIC	70	32	9	13	3	12	139	71%
STRATEGY ERRORS	66	33	9	8	2	4	122	63%
Improper truncation	32	26	8	8	2	4	80	41%
Needed Substance Name or RN	24	2	1				27	14%
Complex phrase instead of ANDed terms	1	11	7				19	10%
Needed CA abbreviation	12	2	2			1	17	9%
OR instead of AND	3	10	1				14	7%
Too broad	3	5					8	4%
Needed Subject Term instead of Substance Name	4						4	2%
Improper syntax		3					3	2%
ANDed identical terms	2						2	1%
TOTAL NUMBER OF SESSIONS	**97**	**42**	**13**	**25**	**4**	**14**	**195**	

Table 2: Analysis of 'Guided Search' strategies. Sampling of subject searches (Jan. 1990 – June 1992)

Another common error noted in subject strategies was the use of chemical names as subject terms rather than as chemical substance names (in a Guided Search, a chemical substance name retrieves the relevant Registry Number(s) for subsequent use in File CA). The use of text terms to search for chemical substances might retrieve some useful references, but would miss some really good articles indexed by Registry Numbers.

On the other hand, some strategies attempted to use subject terms as chemical substance names; for example, 'alkane' and 'paraffin' were entered as chemical substance names. After three tries, this end user did realise the error and entered these terms as subject terms.

About 23 % of the sessions reported in Table 2 had errors that were so serious that they would have resulted in either no references or so many as to have been useless.

However, the percentage of sessions with serious strategy flaws is deceptively high because many of these represented more than one attempt by the same user to retrieve references on a particular subject. Of a total of 22 subject searches employing multiple sessions, nine (41%) resulted in an eventual improvement in strategy. Some examples of serious subject errors are shown below.

Complex phrases	* ultraviolet absorbers for polymers * light stabilisation of thin films * crystal structure of sodium chloride
OR instead of AND logic	* stainless OR steel OR surface OR defect (second attempt at improving this strategy truncated all the terms and still ORed them)
Strategy too broad	* carbon dioxide * gel
Improper syntax	* hindered w amine

Even though some end users attempted to modify faulty strategies, many tried once and apparently gave up. We have no way of knowing if they obtained any useful references. In some cases end users consulted the professional staff after they failed to retrieve useful information, perhaps requesting the searchers to do the search.

Our analysis suggests that most end users at our public searching PC prefer to do simple searches and are content to get a few good references. When they tried to use more complicated strategies, most end users ran into trouble. However, there were a few end users who used fairly sophisticated strategies and probably obtained good results. The most common errors reported here are improper or incomplete truncation and lack of understanding of Boolean concepts.

Before thinking too badly of end-user search strategies, however, we should remember that the searches done at our public searching PC are often done by true novices with no training whatever. After they run into difficulty they may seek an explanation of the system and the best way to use it. However, it is disquieting to think of those who never asked for help, who simply accepted the answer the computer gave them.

The dedicated end user — growth in use of personal IDs

Searching STN in the office or lab

Shortly after we introduced STN Express at the public searching PC, we were requested to make end-user searching more widely available to the scientists in their labs or offices. Some users did not want to walk all the way to the Information Centre — in another wing of the building — to do a search, perhaps finding upon arrival that the public PC was already in use. Now that all scientists had personal

PCs or Macs, perhaps with telecommunications capability, they did not see why they should not use these to do some online database searching.

While we had already been providing advice and passwords to individuals who sought our help, it became clear that we needed to take a more pro-active approach. Some researchers thought our failure to offer courses was an attempt to discourage the scientists from 'stealing' work from us.

We wanted to support end-user searching. While aware of the pitfalls, we had learned from our previous experiences that most end users did not want to do their own patent novelty searches or in-depth studies. Further, we thought that a formal program to support the end users would lead to increased trust and interaction between us. We wanted the scientists to know that they could ask us for help when they ran into difficulties without encountering our disapproval for having tried a search on their own.

We had many decisions to make about what type of program to offer. We wanted to keep it simple by starting with only one system. Which should it be? We ended up choosing STN for a variety of reasons: clients were gaining some familiarity with STN through the use of Express Guided Search on the public PC; STN had nearly all of the databases most used by our clients, including CA *with* online abstracts; STN Messenger had both novice and expert versions of commands, either invocable at any time in the search process; we could get an unlimited number of STN passwords on our existing account; there were many training options, including tutorial diskettes and a helpful manual 'Getting Started in CAS Online'.

Hearing of our plans to promote use of STN by the end users, our STN account representative offered to develop and deliver a very short introduction to the basics of online searching. The tutorial session took about 1 1/2 hours and covered the following topics:

1. Evolution of literature from primary journals to secondary sources and online databases
2. Scope of the CA and Registry Files
3. Format of STN commands
4. Basic commands: FILE, EXPAND, SEARCH and DISPLAY
5. Truncation and character masking
6. Boolean and proximity operators
7. Limiting retrieval to articles by a particular author or company
8. Finding Registry Numbers by searching a common name; and
9. Using Registry Numbers as search terms in File CA.

The lecture portion of the course was followed by individual, half-hour online practice sessions in which the scientists could try to solve one of their own search problems with the aid of the instructor. This private practice session is the most

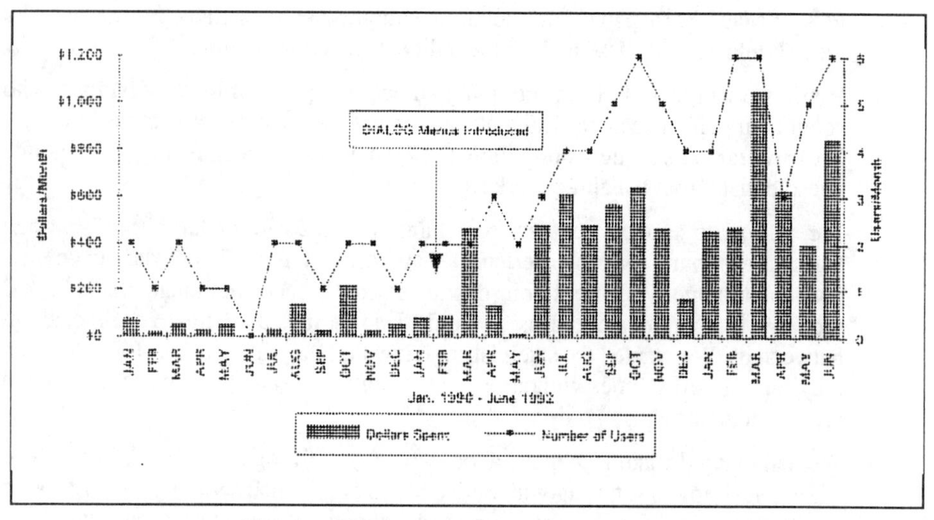

Figure 3: Dialog use in offices or laboratories

popular feature of the course. Some people take the course more than once chiefly to get a second practice session.

This introductory session has been offered at least ten times since it was first developed and offered in 1988. Fourteen people attended the first offering of the course. Twenty-two people attended the second time the course was offered, but we found that the large size inhibited questions and interaction. Later sessions have been limited to 12 or fewer people.

We have made changes and improvements in format, principally by incorporating live online demonstrations into the lecture portion of the course. However, the basic format of a short introduction followed by hands-on practice has been so successful that we have not wanted to tamper with it.

Not all of the attendees become frequent searchers, however. Some leave the practice session with the realisation that the types of searches they want to do are not possible after just a short introduction. We know of at least one course participant who continues to use Guided Search menus at the public PC rather than Messenger commands; even though he searches frequently, he does not think it worthwhile to learn and remember command mode searching.

There are currently 56 scientists with individual STN passwords. However, the average number of STN end users per month is between 10 and 12 — only about 3% of the target population. This number has not varied significantly in the last four years, despite the fact that there are new participants in every introductory session. The average monthly cost per user hovers around $150/month. Is the market for end user searching saturated? Have we reached the limit of the number of people willing to search in command mode?

Personal Dialog IDs

When we introduced Dialog Menus on the public PC, we also made private Dialog passwords available to the scientists. However, we did not widely advertise this availability because we were limited in the number of Dialog passwords that we could have on our main account. Nevertheless, a few of the scientists began using Dialog Menus in their offices.

Figure 3 shows the significant growth in the use of Dialog since the advent of Menus at the beginning of 1990. The average number of Dialog users per month increased from one to five; the average number of dollars per month per scientist quadrupled. In contrast to the STN password owners, most scientists with Dialog passwords use Dialog at least monthly.

Current level of use

Figure 4 gives an idea of the total amount of end-user searching done at our basic research facility. The dollars spent per year, whether at the public searching PC or in private offices, grew slightly upon the introduction of Dialog Menus. We find that this number has levelled off a bit in 1992, perhaps due to a severe budget crunch in the organisation. Use of Dialog, primarily via Menus, is growing. The number of STN users, most of whom do searches in command mode, remains fairly constant.

The amount of money spent by STN end users has been decreasing since 1990. This was surprising to us as the amount of online information that can be retrieved and printed grows significantly every year — and online charges increase annually also.

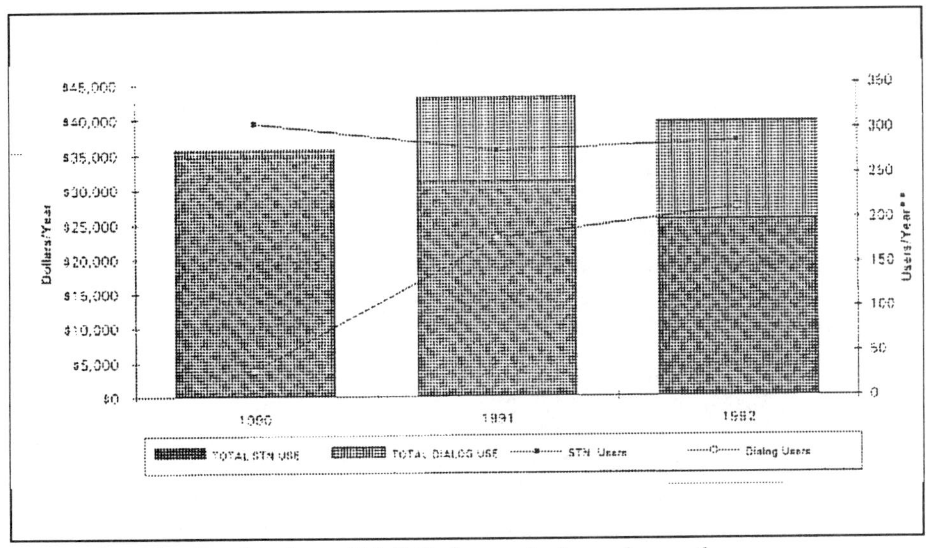

* 1992 projected from Jan.–June data. ** Individuals counted only once in a month.

Figure 4: Total end-user dollars 1990–92. Public PC and personal IDs

Is the decrease then due to increasing proficiency? Increasing cost consciousness? Or is the decrease the result of changes in pricing strategy at STN? This year STN significantly lowered connect-hour charges while dramatically increasing charges for every search term used. The cost of our professional searches has risen under this scheme due to the complexity of our searches and the efforts we make to be comprehensive. The end user, however, who often has little interest in comprehensive retrieval, may have benefited from this change. However, we need more data to be able to answer this question with any certainty.

What databases are the end users searching? Figure 5 shows the databases compiling at least two hours of end-user business in the last 30 months. *Chemical Abstracts* is the overwhelming choice of most of our end users. BIOSIS, Scisearch, INSPEC, and Compendex all show up on the list, but at levels that must be discouraging to their producers. They come in behind Dialog Menus, the pseudo database to which the end user is connected while choosing the file of interest and entering search terms. However, even if the time spent on Menus were divided among the databases actually searched, the use of those databases would still be far, far behind the use of CA.

STN's numeric databases are fairly popular with the end users, especially the C-13 NMR file (now titled SPECINFO). DIPPR and Inorganic Crystal Structures are also used.

The patent files, Claims US Patents and Derwent's World Patent Index, have been accessed by the end users, but at low levels. It is probable that they are searching these files by inventor or company names.

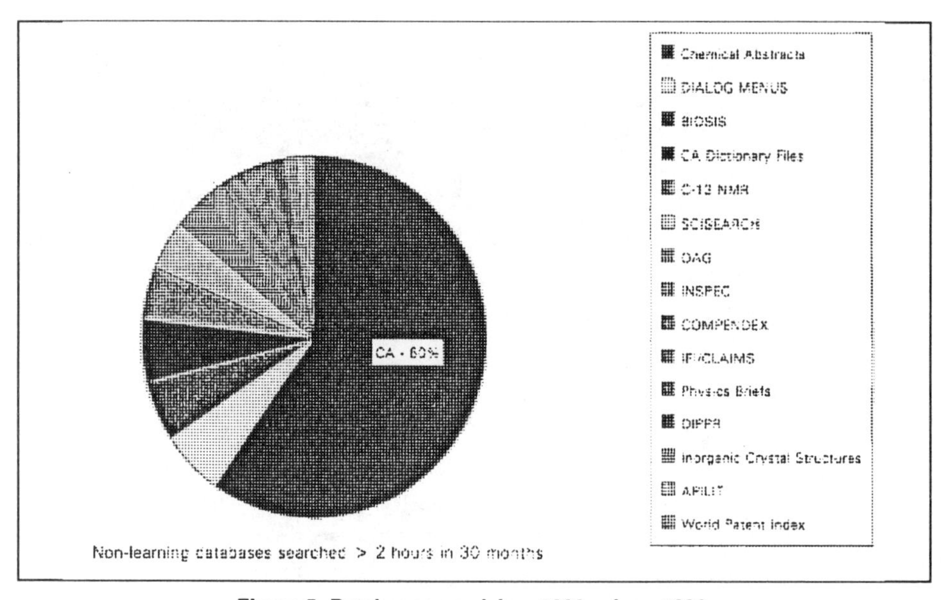

Figure 5: Databases used Jan. 1990 – June 1992

Continuing support and future plans

In addition to offering the introductory courses, we have occasionally offered special sessions on databases like Beilstein, Gmelin, and SPECINFO. These sessions were offered to us by the database producers and so required little effort on our part beyond arranging for meeting rooms and telecommunications facilities. These sessions have been attended by small numbers of interested users, as well as the professional searching staff. Advertising for these sessions keeps our name and our end-user support function in the public eye.

Continuing support is not so much an issue as is 'start-up' support. Some new end users have trouble getting their telecommunications software on their workstations to work with our site modem pool. They may have trouble figuring out which downloading protocols to use. If they have purchased STN Express software, they may have trouble loading it. We generally make 'house calls' to troubleshoot these initial problems, but are not needed after that.

Most of our continuing support efforts are focused on our public searching PC. Many new employees, visiting scientists, and summer interns need a tutorial session on the use of the public searching PC. First we help them decide which system to use (STN Guided Search or Dialog Menus) for the question of the moment. Then we usually sit with them until the first search is successfully completed. This can be somewhat time consuming, but generally results in a client who will be self-sufficient in future.

We recently surveyed a small number of our end users as part of a larger survey to evaluate satisfaction with our services. The questions were designed to find out whether the end users were satisfied with the level of support provided by the professional searching staff and with the balance of doing their own searching versus having it done for them. Most felt satisfied with the level of support although some would like a continuously staffed help line. However, some users felt the need for more advanced training and additional seminars to increase their skills. One user expressed appreciation for the 'enrichment' seminars offered (like those on Gmelin, SPECINFO, and Beilstein), but expressed a fear that something was being missed that could be of benefit to the end-user searcher. In general, the scientists would like the searching staff to be more pro-active about seeking out additional training or enrichment courses.

On the other hand, some end users are tired of taking training courses and would prefer us to develop simple one- or two-page handouts that would make training unnecessary. Just as no one end-user searching option satisfies all users, no training method seems right for all.

Our near-term plans include offering a course in 'Beyond Menus — Dialog Basics' for Dialog users who want to increase their capabilities. We will continue to offer specialised database introductions as they are available; we want to sponsor a course on use of STN's Chemical Properties Data Network. We will also continue to offer periodic introductory courses to STN commands.

Is the end user replacing the professional searcher?

Is the growth of end-user searching decreasing the work for the professional searchers? In one follow-up questionnaire, we asked STN end users how they would have obtained the information they sought prior to learning to search online. Their answers were revealing: 41% would have done the search with the aid of printed index sources; 27% would have done without the information. Only 32% would have asked the professional searching staff to find the information for them. In other words, 68% of the questions asked by end users are new questions; questions that would not otherwise have been searched online, or that may not have been asked at all.

Others have also reported that end-user searching fulfills a need not previously met by mediated searching. In a test of Dialog Corporate Connection (the forerunner to Dialog Menus) at Sandia Labs [Erickson and Pruett, 1990], end-user scientists reported that "51% of the time they searched for information they would not have 'bothered' the Library about." In a study of users of Dialog's Medical Connection at Glaxo Group Research [Boyd and Warne, 1990], 50% of the end-user population said that they were "more likely to seek information because they had instant access and that they are now, as a result, better informed."

At Imperial Chemical Industries, the introduction of end-user searching appeared to lead to an increase in online dollars spent by the information scientists [Warr and Haygarth Jackson]. The number of professional information scientists on staff increased as well. The introduction of end-user searching increased the level of information awareness in the scientist population and resulted in greater information use by both end users and intermediaries.

Statistics on our workload show that the total number of search requests handled by the professional searching staff has remained relatively constant over the last several years. However, the number of shorter, 'quick and dirty' searches has been decreasing over this period. We feel that this is exactly as it should be. Some of our best clients for professional searching services also do some of their own online searching. We strive to achieve this relationship with all of our end users.

In recent years, we have felt that the amount of work we do for our clients is limited by the amount of time we have rather than by the amount of work our clients bring to us. We feel certain that the number of requests that we receive and handle would actually increase, rather than stay constant, if we had the staff to handle more requests in a timely fashion. We do not feel at all as if our jobs are threatened by the amount of searching that our clients do on their own.

Workload for the professional searchers in the Glaxo study appeared to remain rather constant after the introduction of end-user searching; Boyd and Warne speculate that end-user activity absorbed some potential growth in demand. Again, this is perhaps as it should be. In our current environment it would not be possible for us to increase the number of professional searchers on staff. Instead, supporting the end users allows us to 'leverage' our skills to provide the information our scientists need.

Notes and references

1. Boyd, T.; Warne, K., 'End-user searching within Glaxo Group Research Ltd: an evaluation of the Dialog Medical Connection,' 1990, In P. T. Bysouth (Ed.) *End-user Searching*, London: Aslib, The Association for Information Management, pp. 125-134.

2. Bysouth, P. T., 'Evaluating the use of several approaches to online literature retrieval by research scientists,' 1990, In P. T. Bysouth (Ed.) *End-user Searching*, London: Aslib, The Association for Information Management pp. 105-124. Our approach to analysing these data was stimulated by this article.

3. Dedert, P. L.; Johnson, D.K, 'Promoting and supporting end-user online searching in an industrial research environment: a survey of experiences at Exxon Research and Engineering Company' *Science and Technology Libraries* **1990**, *10(1)*, 25-45.

4. Erickson, L. J.; Pruett, N. J., 'End-user searching in the corporate research setting: a planning assessment at Sandia National Laboratories,' *Science & Technology Libraries* **1990**, *10(1)*, 15-23.

5. Landsberg, M. K.; Lorenz, P. A.; Lawrence, B.; Meadows, C. T.; Hewitt, T. T. 'A joint industrial-academic experiment: an evaluation of the IIDA system.' *Proceedings of the American Society for Information Science* (43rd Annual meeting), **1980**, *17*, 406-408.

6. Walton, K. R.; Dedert, P.L. 'Experiences at Exxon in training end users to search technical databases online,' *Online* **1983**, *7(5)*, 70-79.

7. Walton, K. R., 'SearchMaster — programmed for the end-user,' *Online*, **1986**, *10(5)*, 70-79.

8. Warr, W. A.; Haygarth Jackson, A. R., 'End-user searching of CAS ONLINE. Results of a cooperative experiment between Imperial Chemical Industries and Chemical Abstracts Service,' *Journal of Chemical Information and Computer Sciences*, **1988**, *28(1)*, 68-72.

Researching and analysing information on international chemical companies

Sylvia James

'Daymer', Birchen Lane, Haywards Heath, West Sussex RH16 1RY England

Abstract

Information professionals working in any scientific and technical field increasingly need to be able to resource business information on companies and products as part of their work. This applies to the companies and organisations in the chemical industrial sector as they operate, diversify and compete in a global trading environment.

This paper will firstly examine a method for researching the numerous sources of information on the larger international chemical companies; those quoted on stock exchanges, state and partially state owned operations, and large private companies. Some of the more unusual sources of published company information are examined and discussed, which may be unfamiliar to many researchers. There will be an outline of an approach to the systematic collection of all the possible aspects of published information on chemical companies. This has been developed over several years, whilst working with all types of client companies in order to streamline and upgrade the methods of data collection in this important area.

In the second part of the paper there will be a brief rundown of the analysis of chemical company information with some suggestions on how the information professional can 'add value' to the research obtained on basic company data retrieved from the various services and suppliers. Some of the most common reasons for company information analysis will be covered including the principles of company valuation.

Introduction

The information industry, more than any other, is being affected on a broad scale by global business trends. Information professionals and suppliers are having to be more and more international in their coverage as the cross border nature of the businesses, activities and horizons of their users and customers expand. This is apparent in the formation of major economic trading groups, as in Europe with the EEC Single Market and the North American Free Trade Agreement (NAFTA) accord in North America. The breakdown of the structures of the command, state run, communist countries and the global privatisation of industries is also contributing to an international approach.

The investigations necessary for participating in this international expansion are taking place of necessity, exacerbated by the world recessionary climate of the early 1990s, as organisations realise the possibilities that exist in new markets and different countries that were previously unthought of. Only an efficient and effective information industry can support this activity.

All the participants must recognise and be part of this 'globalisation' process and look beyond the traditional boundaries of their operations for the continuation and success of the profession. Whatever subject area or industry sector the information researcher is used to working in, they need to become more business orientated. For information specialists in the chemical sector this requires an understanding and knowledge of the chemical business in which their company operates.

Once this business is appreciated the whole information service can be looked at with the intention of implementing a whole new approach to information provision. This should include a service which takes basic information from sources and really makes it relevant and cost effective for users at all levels.

Reading the most recent *Financial Times* Survey on the Chemical Industry (December 10th 1991) in preparing this paper, the whole business approach to the industry is evident. The supplement states in few articles the essential facts and concerns which should be pre-eminent in the provision of the information service in any chemical company.

This paper will concentrate on one aspect of information research; into chemical companies from a business rather than a technical viewpoint, looking at a method of collecting the information and then at additional ways in which the basic company information can be enhanced to be of most value to business users. It is intended as background detail to the actual presentation at the Conference.

SECTION 1

A systematic approach to collecting chemical company information

Setting up a systematic service of collecting all available published information on a group of companies for a specific project or on a regular basis will require a mixture of desk research and use of automated services. The reasons for undertaking this will be discussed in Section 2 of this Paper.

Company information disclosure falls into four categories:

1. Legal and regulatory requirements
2. Investment/funding
3. Industry/professional
4. Information available to the general public

In each case the company itself discloses some information to a greater or lesser degree, or information is publicly available from other sources. This paper will concentrate on the first two areas of information collection which may be less familiar to information specialists in chemical companies.

Researching chemical company information

Basic methods of researching

Legal & regulatory information

Investment & funding information

The methods of tracking company information are not unique to those researching chemical companies but are worth defining as information on many chemical companies may be missed when in fact there is disclosed detailed information in areas that may not be thought of by the non-specialist.

The most important factor in finding any company information is establishing the status of the company in relation to listing on a stock exchange or a capital market. If this is so, then there will be mandatory disclosure of 'material' information aimed at existing and potential investors. Unfortunately establishing status is not simply a matter of the company being 'public' or 'private'. Today there are additional complications of what is company status in these days of massive state privatisation of holdings in industry.

Company status

A company could be:

- Publicly listed with a full quotation on one or several stock exchanges; this does not necessarily have to be the main stock exchange in a country, nor even the country of the main office or headquarters.
- Publicly listed with a partial listing for just some of the capital or securities that make up the company's financial structure. Again, this does not have to be in the main stock exchange of the country or in the main country of operation.
- State owned with all capital owned by the government of one country
- Partially state owned with divisions, or some part of the capital structure, issued on a stock exchange or capital market.
- In the stage of being privatised; the capital being in the process of being offered to the public, private investors, or another company or companies.
- Public, with a certain number of shareholders defined by domestic company law, but not listed on any stock exchange.
- Private; with some part of the capital structure issued on a capital market but not listed.
- Private with no part of the capital structure listed or available for public investment.

In all but the last category of companies, it is possible to obtain some disclosure of information in virtually any country of the world. The problem for the information specialist is to establish into which category the companies they need to research fall. There are examples of large chemical companies in all these categories. Failure to establish this important listing status could mean missing vital documents.

Chemical companies also often form joint ventures with other companies for specific projects in certain countries under a new company name.

Establishing status and the documents available if listed

Stock Exchanges

In many large companies the corporate treasurer or company secretary will possibly have access to a real time or near real time trading or pricing system. This may be for foreign exchange trading or simply to track investments. A full directory of traded stocks usually accompanies the services which will be invaluable for the information specialist in establishing listing of full companies and securities and the country where listed.

Other sources are stock exchange yearbooks and listing newspapers of price quotations. Newspapers may be issued by the domestic stock exchange or be a section of the main financial newspaper of that city. These can all be acquired at very low cost and are well worth maintaining and updating for countries where company investigations are regularly made.

Many countries have several stock exchanges. There is often one major exchange that dominates trading and listing activity, usually in the capital city. In countries where there is a strong provincial, decentralised culture and government there may well be important smaller stock exchanges where a company important to that region may choose to list securities rather than on the main exchange. These can often be missed, with the researcher assuming the company to be private when it is not and missing important published information.

The process of listing itself will produce interesting and informative published documents. The initial listing document will be very detailed and may give information that will never be disclosed again. Watch out for subsequent listing documents on one or multiple 'foreign' exchanges (International Public Offerings; IPOs) and for further offerings of securities which will require considerable disclosure. Commentary about these offerings and the industry involved from analysts of the underwriting banks and brokers will also be desirable, although not so easy to obtain.

During the life of the company when listed, each stock exchange requires different degrees of disclosure, but detailed documents are generally available because of the increasing international pressures of protecting shareholders.

Documents that can be expected are

- financial statements; annual, semi-annual, quarterly
- directors details, interests, salaries
- large shareholding stakes and changes
- announcements on significant changes affecting market price
- press releases.

Updating is generally very rapid with documents usually being required to be available to the public on the day they are lodged by the company.

There are moves to harmonise listing and disclosure requirements for stock exchanges belonging to various national groups and associations.

Mergers and acquisition documents are available if two listed companies are involved. These are still very rare outside the UK and the USA.

Information provided on listed companies by analysts in brokers and banks can often be very informative. Tracing the brokers in the US, UK and Europe who are covering the chemical industry can be done by consulting surveys such as the US and international editions of *Institutional Investor* magazine.

Privatisation

Tracking state owned companies and their divisions and subsidiaries can be very difficult. Global government directories such as 'Europa World Year Book' or 'International Year Book and Statesman's Who's Who' are useful as a starting point, for establishing the state structure of the chemical industry at a fixed date. Then, press comment sources and governmental libraries can be more reliable for establishing just where the company is along the privatisation process. This is a very common area for information to be missed by inexperienced researchers. Parliamentary proceedings, draft legislation and advisory papers prepared before and during privatisation can also be surprisingly revealing about the companies actually under the denationalisation process as well as those planned for privatisation.

Chemical companies are very often owned by governments as one of the strategic sectors of industry within a country. They are now often the targets of privatisation. It is important to also check information about large state 'holding companies' where units and divisions of chemical production of particular interest may be lumped together in a structure with companies with non-related activities.

Other capital markets

Checking if securities are quoted on other capital markets requires a specialist knowledge of the main markets that exist for international investment, which should be acquired by the company information specialist. They will soon get into the routine of ensuring that information supplied to these markets by companies raising capital on them are investigated.

The most common markets used by companies for such capital raising are:

- US stock exchanges where non-US companies can raise capital by means of American Depository Receipts (ADRs). Securities and Exchange documents (SEC) are required to be filed.
- The International Capital Market (commonly known as the Eurobond market). Deals are announced daily in *The Financial Times*, with a summary of the. previous week's deals in the Monday issue in the second section on 'Companies and Markets', in a table on the 'International Capital Markets' page. Public

prospectuses are available from the managing bank listed in the table. Many chemical companies use this method of raising funds and regular collection of these documents on all or selected chemical companies is recommended.

- Swiss Foreign Bond Market. Deals in this market are also listed in the table mentioned above published in the *Financial Times*. Managing banks are always distinguished as Swiss.
- Foreign bonds issued in the country of interest. If a major project is being carried out in a country by a chemical company or group of chemical companies it is possible that the consortium or individual company could raise funds for the specific project by issuing 'foreign bonds' on the local stock exchange. Similarly all the other international markets could be used. Researchers should be aware that a specific company may be set up as a funding vehicle which will not necessarily be the same name as the project, consortium or individual company undertaking the deal. Often documents from such funding are prepared in such detail that they may have relevance long after the project is over or the funding repaid.

Private, unlisted companies

The majority of companies are unlisted and private. This means that there will be very little publicly available information about the company. If the company has any form of limited liability status and has been formally incorporated as a business under the company law of the country of operation, then it is likely that there is some disclosure in a public registry or in an official gazette and this should always be obtained as primary source data.

Unfortunately, most countries have de-centralised registries operating in various regional centres, so it is necessary to know the full address of the office of the company and then locate the office of the registry in that area. In countries like the UK with a centralised company registry, it is very much easier to track down a company. Official gazettes are very authoritative and excellent sources for private company information, but are published daily and are poorly indexed and not very accessible.

Information produced by these registries and gazettes is rarely automated and information must be sought by basic desk research methods. This means that many automated company information services that attempt to cover many countries have very incomplete data.

Family businesses may never be incorporated, yet may be very large concerns operating simply under the family name with no requirement or need to register anywhere and therefore disclose any public information. In many countries there are many other official forms of non-limited liability registration, such as partnerships, which will have few disclosure requirements apart from registration documentation.

There are significant chemical companies in the private company sector in all the developed world countries.

Information services on companies

Apart from undertaking traditional desk research into companies, most chemical information specialists will today turn to online and other automated information services as their main source of company information.

Most online information services have been very effective at actively soliciting sources of company financial data in the past five years, and the large composite Host systems such as Data-Star, Dialog and Profile have a considerable variety of services covering company information from individual and groups of countries.

In European countries, some of the most detailed company financial information sources for any individual country will be provided by the Host service or services of that country.

Two major providers of 'global' company financial data also act as individual Hosts with direct access for users; Dun & Bradstreet and Kompass. These compiler/hosts are also providing the first reasonably reliable attempts at standardised databases of company information across geographical regions in the Kompass 'EKOL' (online) and 'EKOD' (CD-ROM) and Dun and Bradstreet 'Europa' (diskette or CD-ROM) products.

Credit information on the payment records of companies is another type of information which is widely available on online databases. Researchers should always be extremely wary about collecting this type of financial information masquerading as full financial information on companies.

From the methods of researching company information described above, the sources of information on funding and mergers and acquisition could be used very effectively to highlight possible deals where fuller documents could be obtained. Many of these services are extremely specialist and aimed almost exclusively at the financial end users working in the funding area. There are, however some services available on the composite Hosts which are worth mentioning. The IDD services on Data-Star and Dialog are well worth using for deal information both in Mergers and Acquisitions, and Funding. Profile has a similar service in the 'FT Mergers and Acquisitions' database and the Predicasts databases have good event codes which can be used to identify deals. Standard & Poor's and Disclosure Services on Dialog are also useful for identifying companies who have securities quoted on US Stock Exchanges.

Many of the Hosts have also grouped the company services available into small cross-file groups for easier identification and searching where these more unusual services are often grouped.

The automated services are generally very poor at covering information on state owned and partially privatised companies. The only services where information is covered routinely on these organisations is in the press databases.

The chemical industry is one of the only industrial sectors that boasts an actual business information database devoted to the sector in the 'Chemical Business NewsBase' produced by the Royal Society of Chemistry in the UK. The database concentrates on Europe but covers the business aspects of the chemical industry in the main markets of the USA & Japan.

Researching company information

Industry/professional information

Information available to the general public

Technical information specialists in chemical companies will be well aware of the information sources and services available for collection of company information from the industrial and professional areas. These services are not necessarily indexed or abstracted by company and may be difficult to research purely for company information.

For completeness these will be summarised here but not discussed in any detail.

Company information will be available from:

- Patents and trade marks
- Licences
- Catalogues and manuals
- In-house magazines and journals
- Conference and seminar papers
- Product reviews.

In the industrial and professional field there will also be sources of information from:

- Suppliers to the company
- Customers of the company
- Contracts.

which may be reported publicly in:

- Trade journals
- Newspapers.

but may be much more effectively resourced from in-house information collected in some effective way from company representatives, reports from external meetings and correspondence.

Company information can also be gleaned from sources available to everyone and often forgotten by information professionals in the pursuit of information from ever more sophisticated online services which simply ignore many of these sources.

These can be summarised as those generated by the company:

- Advertising; general and recruitment
- Press releases
- Speeches and papers.

and those generated by others:

- Books, company histories

- Journals; business and popular
- Case studies
- Market research
- Unions & employee groups
- Pressure groups
- Biographical profiles of directors and staff.

SECTION 2

Analysing company information & adding value

As information professionals, the chemical information specialists will be concerned about their position in the organisation and will wish to enhance their service as much as possible to the benefit of users.

In collecting company information in the business environment described in the introduction to this paper, there are several identifiable areas where the information is most likely to be used:

The most usual reasons are:

- Full Mergers and Acquisitions of complete companies
- Competitive intelligence
- Purchase/sale stakes in companies
- Purchase/sale brand name, licence, trademark, patent.

Others:

- Forced liquidation
- To discover synergies between two companies
- Diversification
- Legislation or professional practice
- Find hidden potential in a company.

In each case it is most likely that the company information will be used to arrive at a company valuation which can be defined as a formal assessment of the worth of the company. The valuation of a company is the first stage in the bargaining and negotiation process for the merger, acquisition, or purchase of assets and strategic stakes.

There are considerable differences in valuing a listed (quoted) company from valuing a private company. It is usually very difficult if not impossible to value a private company without some information provided by the company itself.

Information gathering is the most important part of the company valuation process and information specialists should be aware of the most common company valuation methods and the information requirements for each.

Methods of Company Valuation

There are many methods and practices of valuation. They can be very complex incorporating detailed theoretical mathematical formulae. One method of valuation is not usually sufficient, therefore a combination of methods is normally used to establish a fair price.

Main methods of company valuation

- Book value or underlying net asset value
- Market value
- Discounted Cash Flow (DCF)
- Capitalised earnings
- Dividend yield basis.

Users actually undertaking the valuation will need as much information as can be collected as described in Section 1 of this paper plus some additional requirements:

- Formal valuations from comparable companies who have undertaken deals in the past. These valuations can be many years old, especially if they are for a particularly relevant company. Deal databases mentioned in Section 1 are very helpful in pinpointing these.
- If the companies are quoted; share price information, dividend information and relevant share price ratios quoted in summaries of earnings estimates published by services like I\B\E\S (International Brokers Earnings Estimates) or the US Value Line.
- Many years of historical financial statements, particularly for the DCF method.
- Industry and sector indicators and measures.
- General and specific economic and financial indicators.
- Specific details of company management.

Issues concerning the use of databases for researching financial information

Information professionals can also add value to information services by including explanations of the sources. When using online services which are not primary sources of company information as described in Section 1 they should be able to inform users of the issues involved in the data collection. As chemical information specialists may not be familiar with company information suppliers who are not usually same suppliers of technical information, they should investigate these areas of concern particularly carefully.

In using any database to research any information, the information specialist must be sure that the database compiler has prepared the data offered to the highest standards.

There are several criteria that should be considered before a user regularly relies on any database for company information:

- The position of the database and access
- Quality of information supplied
- Quantity of information supplied
- Format
- Updating schedule
- Sources used in compilation
- Cost.

The position of the database and access

Do you know about the operations and services of the database supplier? Is the database being supplied to a Host system(s) or is it being compiled and offered direct to the user by the supplier? Which is the best in terms of the standards of service? Is it a problem in your organisation to have many different contracts with individual databases for occasional use?

In the company information field there are a mixture of both types of database on offer, with many services only being available direct from suppliers. This can be an important management decision for many organisations and may lead to good databases not being chosen because of the difficulty of access.

Quality of information supplied

is of vital importance in accessing company information. Users have to be convinced of the high quality and accuracy of the information available on the database. Compilers should have a statement of the quality controls employed to ensure accuracy in their introduction to the service actually stated online, as well as in their sales literature. It is also essential that the user be quite clear as to what additional calculations/input have been added to database numerical data by the compiler, and what quality standards have been applied.

Quantity of information supplied

Much of the company information supplied on databases is incomplete in some way, generally because the information has not been disclosed in full. Often there is no way the user can know how well covered his search will be and some compilers give no indication of the completeness of their data sets. The main type of information where this is particularly common is for company financial data where company status and the disclosure policies and filing practices of individual countries can dramatically affect the quantities of data on the service. Again, some statement of collection policy should be stated in the database preliminaries and in product literature.

Format

Company information sources can be accessed in a wide variety of automated formats. To take a fair selection of databases would involve a considerable outlay in computer hardware and software, and to make all the services compatible to link with in-house systems would also involve considerable resources in time and cost. This may have to be done with essential sources even though the formats are not ideal.

Updating schedule

Company information obviously has to be very up-to-date to be useful. The primary sources detailed in the methods of researching company information are often very up-to-date for listed company information but woefully out-of-date for private company information. Information suppliers providing databases made of a mix of the two types of company often intimate that the service is up-to-date, neglecting to inform users of the difficulties that have been encountered in really updating the private company component.

Sources used in compilation

Primary sources of financial information used in databases must be acknowledged in the database, not just in the sales literature. It is then easy for the user to go back to the primary source if necessary, to do the appropriate checks. This element of database compilation is becoming more and more important as databases are launched with no equivalent paper publication, (with which the supplier has built up a publishing reputation).

Cost

Company information sources are expensive in comparison to abstracting databases. Users must expect to pay more assuming the additional controls and standards that have to be used to ensure accuracy in the compilation of such databases

Conclusion

This paper has covered a basic summary of researching company information in the business environment of the chemical company. It has been specifically orientated towards the technical information specialist who has little knowledge of the detail of providing in depth company information for increasingly demanding users.

Appendix: contact points for suppliers

Data-Star

RadioSuisse Services, D-S Marketing Ltd, Plaza Suite, 114 Jermyn Street, London SW1Y 6HJ UK. Fax +44 71 930 2581

Dialog Europe

PO Box 188, Oxford OX1 5AX UK. Fax +0865 736354

Dun & Bradstreet Offices in most capital cities

Europa World Year Book

Europa Publications Ltd 18 Bedford Square, London WC1B 3JN. Fax +71 636 1664

International Year Book & Statesmen's Who's Who

Kompass, EKOL & EKOD, Reed Information Services, Ltd., Windsor Court ,East Grinstead House, East Grinstead, West Sussex RH19 1XA UK. Fax 0342 335612

FT Profile

P O Box 12, Sunbury on Thames, Middlesex TW16 7UD UK. Fax +932 781425

Enhanced access to information via personal bibliographic databases

Thomas E. Wolff

Amoco Research Center, Naperville, Illinois, USA

The Information Research and Analysis group provides scientific, technical and business information to customers within Amoco Corporation. In general, this information is retrieved from databases on the major online systems. Since members of our group have considerable and varied experience working in Research and Development within Amoco and elsewhere, we understand our customers' businesses and concerns and are well situated to carry out searches in accordance with their needs. We add value to our services by analysing the search results for relevance and editing the output for clarity and improved accessibility. Each report includes a cover letter describing search strategies used and results obtained. On request, additional technology assessment will be provided. However, our reports are generally collections of database citations with abstracts. The customer can then order the full documents as they deem necessary, doing his or her own assessment of the subject area.

We place considerable emphasis on providing search reports which can be most effectively used by our customers, and we look to add this value at minimal cost. Traditionally, our efforts have focused on printing reports with only the most appropriate amount of bibliographic or indexing information, or on organising reports into useful subject categories, occasionally adding a Table of Contents. But no matter how well the information is presented, reports are almost always used just once and just by the original requestor. It is clear to us that this is not an effective use of the company's resources. Information is too expensive to be used in such a short-term and short-sighted manner. Therefore, we have begun to provide our search results as bibliographic databases using selected personal computer programs. As will be discussed further, these databases provide our customers with a convenient means to evaluate the search results initially, to revisit these reports, and in some cases, to share that information with colleagues.

The development and usage of these bibliographic databases involves numerous process steps of varying complexity. As these are discussed, reference will be made to Figures 1 and 2 at the end of this article, which show patent and journal article citations, respectively. Each figure is divided into: (a) information downloaded from the online system; (b) personal computer 'screen' images of the information in a Library Master database; and (c) the citation information reformatted and output from the Library Master database and automatically converted to a word processor document.

What are personal bibliographic databases?

Two years ago, the moniker Personal Bibliographic Databases seemed perfectly appropriate for the products we were developing. Individuals would request a search and the information was provided as a personal computer database along with a hard copy report if requested. So 'personal' referred to the personal requestor as well as to the PC. The PC was the medium of choice because no appropriate mainframe (IBM or VAX) database programs were available at Amoco and the purchase price for new ones was generally over $10,000, well beyond the means of any individual users. Now we find that research groups are requesting searches for the whole team to use. An individual's PC is too restrictive. We are again looking at mainframe programs, but more likely we will follow the distributed computing path and load the databases on local area networks. At that point, these may well be 'Team Bibliographic Databases' or 'Research Project Bibliographic Databases.'

The designation 'bibliographic' may not be appropriate in the future either. Right now we have difficulty incorporating chemical structures or numerical data, as from Beilstein, in our personal bibliographic databases. In addition, we look forward to adding graphics, either from CD-ROM records or from the online patent databases of the future. Finally, the addition of hypertext capability would further enhance the value of the customers' databases. We anticipate that software packages will improve considerably and require that any data imported into today's databases must be exportable in formats appropriate for tomorrow's improved programs.

The personal bibliographic databases of today are requested by individual researchers to augment or replace 'traditional' searches. Databases are particularly appropriate for broad searches of hundreds or thousands of citations or for periodically updated searches which gradually build up the database. Since the database software is well suited for citation categorisation and sorting, and for putting out reports in word processed formats, formation of the database can be a means to produce a better hardcopy report. In this case, the database may be a side benefit.

Search results from numerous databases and online systems are often combined into a single personal bibliographic database. Output from ORBIT, QUESTEL and STN is well-suited for importing because of the field-delimited formats with unique field tags and well-defined text formats. Information from Dialog Information Services may also be used, either in field-delimited 'tagged' or in native, untagged format. However, problems arise in either case. Tagged format is preferred, although one must deal with non-unique field designators and inconsistent formatting. Untagged information is probably best suited for importing into structureless personal databases, which are not further considered here, although it may be imported into structured databases using the special untagged-format Biblio-Links import/conversion program used with Pro-Cite bibliographic software.

Personal databases will soon be drawing upon other sources of information as well. Already, diskettes and CD-ROM are replacing many hardcopy sources, such as the Science Citation Index from the Institute for Scientific Information, which is available on CD-ROM and includes abstracts.[1] The CABI (Commonwealth Agricultural Bureaux International) now provides 47 printed abstract journals on diskette.[2] Many 'full-text' journals are available online, such as the extensive Chemical Journals Online (CJO) on STN. These online journals suffer from lack of tabular and image information, which both the online services and personal bibliographic database software will have to address. Eventually, many journals will be published exclusively online. While this may seem unlikely, The Online Journal of Current Clinical Trials, a peer-reviewed journal, is the first of a series of anticipated publications from Primary Journals Online. Although many issues must be dealt with, including fees, copyright and distribution, a level of acceptance has been reached with BIOSIS now indexing articles from the Online Journal of Current Clinical Trials.[3]

The value of personal bibliographic databases

The use of personal databases may be characterised by the following: convenience, communication, idea generation and cost.

Convenience

Search results are in an accessible format with many of the search techniques available online and with enhanced sort and output capability. Information has extended usable lifetime, well beyond the 'read-once-and-file-away' lives of most hardcopy reports. Multiple searches may be combined into larger databases, or subsets may be conveniently separated into narrow-focus databases. The value of individual records may be enhanced by annotation, cross-referencing or combination of related records.

Communication

Databases can be used to create custom bibliographies and topical reports. Shared databases can be annotated by individual users to communicate key features of the cited references.

Idea generation

Convenient searching and browsing through information should lead to generation of new ideas. One's perspectives change as projects progress, so each time questions are asked and the data evaluated, there are new opportunities to learn and generate new search questions. When information in a personal database is found to be incomplete, or when relevant information cannot be found because of limitations in the database indexing or search software, new queries are made in the source online databases, which may then lead to an expanded personal database.

Cost

Some of the value in maintaining information in an accessible personal database may lead to actual cost savings. Certainly, many 'trivial' questions could then be answered on one's own PC. In addition, some queries of the source database may become 'unnecessary' because the answers have already been obtained in prior searches. On the other hand, the more 'information-aware' researchers should continue to generate more search questions as their knowledge of a field increases.

The net search costs may decrease or increase, but the savings in doing better research should be substantial when search information is well used.

Database packages

Many database software packages appropriate for developing personal bibliographic databases are available and have been reviewed [4-6]. Our experience is with IBM-compatible personal computers, for which we have evaluated four programs in detail:[7] Library Master; Notebook II; Papyrus; and Pro-Cite with Biblio-Links. Two of these have recently been upgraded, Library Master to a local area network version with improved functionality, and Pro-Cite to a more efficient version. Lately, we have also considered three other programs: EndNote, recently translated from the Macintosh version[8]; ideaList[9]; and STN Personal File System, essentially STN Messenger language for the PC with very efficient importing for downloaded STN records. On the basis of our evaluations and our customer acceptance, we have settled on Library Master and Pro-Cite as our recommended software. However, as stated earlier, we will be continually looking for improved ways to bring this information to our customers. For example, our department is evaluating document management software, such as Verity or Excalibur, for managing Amoco's internal documents. Many of these have sophisticated search capabilities. When Amoco moves from the mainframe environment and adopts one of these programs, we may find that they will also be well suited to our bibliographic databases.

Principal concerns regarding personal database development

User-definable Record Formats

The patent user community is not a significant constituent of the bibliographic software market. This is most clearly shown by the patent record type available with the software. Some programs list patent types as standard but the format is so limited as to be almost worthless. For example, Papyrus has a patent record type with only nine simple fields, and it is impossible to search or sort on assignee. In Pro-Cite, the patent record format, which is available as a supplemental workform, is acceptable in its simplicity and, even better, modifiable through standard Pro-Cite procedures. However, multiple paragraph abstracts, multiple entry fields (e.g., abstracts from multiple equivalents), and lists of applications or patent family equivalents cannot be easily imported or handled properly in Pro-Cite. These specially formatted fields are all 'word-wrapped' into single, continuous, difficult-to-read paragraph fields. Other database program producers have not considered patents at all. However, for some, the program's flexibility and power can enable creation of a useful patent record type.

The first two screens of a patent record as created by Library Master are shown in Figure 1b. These 'data input forms,' as the screens for editing and browsing are called, are fully user-customisable. All bibliographic information fields are shown on the first screen including the first eight lines of the abstracts. Multiple-element list fields, such as author, equivalent patent and applications, are fully viewable by moving the cursor to them, but only the first two lines are shown initially. The remainder of the abstract is also available through the expand-field option. The second and subsequent screens in our Library Master patent records include a comment field, indexing, other abstracts, and patent claims. The first screen of a journal article record in Library Master (Figure 2b) shows the similarity we have maintained between the layouts of patent and article screens. For example, in each record type the first screen is headed by the title, followed by bibliographic information and then the abstract.

The patent record screen also shows new fields generated by parsing fields in the imported data or by modification of the downloaded information prior to importation. One useful conversion is the separation of patent number from its date, required since most patent files combine them in one information field. Another is the creation of the earliest priority number and date fields, information that is usually buried in other application fields in online records. These separate, searchable earliest priority application fields are valuable to simplify the identification of patents from the same family. We make substantial use of KEDIT macros to reorganise downloaded records, to make the information both more accessible in the personal bibliographic database and more consistent from source to source. For example, the information to be imported is reorganised to take advantage of the Library Master feature which allows any field to be designated as a date-format field. These fields can also have searchable text following the date. The application and patent number fields have dates first, which allows both dates and numbers to be searched appropriately.

Importing

As in most computer software applications, a balance must be struck between program flexibility and ease-of-use. The broad variety of information sources makes critical the program's flexibility in importing process. At the same time, this process should be straightforward and efficient, since users generally wish to pass quickly through the importing stage and on to using the information. Software producers have met the importing challenge various ways. Most create separate importing modules or programs. For our purposes, EndLink, the importing program for EndNote, errs on the side of simplicity because it is a 'black box,' as described by a sales representative, which allows for no alteration of the downloaded information. At the other end of the scale, Papyrus requires development of a complicated import template which must account for all possible variations in the imported information. Although the Papyrus producer will create customised importing programs for the users, we have found them to be generally ineffective, as for importing Chemical Abstracts information from STN, for example. Developing import templates can be much more straightforward. The Convert program with Notebook II is extremely simple to customise, and the import facility for Library Master is nearly as clear-cut. The balance is probably best struck by Pro-Cite, which provides separate Biblio-Links for each online systems and many CD-ROM products. The Biblio-Links can also be modified readily. However, until recently, the pricing policy for the Biblio-Links made them considerably more expensive than the database software Pro-Cite itself; with the newly created package sets, the import and database software are more nearly matched in cost.

Responsibility for importing difficulties also lies with the database producers. Personal computer importing programs must be powerful because online information is so complex and inconsistent. Some databases have dozens of document types. For example, at last count, Compendex has 26 document types on STN. Of more concern, fields frequently contain more than one type of information, necessitating the use of sophisticated field parsing or editor programs. For example, in APILIT, the source field has almost no consistent format at all and can contain CODEN or ISSN numbers or references to other sources such as Petroleum Abstracts. Similarly, the source field in CA file article citations contain all the bibliographic information except year of publication (see Figure 2a). CA file patent record source (SO) field does not contain the application or patent numbers or issue date, as might be expected, but rather the 'country' information, number of pages, and, for Polish patents only, a phrase which describes that the abstract was taken from the application. Rather, the actual 'patent information' is found in the PI field.

Online database information can also be presented badly. For example, the tagged formats for Dialog Information Services appear to be just an afterthought created out of the 'normal' format. Dialog field tags frequently are not unique, and they may contain subfield descriptors in angle brackets at the beginning of the field's text area (see Figure 1a of a Derwent World Patent Index record). Dialog also uses the virgule or vertical bar to designate end-of-field, but this mark is not applied consistently throughout records or between files. Our KEDIT macros make major changes to Dialog tagged output.[7] In addition, we have recently written macros which convert untagged Derwent output (e.g., Dialog format 7) to the equivalent tagged format. This allows us to use Derwent citations for importing, even if we had not planned on it when the search was originally downloaded . The whole issue of having to remember to use tagged format does not exist when downloading information from ORBIT, Questel, or STN. These online systems only produce tagged output, which is generally more consistent than that from Dialog.

The customer-oriented approach of the CABI seems particularly enlightened.[2] Subscribers to CAB Abstract Journals on diskette requested search and retrieval software. Other database providers have responded by including specialised retrieval software with their data, ISI (Science Citation Index on diskette) or Ziff-Davis (Computer Library on CD-ROM). However, CABI recognised that subscribers had their own preferred database software which could be used for information from varied sources. To accommodate this need of the majority of their subscribers, CABI publishes their abstracts journals in both comma-delimited and Pro-Cite proprietary formats and provides information on potentially useful database software. The lesson to be drawn is that information should be formatted consistently and thoughtfully in consideration of importation into the personal databases of the customer's choice.

Browsing, searching and general ease of use

The forecasted demise of hardcopy information is often regretted because of the value and pleasure in browsing and the serendipity in finding unsought-for information of interest. This ability to browse information conveniently is critical to idea generation but need not necessarily be lost with computer databases. But software designers must remove barriers to information access. For example, consider the situation in which a search has been carried out in a personal database and the user wishes to browse the results. One of three approaches is taken by software producers. The first is to provide just enough information to enable a judgment about whether the 'hit' citation should be further reviewed. This information might be a page of author and title information. The problem with this implementation is that it generally takes two or more keystrokes to move from abstract to abstract, whether the user looks at all retrieved records or selects on the basis of the limited scan information. The second approach is to have the user move from citation to citation. However, if the screens are not well designed, one may again need multiple keystrokes to go from abstract to abstract. A third approach would be browsing actual occurrences of hit terms wherever they occur in the 'hit' record. This approach is available in some document management software (e.g., Magellan or Verity) but not any on the personal database software evaluated by us.

The ability to browse in a database generally seems to parallel other functions of the database software. That is, if it is straightforward to get around the data, it is often easy to get around the software. Most of the programs we have evaluated have menu systems and short-cut keystrokes to skip through, but not avoid, the menus. With few exceptions, our customers would prefer good menu-driven programs to command-line programs. However, with experience, menus can become tedious. The ease of database searching is another tell-tale area. Boolean operators are generally available, but each software program has its own rules about operator order, use of parenthesis, and other syntax, e.g., the use of 'not' vs. 'and not' vs. the caret symbol (^). Similarly, truncation operators are generally straightforward, although the specific characters used vary considerably. Proximity operators are generally not available. Also of concern is the layout of the search information screen, the format for entering field codes (e.g., from a menu of useful abbreviations vs. unrecognisable field code numbers) and the convenience of accessible lists of database index terms for searching. Other database functions, such as editing, sorting and set handling, frequently follow conventions already used in the browse and search capabilities.

Report generation

Database software should integrate well with other personal computer operations, especially for reports generated for importing into word-processing programs. Most personal bibliographic database software is designed to produce bibliographies for incorporation into published works, and most leading word processing programs have been accommodated. In some cases, the database program can also scan an article in word-processor format, find reference indicators and create properly ordered and formatted reference lists. The database programs may have options to conform with dozens of bibliography syntaxes required by journals, which relieves the author or secretary from much tedious footnote reformatting.

For our work as industrial scientists and engineers, we need to create subject bibliographies or reports based on online search results. The output syntax is seldom specified by the recipient, who only needs the information complete and in easily-readable format. Therefore, we have created custom output formats which resemble the 'CBIB AB' display format of the CA file on STN (see Figure 1c and 2c). Our output formats always have the title first, in bold-face, followed by pertinent bibliographic information, some of which is italicised for ready identification, then lists of equivalent patents when available, and the abstract(s). When the basic patent is non-US but a US patent abstract is available, both the basic and US-equivalent abstracts will be printed. Index terms are rarely printed, except when the 'hit' terms, such as Registry numbers, would be useful to the recipient. This bibliographic information is usually output in word-processor format for generation of readable hardcopy reports. However, for some of our reports which are viewed or transferred electronically, ASCII text is preferred.

Cost

One of the first concerns of the potential customer is about the cost of the personal bibliographic software program. After all, they think, if the software is expensive, there is no point in considering further the development of a personal database. Even after hearing that the database program cost is a very small fraction of the total cost of generating the database itself, customers still want to purchase inexpensive programs to keep their initial investment down. Fortunately, the effective personal computer database software is inexpensive, costing at most a few hundred U.S. dollars. The STN Personal File System is one of the more expensive personal computer programs at $600, which is enough for most of our customers to look elsewhere. But this price is still very reasonable compared to mainframe programs, which may cost $10,000 to $20,000 and more depending on the computer platform. The exception is Papyrus, whose VAX-based version is equivalent in cost ($99) and functionality to the personal computer edition, at least up to version 6.0. This low cost has attracted several of our customers, but they have never been satisfied with the program for reasons described previously.[7] The high cost mainframe (VAX and IBM) programs are completely out of the reach of our employees, until the usage of personal databases reaches the critical number required to justify the software purchase. By this time our use of the PC-based programs will probably be entrenched.

As already suggested, the cost of developing personal bibliographic databases involves far more than the software purchase. The principal costs are: downloading information, or typing it in manually; editing or reformatting the downloaded information; importing the information into the personal database; and copyright or licensing fees. The most expensive portion is for information downloading, which includes search strategy development, online time and citation print charges. The cost per record ranges generally between two and five dollars, depending on database and online service. Citation reformatting and importing are carried out as automatically as possible using editor macro programs and standardised import routines.[10] The incremental cost for reformatting and importing is kept to a minimum, but can still run at one to two dollars per record. Finally, licensing fees are frequently required for storing the information electronically, as will be discussed later. These fees can be another one dollar per record per year. Thus, the total cost of personal bibliographic database development seems to average about $6 per record, or thousands of dollars for hundreds of records. As stated earlier, the cost for the personal database software itself is a hardly significant fraction of the total personal database development cost.

Local area network (LAN) access

Convenient sharing of information among group or team members will be a critical requirement of bibliographic databases in the future. Our customers are thinking about this cost-effective approach to information usage, but both the hardware and software are holding us back. Many of our employees are still working on mainframes, and the conversion to personal computers is taking place slowly. Those fortunate individuals with personal computers are not yet connected to their own local area networks. We are considering loading our customers' bibliographic databases on our LAN, at least until their own networks are established. For this, we would use Library Master, whose LAN version is undergoing extensive beta-testing prior to release in late summer. Multiple users have full, simultaneous access to any database, except that no two people can have simultaneous write access to the same record in a database. We have been very impressed with the implementation of the LAN version of Library Master. Since the licensing cost per user is comparable to the cost of individual copies, we anticipate having our whole group using the LAN version. This will facilitate purchase and installation of database software upgrades.

Non-textual information

Bibliographic database software will need to accommodate more than simple ASCII textual information, as our sources become more complicated. Even importing chemical structural information in text mode (as opposed to the graphics mode available with 'type 3' terminals) from the Registry file on STN cannot generally be imported without 'word-wrapping' the characters into meaningless strings. Other text information frequently handled improperly include lists, such as patent equivalents, tables as output with the Dialog report format, or numerical data. The software packages that we currently use have no means of incorporating images in binary formats. As images become more prevalent on CD-ROMs or online, e.g. Registry file structures or the online Derwent abstracts of the future, we will probably want some hypertext pointers in our databases.

Other significant issues

Comprehensiveness

Personal bibliographic databases are not comprehensive. This is an important principal for all to understand. Search professionals often search multiple sources to increase the likelihood that "all relevant" information is retrieved. However, cost-benefit considerations usually prevent searching every conceivable source. Even if the imported information is 'complete,' the personal bibliographic database must be continually updated as the source files are. Furthermore, all search techniques available online are not available on personal computers, e.g. proximity operators or online database coding. So, 'relevant' citations in the personal database may be missed on subsequent searching. Personal databases are valuable information repositories which may reduce, but can never replace, searching the source literature and online files.

Copyright, licensing agreements, and cost

Most of the information we use in our personal bibliographic databases is copyrighted, and appropriate licensing agreements have been established with the information producers. It is easy to empathise with the headline that copyright and licensing agreements are 'Making Life Complicated.'[11] This results from the conflicting views of information users and producers. The former tend to think of information as free, while the latter require fair compensation for creating and assembling the information. We search intermediaries are caught in the middle, with particularly strong influence from corporate policy requiring full compliance with the law. However, the latest major revision of American copyright law, the Copyright Act of 1976, "was not designed with information brokers [or] rapid computer transmission of information . . . in mind."[12] Personal computers were not even invented then, so what do we do about sharing electronic information on local area networks? The answer seems to be that we work out individual licensing agreements with every information provider. Even having source files on the same online system is no benefit, because the online systems are just conduits of the information and the licensing policies. In the near term, the best goal to aim for might be having similarly structured licensing policies which would differ in cost but not billing structure.

The factors influencing database storage licensing costs should be: (1) lost revenue from reduced online searching or journal subscriptions, and (2) recovering fees for multiple distribution of information. Information producers also suggest that information users should pay for the convenience of electronic storage of information, although I believe this is a red herring. One should not be penalised for electronic storage for information when hard copy storage is free. However, if electronic storage means reduced revenue for the producer, then the producer may be entitled to make up this revenue via licensing fees. Similarly, fees are paid for photocopying hard copy materials via the Copyright Clearance Center in the United States, but there is no parallel organisation or payment mechanism for copying electronic information. Note that the difficulty of monitoring what documents are photocopied is magnified when translated to monitoring electronic access of shared database material or electronic distribution beyond the initial group of recipients. For example, very few database programs have any sort of access monitoring or copy-protection schemes. Some producers have suggested that any fees be paid on the number of potential information users, i.e., all personnel connected to a local area network, rather than on 'actual' users. In one such case, a licensing fee algorithm for 5000 citations available to 8-10 people calculated a fee of about $15,000 per year. After we got this proposal we found that there might be 50 interested users. Even the information provider realised that the algorithm broke down as the whole database could be licensed for $25,000 per year.

The most manageable licensing fee schedule would be based on the amount of information, i.e., the number of database records, rather than on the number of users with either potential or actual access. The latter requirement is unworkable and untenable with today's software. Furthermore, we already spend considerable time just complying with the simpler licensing fees based on number of records. We would prefer agreements which accept annual payment based on actual records in personal databases at the beginning of the agreement year. Additional records added during the course of the year would be exempt from payment until the next year. Some of our current licensing agreements have this structure.

To ensure that proper copyright credit is given in personal bibliographic database records, database producers and online systems should have the copyright statement as an integral part of each record. Copyright information is frequently found in the header information when the database is first entered or as a separate line or field before or within each record. This information is seldom incorporated into the personal database or subsequent output. We use KEDIT macros to append copyright statements to accession numbers, which are always printed as part of the customised record format (see Figures 1c and 2c).

Conclusions

- Our customers will continue to look for cost-effective ways of handling information, and we need to be at the forefront of this issue.
- Personal bibliographic databases will gain in acceptance as long as the cost of developing and maintaining them is reasonable compared to producing hard-copy searches.
- Customers are willing to pay a 'fair' price for access to information in their personal databases. Producers will have to work with their customers to determine this fair price, or they will lose this portion of their market.
- Electronic storage licensing fees should be based on the amount of information and not require an exact counting of users of that information. Fee scheduling over broad numerical ranges of users would be tenable.
- Straight bibliographic database software will satisfy most of our customers' needs in the near term. However, as the computerised information environment evolves, so too will the database programs of choice. Therefore, it is critical that we be able to export data from the currently-preferred software if that software does not continue to meet customer needs.
- Database producers and online systems must appreciate that searchers are using advanced computer technology to obtain and process downloaded information. The format of that data should be improved to allow everyone to readily take advantage of the technology. Specifically, database fields should be rationalised and simplified so each contains only one data element. Format consistency should be a hallmark of the database.
- Personal software developers need to continue to improve the process of browsing information so that users will feel as comfortable just 'cruising'

through the database as they would a hardcopy reference. Similarly, searching should be improved, at minimum by avoiding symbols where words should work, by adding proximity searching and possibly by including non-Boolean search techniques such as frequency searching or relevance ranking, or hypertext, techniques often found in document management programs.

References

[1] Garfield, Eugene. 'New Chemistry Citation Index on CD-ROM Comes with Abstracts, Related Records, and KeyWords Plus.' *Current Contents,* **No. 3** (January 20, 1992): pp. 5-9.

[2] Powell, Andrea. 'Making the Most of an Electronic Journal.' C.A.B. *International Database News*, **No. 10** (Nov. 1991): p. 2.

[3] 'BIOSIS to Index Articles from The Online Journal of Current Clinical Trials.' *Database* **15**, *No. 3* (June 1992): p. 12.

[4] Lundeen, Gerald W. 'Software for Managing Personal Files.' *Database* **13**, *No. 3* (June 1989): pp. 36-48.

[5] Lundeen, Gerald W. 'Bibliographic Software Update.' *Database* **14**, *No. 6* (Dec. 1991): pp. 57-67.

[6] Stigleman, Sue. 'Bibliography Formatting Software: A Buying Guide.' *Database* **15**, *No. 1* (Feb. 1992): pp. 15-27.

[7] Wolff, Thomas E. 'Personal Bibliographic Databases: An Industrial Scientist's Perspective.' *Database* **15**, *No. 2* (April 1992): pp. 34-40.

[8] Mead, Thomas. 'Making the Link: Importing Downloaded Bibliographic References to Pro-Cite and EndNote on the Macintosh.' *Database* **14**, *No. 1* (Feb. 1991): pp 35-41.

[9] Stigleman, Sue. 'ideaList: Flexible Text Storage and Retrieval Software.' *Database* **15**, *No. 3* (June 1992): pp. 50-56.

[10] Wolff, Thomas E. 'KEDIT: Text Editor for Post-processing Searches.' *Database* **15**, *No. 3* (June 1992): pp. 43-49.

[11] Collier, Harry (Ed.). 'Making Life Complicated.' *Library Monitor* **33** (July 1991): pp. 1-3.

[12] Berring, Robert C. 'Copyright: Online Abstracts Raise New Issue.' *Database* **11**, *No. 3* (June 1988): pp. 6-9.

Information on packages

Library Master version 1.3: Balboa Software, 61 Lorraine Drive, Willowdale, Ontario M2N 2E3 Canada (416/730-8980; e-mail address hahne@epas.utoronto.ca), $199.95, $599 for a 5-node Local Area Network version, $5.00 for a demo version.

Pro-Cite version 2.0 for IBM-compatible PC or Macintosh with Biblio-Links: Personal Bibliographic Software, Inc., P.O. Box 4250, Ann Arbor, MI 48106 (313/996-1580), Pro-Cite $395, Biblio-Links at various prices.

KEDIT version 5.0: Mansfield Software Group, Inc., P.O. Box 532, Storrs, CT 06268 (203/429-8402), $150 (DOS only), $175 (DOS and OS/2), free demo version. Technical support on MSG bulletin board at 203/429-3784; CompuServe (GO PCVEN); BIX (JOIN MANSFIELD). KEDIT macros, including many alluded to in this paper, may be found in the PCVEN library on CompuServe and on the MSG bulletin board.

43/4/24
AX- 91-193135/26I
AX- <XRAM> C91-083597I
TI- Prodn. of aromatic anhydride(s) and ester(s) with superior colour — by treatment of crude prod. with activated boric acid and fractionation, useful in polymer prodn.I
PA- (STAD)_AMOCO CORPI
AU- <INVENTORS> PARK C M; COATES R; HOLZHAUER J K; PETERSON J VI
NP- 002I
PN- <BASIC> WO 9108204_A_910613_9126I
PN- <EQUIVALENTS> EP 455802 _A_911113_9146I
AN- <PRIORITIES> US 606603 (901031; US 443564 (891129)I
AN- <APPLICATIONS> WO 9OUS6938 (901128); EP 90901071 (901128)I
LA- EnglishI
CT- US 2971011; DE 1948374; US 3888921; US 4794195; 3 Jnl.REFI
DS- <NATIONAL> JPI
DS- <REGIONAL> AT; BE; CH; DE; DK; ES; FR; GB; GR; IT; LU; NL; SEI
AB- <BASIC> WO 9108204
Prodn. of aromatic anhydrides and esters with improved colour comprises treatment with activated boric acid followed by fractionation at 200—275 deg C and 25–1 mmHg. The boric acid is activated by heating with an organic hydrocarbon acid or anhydride. Specifically claimed is treatment of trimellitic anhydride (TMA) and dimethyl-2,6-naphthalene dicarboxylate.
USE/ADVANTAGE — TMA is used an an intermediate in the prodn. of quality plasticisers and ployester resins. Dimethyl-2,6-naphthalene dicarboxylate is a monomer used in the prepn. of high performance polyesers, esp. poly(ethylene-2,6-naphthalene) (PEN) which is used in 'hot- filled' food and beverage containers, tyre cord and magnetic recording tape. The delta E colour of TMA is improved from, e.g. 2.69–0.44 in a process which does not require expensive recovery and regeneration of dehydration agents. @22pp Dwg.No.0.0)@I
FS- CPII
DC- A41; E13; E14; I
IC- C07C-051/42; C07C-067/48; C07C-069/76; C07D-307/77I
MC- A01-E11; A08-P03; E06-A02A; E10-G02AI
DR- 1894-U; 1924-UI

Fig. 1a: Derwent World Patent Index (Dialog) citation as downloaded in tagged format

```
BROWSE      DB: T:AMOCO      TOT: 1571    REC: 20       SC: 1
                         RECORD TYPE: PATENT
TITLE:        Prodn. of aromatic anhydride(s) and ester(s) with superior
              colour - by treatment of crude prod. with activated boric acid
              and fractionation, useful in polymer prodn.
AUTHORS:      PARK, C M              | ASSIGNEE: |(STAD ) AMOCO CORP
              COATES, R              |
PATENT NUM:   WO 9108204 A              | DATE: |13 Jun 1991      |
APPL NUMBER:  28 Nov 1990  EP 90901071 |
APPL PRIORIT  US 443564               DATE PRIORIT: |29 Nov 1989      |
EQUIV PATENT  13 Nov 1991  EP  455802 A  ALL PRIORITY: |29 Nov 1989  US   443564
              31 Mar 1992  US 5101050 A               |31 Oct 1990  US   606603
ACCESS. NUM:  91-193135/26 - Copyright | LANGUAGE: |English         |
USE:          3  | TYPE: |9202 92-| LOCATION: |
                              ABSTRACT:
BASIC:  WO 9108204
    Prodn. of aromatic anhydrides and esters with improved colour comprises
treatment with activated boric acid followed by fractionation at 200-275 deg
C and 25-1 mmHg. The boric acid is activated by heating with an organic
hydrocarbon acid or anhydride.
    Specifically claimed is treatment of trimellitic anhydride (TMA) and
dimethyl-2,6-naphthalene dicarboxylate.
    USE/ADVANTAGE - TMA is used as an intermediate in the prodn. of quality
1HELP 2PRVIEW 3FLD CONT 4DUMP 5PRINT 6BRIEF 7ED 9EXPAND PGDNNXT PGUPPRV ESCQUIT
```

Fig. 1b: Derwent WPI citation as a Library Master database record (Screen 1)

```
BROWSE    DB: T:AMOCO    TOT: 1571    REC: 20    SC: 2
SECTION:
                         COMMENTS:
Comments by G. E. Kuhlmann (from February 1992):
A very interesting Amoco US patent concerning the use of activated boric
acid to improve the color of 2,6-NDC during distillation.  Good color
properties for Amoco 2,6-NDC are necessary to pass customer heat stability
tests.
                         INDEX TERMS:
C07C-051/42
C07C-067/48
C07C-069/76
C07D-307/77
A01-E11
A08-P03
E06-A02A
E10-G02A
1363-P
1894-U
1894-U
1HELP 2PRVIEW 3FLD CONT 4DUMP 5PRINT 6BRIEF 7ED 9EXPAND PGDNNXT PGUPPRV ESCQUIT
```

Fig. 1b: Derwent WPI citation as a Library Master database record (Screen 2)

Prodn. of aromatic anhydride(s) and ester(s) with superior colour — by treatment of crude prod. with activated boric acid and factionation, useful in polymer prodn.

Park, C.M.; Coates, R.; Holzhauer, J.K.; Peterson, J.V. ((STAD) AMOCO CORP). *WO 9108204 A*, 13 Jun 1991 (Appl. US 443564, 29 Nov 1989) (English) [91–193135/26 — Copyright (C) 1992 Derwent Publications Ltd. — For Internal Use Only].

BASIC: WO 9108204

Prodn. of aromatic anhydrides and esters with improved colour comprises treatment with activated boric acid and followed by fractionation at 200–275 deg C and 25–1 mmHg. The boric acid is activated by heating with an organic hydrocarbon acid or anhydride.

Specifically claimed is treatment of trimellitic anhydride (TMA) and dimethyl–2,6–naphthalene dicarboxylate.

USE/ADVANTAGE — TMA is used as an intermediate in the prodn. of quality plasticisers and polyester resins. Dimethyl–2,6–naphthalene dicarboxylate is a monomer used in the prepn. of high performance polyesers, esp. poly(ethylene–2,6–naphthalene) (PEN) which is used in 'hot- filled' food and beverage containers, tyre cord and magnetic recording tape. The delta E colour of TMA is improved from, e.g. 2.69–0. 44 in a process which does not require expensive recovery and regeneration of dehydration agents.

@(22pp Dwg.No.0.0)@

Comments by G.E. Kuhlmann:

A very interesting Amoco US patent concerning the use of activated boric acid to improve the colour of 2,6–NDC during distillation. Good colour properties for Amoco 2,6–NDC are necessary to pass customer heat stability tests.

Fig. 1c: Derwent WPI citation as output by Library Master in Word Perfect 5.1 format

```
L35 ANSWER 31 OF 45 COPYRIGHT 1992 ACS
AN CA116(12):108732t
TI Deactivation mechanisms in liquid phase oxidations caused by carboxylic acids
AU Partenheimer, W.; Kaduk, J. A.
CS Amoco Chem. Co.
LO Naperville, IL 60566, USA
SO Stud. Surf. Sci. Catal., 66(Dioxygen Act. Homogeneous Catal. Oxid.), 613-21
SC 45-4 (Industrial Organic Chemicals, Leather, Fats, and Waxes)
SX 25
DT J
CO SSCTDM
IS 0167-2991
PY 1991
LA Eng
AB The addn. of selected arom. acids to the Co/Mn/Br-catalyzed, homogeneous liq.-phase autoxidn. of 1,2,4-pseudocumene in aq. HOAc
   reveals two different types of catalyst deactivation. One type occurs without catalyst metals pptn. and it is suggested that this is caused by
   Co(III) decarboxylation of the arom. acids. The second type is caused by the decrease of the catalyst concn. due to the pptn. of the metals
   as their arom. acid compds. The x-ray crystal structures of Co and Mn pyromellitate suggest that the driving force for the pptn. may be H
   bond formation.
KW pseudocumene oxidn catalyst deactivation; arom acid deactivation oxidn catalyst
IT Hydrogen bond
(in manganese pyromellitate, autoxidn. catalyst deactivation in relation to)
IT Oxidation, aut-
     (liq.-phase, of pseudocumene, in presence of cobalt acetate-manganese acetate-sodium bromide, effect of arom. acids on)
IT Crystal structure
     (of manganese pyromellitate, deactivation of pseudocumene oxidn. catalysts in relation to)
     ... ...
IT 7647-15-6, Sodium bromide, uses
     (catalysts contg., for oxidn. of pseudocumene, deactivation of, mechanism of)
IT 71-48-7, Cobalt diacetate 638-38-0, Manganese diacetate (catalysts, for oxidn. of pseudocumene, deactivation of, mechanism of)
IT 56004-36-5
     (crystal structure of, deactivation of pseudocumene oxidn. catalysts in relation to)
```

Fig. 2a: Chemical Abstracts file (STN) journal citation downloaded

```
BROWSE   DB: T:AMOCO     TOT: 1550   REC: 14     SC: 1
                    RECORD TYPE: JOURNAL ARTICLE
TITLE:          Deactivation mechanisms in liquid phase oxidations caused by
                carboxylic acids

AUTHORS:        Partenheimer, W.            | ASSIGNEE: |Amoco Chem. Co.
                Kaduk, J. A.                |           |
TRANS TITLE:
PERIODICAL:     Stud. Surf. Sci. Catal., 66(Dioxygen Act. Homogeneous Catal.
                Oxid.), 613-21
VOLUME:              | ISSUE: |        | DATE: |1991              |
PAGES:                      | ABBREVIATION: |
ACCESS. NUM:    CA116(12):108732t - Copyr| LANGUAGE: |English      |
USE:            11 | TYPE: |92-0574 | LOCATION: |Naperville, IL 60566, USA
                          ABSTRACT:
The addn. of selected arom. acids to the Co/Mn/Br-catalyzed, homogeneous liq.
-phase autoxidn. of 1,2,4-pseudocumene in aq. HOAc reveals two different
types of catalyst deactivation. One type occurs without catalyst metals
pptn. and it is suggested that this is caused by Co(III) decarboxylation of
the arom. acids. The second type is caused by the decrease of the catalyst
concn. due to the pptn. of the metals as their arom. acid compds. The x-ray
crystal structures of Co and Mn pyromellitate suggest that the driving force
for the pptn. may be H bond formation.
1HELP 2PRVIEW 3FLD CONT 4DUMP 5PRINT 6BRIEF 7ED 9EXPAND PGDNNXT PGUPPRV ESCQUIT
```

Fig. 2b: Chemical Abstracts file (STN) journal citation as a Library Master database recorrd (Screen 1)

```
BROWSE   DB: T:AMOCO   TOT: 1571   REC: 14   SC: 2
SECTION:   45-4 (Industrial Organic Chemicals, Leather, Fats, and Waxes)
------------------------------ COMMENTS: ------------------------------



------------------------------ INDEX TERMS: ------------------------------
pseudocumene oxidn catalyst deactivation
arom acid deactivation oxidn catalyst
Hydrogen bond (in manganese pyromellitate, autoxidn. catalyst deactivation in
Oxidation, aut- (liq.-phase, of pseudocumene, in presence of cobalt acetate-ma
Decarboxylation (of benzenecarboxlyic acids formed in autoxidn. of pseudocumen
Crystal structure (of manganese pyromellitate, deactivation of pseudocumene ox
Benzenecarboxylic acids (pseudocumene oxidn. in presence of, catalyst deactiva
Oxidation catalysts (aut-, cobalt acetate-manganese acetate-sodium bromide, fo
7647-15-6, Sodium bromide, uses (catalysts contg., for oxidn. of pseudocumene,
71-48-7, Cobalt diacetate   638-38-0, Manganese diacetate (catalysts, for oxid
56004-36-5 (crystal structure of, deactivation of pseudocumene oxidn. catalyst
1HELP 2PRVIEW 3FLD CONT 4DUMP 5PRINT 6BRIEF 7ED 9EXPAND PGDNNXT PGUPPRV ESCQUIT
```

Fig. 2b: Chemical Abstracts file (STN) journal citation as a Library Master database record (Screen 2)

Deactivation mechanisms in liquid phase oxidations caused by carboxylic acids

Partenheimer, W.; Kaduk, J.A. (Amoco Chem. Co., Naperville, IL 60566, USA) *Stud. Surf. Sci. Catal., 66(Dioxygen Act. Homogeneous Catal. Oxid.) 613–21* (1991) (English) [CA116(12):108732t—Copyright (C) 1992 ACS].

The addn. of selected arom. acids to the Co/Mn/Br-catalyzed, homogeneous liq.-phase autoxidn. of 1,2,4–pseudocumene in aq. HOAc reveals two different types of catalyst deactivation. One type occurs without catalyst metals pptn. and it is suggested that this is caused by Co(III) decarboxylation of the arom. acids. The second type is caused by the decrease of the catalyst concn. due to the pptn. of the metals as their arom. acid compds. The x-ray crystal structures of Co and Mn pyromellitate suggest that the driving force for the pptn. may be H bond formation.

Fig. 2c: Chemical Abstracts file (STN) journal citation as output by Library Master in Word Perfect 5.1 format

Parallel distributed computing for molecular dynamics: simulation of large heterogeneous systems on a systolic ring of transputers [1]

H. Heller[2] and K. Schulten

Beckman Institute and Department of Physics, University of Illinois, 405 N. Mathews Avenue, Urbana, IL 61801 U.S.A.

Abstract

For the purpose of molecular dynamics simulations we have built a parallel computer based on a systolic double ring architecture, with Transputers as computational units, programmed in **occam II**. *The design is very compact and achieves high reliability even on continuous duty cycles of several months. A parity check logic ensures the correctness of data stored in the external RAM. Our program, EGO, designed for simulation of very large molecules (up to 50,000 atoms), is compatible with the programs CHARMm and X-PLOR. EGO increases its computational throughput nearly linearly with the number of computational nodes. Results of simulations on the Transputer system of a lipid bilayer are presented.*

1 Introduction

A principal aim of molecular biology is to understand the relationship between structure and function exhibited by the molecules of life. Static structures of several of these important molecules have been resolved using experimental techniques such as X-ray crystallography and nuclear magnetic resonance. Molecular dynamics simulations can be used to refine these structures and to determine their

---

[1] This text will also be published in: Proceedings of the Fifth Conference on Neural Networks and Parallel Distributed Processing, editor: Samir I. Sayegh. Department of Physics, Indiana University — Purdue University at Fort Wayne, Indiana, U.S.A., Indiana University — Purdue University at Fort Wayne, 1992. and in: *Chemical Design Automation News* (CDA News), 7, 1992

[2] Electronic mail: heller@lisboa.ks.uiuc.edu, NeXT-mail welcome

stability. For a long time, it was assumed that the static structure along with knowledge of the chemical bonds and the typical atomic charges involved would be sufficient for a full explanation of a molecule's function. However, in the recent decade it became increasingly clear that the *motion* of the individual atoms plays a crucial role in determining many aspects of molecular properties and interactions. The details of these processes are often very difficult to measure experimentally, if at all. The information of interest can then only be obtained by computer simulations. Reviews of this exciting field, and examples of many applications, can be found in [24, 3, 31, 28].

Currently, many groups are developing approaches to allow an increasingly faithful representation of the dynamics of biological macromolecules in computer programs. One effort is to include in the simulation as much as necessary of the natural environment of the protein under investigation, e.g. membrane and water (see Fig.1).

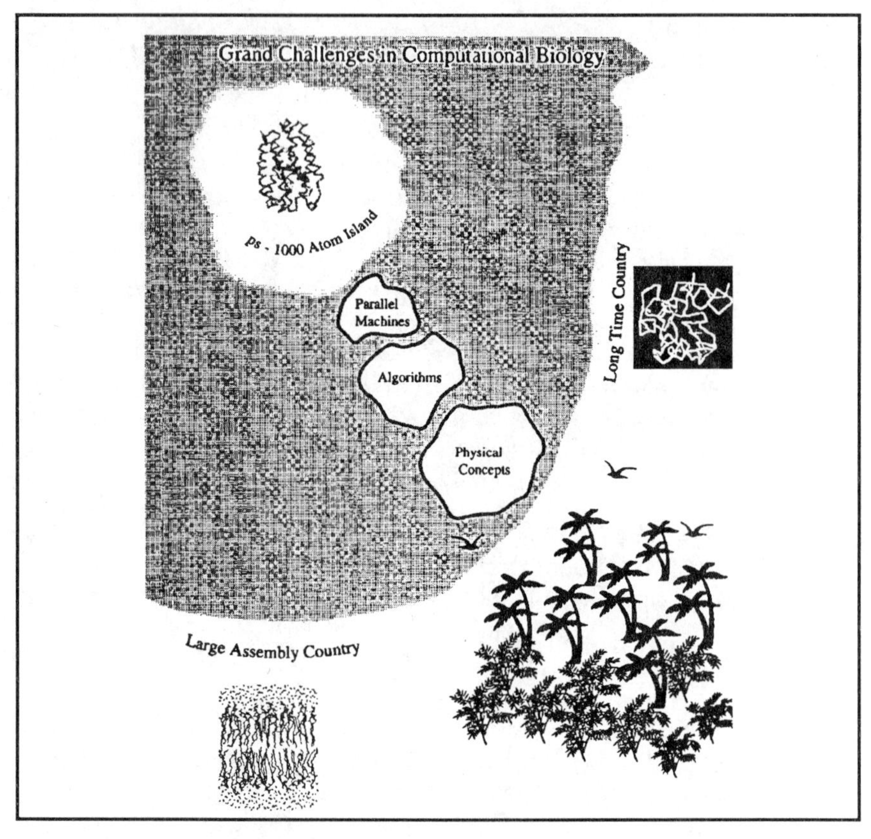

Figure 1: The grand challenge in computational biology is to simulate real world systems. Currently only systems on the small island of short simulation times (pico seconds) and small system size (1,000s of atoms) can be simulated. With the help of the stepping stones "parallel machines", "faster algorithms", and "new physical concepts" we hope to bridge to the real world. These real world systems consist of 10,000 to 100,000 atoms and have to be simulated over time spans in the range of seconds.

In the near future, this will make it possible to leap beyond simple analysis of structure-function relationships and allow significant contributions to biology, biotechnology, and medicine by guiding the synthesis of new materials and drugs and predicting their respective effects. A long-range goal of computer simulations includes the prediction of 3-D protein structures from their respective amino acid sequences. With the introduction of parallel processing techniques into the field of molecular dynamics simulations these goals become more and more achievable. As indicated in Fig.1, parallel computers are one of the stepping stones on the way to the simulation of real world systems.

2 What is Molecular Dynamics?

The exact time evolution of a system consisting of many atoms is described by the pertinent quantum-mechanical Schrödinger equations whose complexity — even with only small molecules — renders this description unfeasible for the task of dynamics simulations. Therefore, a classical approximation is used, where the computational demand increases at most like N_a^2, where N_a is the number of atoms. In this approximation, atoms are represented as points of mass with a certain charge. The motion of each atom is then determined by summing up all forces acting on it and integrating Newton's equations of motion.

$$\left.\begin{aligned} m_i \frac{d^2 \mathbf{r}_i(t)}{dt^2} &= \boldsymbol{F}_i(t) \\ \boldsymbol{F}_i(t) &= -\nabla_i E(\mathbf{r}_1(t), \ldots, \mathbf{r}_{N_a}(t)) \end{aligned}\right\} \quad (1)$$

Here, m_i and $\mathbf{r}_i$ are used to denote the mass and position of the i-th atom, respectively. ∇_i represents $(\frac{\theta}{\theta_{x_i}}, \frac{\theta}{\theta_{y_i}}, \frac{\theta}{\theta_{z_i}})^T$, and N_a is the total number of atoms in the system. E is the empirical energy function, the quality of which is critical for a realistic description of molecules. In the widely used molecular dynamics programs CHARMm [2] and X-PLOR [4], and in the parallel program EGO [16], which we have developed for our Transputer system, this energy function is of the form

$$E = \underbrace{E_{bond} + E_{angle} + E_{dihedral} + E_{improper}}_{E_{bonded}} + \underbrace{E_{el} + E_{vdW} + E_{hbond}}_{E_{nonbonded}} \quad (2)$$

where E is defined as the total energy of the molecule. The individual contributions correspond to the different types of forces acting in the molecule. The first contribution, E_{bond}, describes the high frequency vibrations along covalent bonds. The second contribution, E_{angle}, represents the bending vibrations between two adjacent bonds, and the third contribution, $E_{dihedral}$, describes torsional motions around bonds. $E_{improper}$ relates to the motion of one atom relative to a plane defined by three other atoms. All of the previously mentioned energy contributions portray interactions due to a direct chemical bond between atoms. They are often grouped as so-called *bonded* interactions. The remaining three terms describe the *non-bonded* interactions between atoms that are not chemically bonded. E_{el} accounts for the electrostatic (*Coulomb*) energy between the individual atomic charges, E_{vdW} models the *van der Waals*-interaction and E_{hbond} represents the hydrogen bonds.

The method most often used to integrate the Newtonian equations of motion (1) is the Verlet algorithm [32]. The position $\mathbf{r}_i(t+\Delta t)$ of atom i at the instant $t+\Delta t$ is then determined according to the formula

$$r_i\,(t+\Delta t) = 2r_i(t) - r_i\,(t-\Delta t) + F_i(t)\,\frac{(\Delta t)^2}{m_i}\ , \qquad (3)$$

where $\mathbf{F}_i(\mathrm{t})$ as defined in Eqs. (1) represents the sum of all forces acting on the i-th atom at time t. The time Δt step is determined by the fastest degrees of freedom in the molecule. These are typically bond vibrations of the light hydrogen atoms, and the time step must be no larger than $1.0\,fs$ to insure correct integration of Eqs.(1).

Because of the many interactions that must be considered, and the small size of the time step, the simulations one would like to carry out are severely hampered by the availability of suitable computer resources. In fact, dynamics calculations up to now have been limited to short simulation periods of a few nanoseconds and to biopolymers of at most a few thousand atoms (see Fig.1). Furthermore, biological macromolecules are usually embedded in an environment of water or lipids, and a realistic description of such molecules should include the surrounding material. The combination of these factors demands the fastest computers available today, which are parallel machines.

2.1 Algorithmic improvements to speed-up computations

In addition to parallel machines we need improved and faster algorithms for molecular dynamics simulations if we want to advance on our way towards realistic simulations (see Fig.1).

In the integration of the Verlet equations (3), most computer time is spent on the evaluation of the non-bonded interactions. These are composed of the electrostatic Coulomb interaction and the van der Waals interaction embodied by a Lennard-Jones-potential. For each integration step, due to the long range nature of the Coulomb interaction, $N_a(N_a - 1)/2$ pair interactions must be calculated. The programs CHARMm and X-PLOR avoid the corresponding prohibitive computational effort by introducing a cut-off for these interactions, which for long-range Coulomb forces is not satisfactory. The ensuing error is especially serious for large biopolymers for which inhomogeneous charge distributions are thought to play an important role for stability and function (see Fig.2).

In addition to being of questionable value for the faithful representation of biopolymer dynamics, the introduction of a force cut-off also creates technical problems. The necessity of switching functions to maintain the differentiability of the potential energy function, as well as the need to maintain pair lists to account for the proximity of atoms, entails additional computation and memory requirements which partially offsets the advantage gained through the reduced count of interacting pairs.

Alternative approaches for calculation of non-bonded pair contributions without sacrificing accuracy are the *Fast Multipole Algorithm* developed by Greengard and

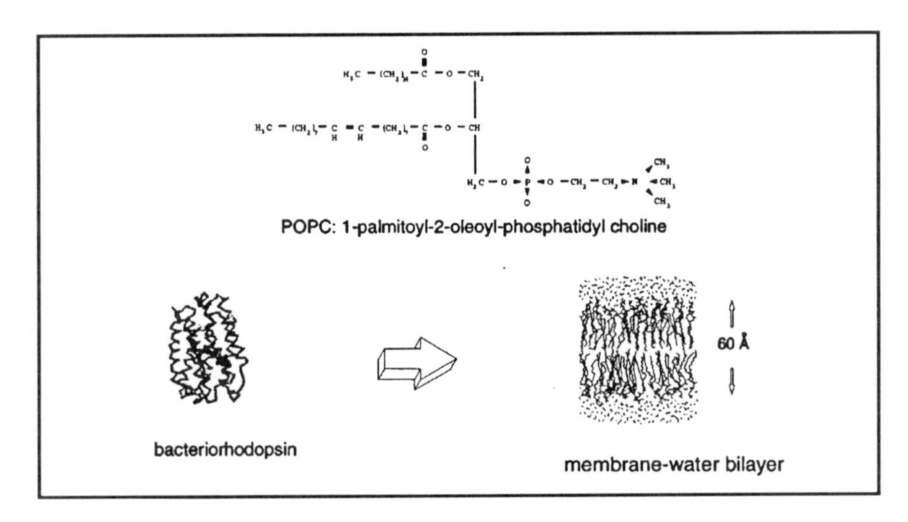

Figure 2: The membrane protein bacteriorhodopsin, *which has so far only been simulated in vacuum, has been embedded into a membrane-water system. The membrane had been built from 200 POPC lipid molecules [17]. These molecules have charged headgroups with a strong dipole moment. Since the headgroups of the two layers of the membrane are about 60 Å apart, a simple force cut-off of the electrostatic interaction at a distance of 10 Å to 15 Å, as otherwise widely used, would completely neglect that important contribution from the headgroup interaction. Therefore our program, EGO, uses a special distance class algorithm [13] to incorporate long range electrostatic effects.*

Rokhlin [11,9,10,12] and a hierarchical distance class approach suggested in [34]. The latter method has been implemented in our parallel molecular dynamics program EGO and is described in detail in [13,14]. The method is based on a spherical subdivision of interatomic distances into several distance classes. The relative motion of atoms that are close in space greatly influences the electrostatic energy, so that these Coulomb force contributions must be calculated at each integration step. Relative motion of atoms separated further apart does not change their Coulomb force contributions as much, and, therefore, these interactions can be evaluated less frequently. The resulting algorithm speeds up computation by a factor of six to ten without a significant loss of accuracy [13].

3 The transputer program

As explained in Section 2, the computationally most expensive task in molecular dynamics calculations of biopolymers is the evaluation of Coulomb (and van der Waals) forces. Due to the long range character of these forces, each atom interacts with every other atom, leading to a total of N_a $(N_a - 1)/2$ force computations for each integration step. The parallelization strategy for our program and hardware had been adopted to be most suitable to evaluate pair interactions like Coulomb forces. However, all other contributions to the energy function (Eq. (2))(and hence the total force) have to be calculated as well and must be integrated in the parallel computational scheme.

3.1 Topology

In our computational scheme, the atoms of the molecule are assigned to different computational nodes irrespective of the atoms' positions in space or within the molecule [1] (see Fig.3). All information about these atoms, such as mass, charge, type, and coordinates, is stored in the associated nodes. The evaluation of the forces acting on these atoms also takes place at the Transputer of the node to which the atoms are assigned. In the following discussion, all atoms allocated to a specific Transputer are referred to as the 'own' atoms of that Transputer, all other atoms are the 'external' atoms.

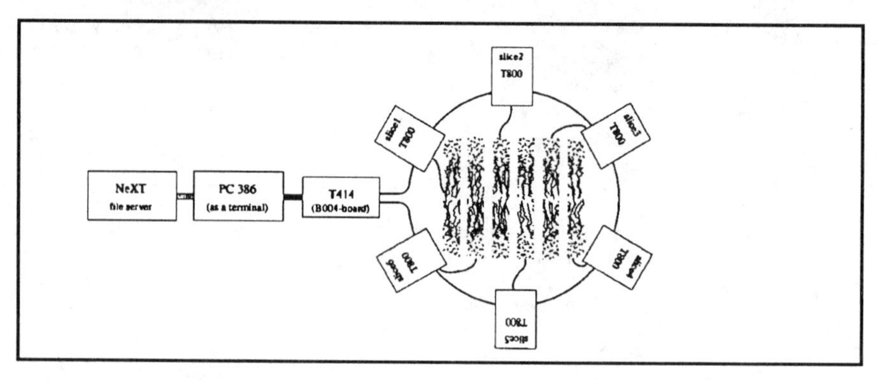

Figure 3: Schematic diagram of the systolic double ring network used in our application. The system, here the membrane bilayer covered with water, is sliced in as many pieces as there are computational nodes in the ring. Then the atoms in each slice are assigned to a node, illustrated by the little curved lines, as the 'own' atoms of that node.

The computation of Coulomb forces is now separated into two terms

$$F_i = \sum_{\text{'own' atoms } j} \frac{q_i q_j r_{ij}}{4\pi\varepsilon r_{ij}^3} + \sum_{\text{'external' atoms } k} \frac{q_i q_k r_{ik}}{4\pi\varepsilon r_{ik}^3} \quad . \tag{4}$$

Here $\mathbf{F}_i$ is the force which acts on atom i, q_i is the partial charge of this atom, $\mathbf{r}_{ij} = \mathbf{r}_i - \mathbf{r}_j$ is the vector joining the atoms at positions $\mathbf{r}_i$ and $\mathbf{r}_j$, and ε is the dielectric constant. The first term in (4) can be evaluated by all nodes in parallel, since it involves only information of 'own' atoms i and j, which is always known locally at each node. The second term in (4), however, involves 'own' atoms i as well as 'external' atoms k and therefore requires some communication.

The systolic double ring topology [18], which is depicted in Fig. 4, provides the communication channels needed. The **network Transputers** (T800), one of which can be entirely devoted to the calculation of hydrogen bonds [1] (**h-bond node,**

[1] Nevertheless it is advantageous to distribute them over the ring according to the order introduced by the protein backbone, as done in EGO by following the sequence introduced by the pdb file.

shown in light grey in Fig.4), and a **host Transputer**, are joined by a bidirectional Transputer link (two antiparallel, unidirectional **occam II** channels). This forms the **first ring**. A **second ring** is established using the two remaining links on each network Transputer. It is neither desirable nor necessary (and without a fifth link not possible) to include the host Transputer in this **second ring**. There is also no need for the **h-bond node** to be tied into this **second ring**.

Using the **first ring**, each Transputer can send a copy of the information it has about its 'own' atoms (coordinates) to its neighbour in clockwise direction (on the dark black channel in Fig. 4). After the coordinates have been received, each Transputer can calculate that part of the sum in Eq. (4) concerning the 'external' atoms it just received from its neighbour. In the next step, the information about the 'external' atoms is passed to the next neighbour in clockwise direction by each node. Knowing a new set of 'external' atoms, each node can calculate another part of the 'external' sum in Eq. (4). This process is repeated until all pair interactions have been computed. By then, each node will have received its 'own' atoms back, which have circulated one full revolution on the ring. Using the total Coulomb force component, along with the bonded force contributions of its 'own' atoms, the entire force $\mathbf{F}_i$ acting on each atom is known, and all nodes can integrate one step according to Eq. (3).

By applying Newton's law $\mathbf{F}_{ki} = -\mathbf{F}_{ik}$ it is possible to cut the computation time to about one half, avoiding redundant calculations. This is achieved by sending partial force arrays, $\mathbf{F}_{ki}$, along with the coordinates on the **force.array** channel (medium grey channel in Fig. 4). One drawback of this method is that the strategy can effectively only use an odd number of computational nodes. Since we have six nodes on each board, i.e. always an even number of nodes available, one node would remain idle throughout the simulation [16, 15]. However, making use of the multiple instructions, multiple data (MIMD) structure of the Transputer network, this node can be devoted to the calculation of hydrogen bonds (see [1]) without significantly slowing down the other calculations.

3.2 Deadlock-free force communication

Two kinds of forces are produced: force items and force arrays. All the pair interactions can be evaluated immediately after coordinates of external atoms have arrived at a node, as they involve only one external atom. The forces produced during those computations can be sent as arrays, which is much more efficient than sending them as single force items for each atom. For this purpose we use the **force.array** channel depicted in medium grey in Fig. 4. Using a separate channel for force arrays and for coordinate arrays speeds up communication since the

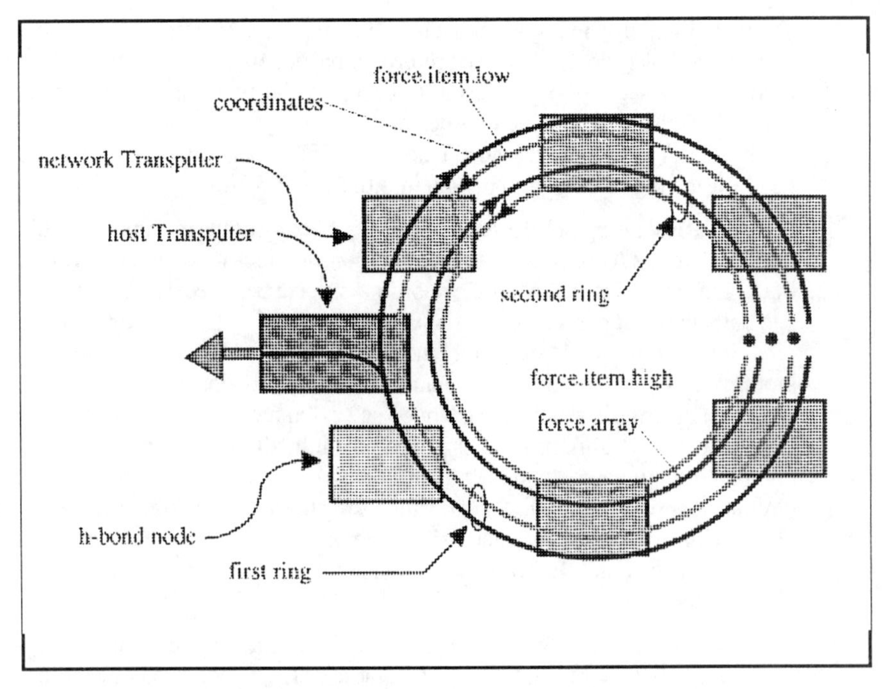

Figure 4: Communication channels for the double ring topology used in our parallel computer. See text for details.

channels reside on different links that can be operated in parallel.[1] Different systolic loop methods are described in [27].

The other type of forces stem from interactions between three (angle interactions) and four (dihedral and improper interactions) particles. For their computation, coordinates from up to four different nodes need to be available. For this purpose the coordinates of atoms involved in many-body interactions are saved while they are passed around in the ring. When all the necessary information to compute a many body interaction is available to a node, the computation is carried out and the forces are sent one by one as force items to the node that owns the respective atom, using the **force.item** channels (light grey in Fig. 4).

All new force items are multiplexed into the **force.item.low** channel. Each node listens on this channel for items targeted to it. This raises a deadlock problem, since a situation is possible where all the nodes might want to forward a force item but can not do so because there is already an item stuck in the channel. To avoid this problem a second force item ring, **force.item.high**, is introduced. No node is

[1] Technically, the force arrays are out of phase with the coordinates by exactly one transfer step. This reflects that the coordinates have to be known *prior* to the computation, while the forces are known only *after* the computation.

allowed to multiplex items into this ring. The only allowed operations are forwarding of items and removal of items addressed to this very node.

However, there is exactly *one* node that passes information from the **force.item.low** channel to the **force.item.high** channel (as shown in Fig.4); thus the rings become a spiral. The **force.item.high** ring can never deadlock since the only allowed operations are removal and forwarding for force items. This implies that there is always one node on the **force.item.low** channel that cannot deadlock (namely the one that crosses both rings). This guarantees the deadlock freedom of the whole system [5].

3.3 Compatibility with other programs

Compatibility with standard molecular dynamics packages was a key issue in the development of our program. The input files used (containing the atomic coordinates, the structure information, and force field parameters) can easily be obtained from the file format used by X-PLOR. The output files containing the trajectories of the individual atoms can also be converted into X-PLOR format[1] for detailed analysis with X-PLOR.

3.4 Dissemination

Besides the installation at the UIUC, several other universities and research institutions installed EGO on their systems (see Fig.5). Recently M. Schaefer and H. Heller ported the approximately 20,000 lines of **occam II** code to the C-language to make EGO available also on parallel non Transputer-based machines, such as the CM5 of Thinking Machines Corporation (TMC), the Paragon/XP of Intel, or networks of workstations. The latter effort builds on the coordination language CHARM [22,7,29] which was developed by Prof. Kale at the UIUC. This will enable many researchers who do not have access to a Transputer system to use EGO for their research.

To try out the EGO program and/or a Transputer system, the NIH Resource "Concurrent Biological Computing" has made available a Transputer guest account on one of its machines. This guest account allows access to a 6 Transputer network which allows for the simulation of small molecules (up to 1,500 atoms) using the program EGO. Further details on how to use this service are provided in the appendix A.

4 Hardware development

In November 1987, when we began our efforts to develop molecular dynamics algorithms for Transputers, to our knowledge no commercial hardware meeting our specifications (large, parity checked DRAM on each node, status indicator, compact design) was available. Therefore, we designed and built our own boards.

---

[1] This feature was implemented by M.Tesch

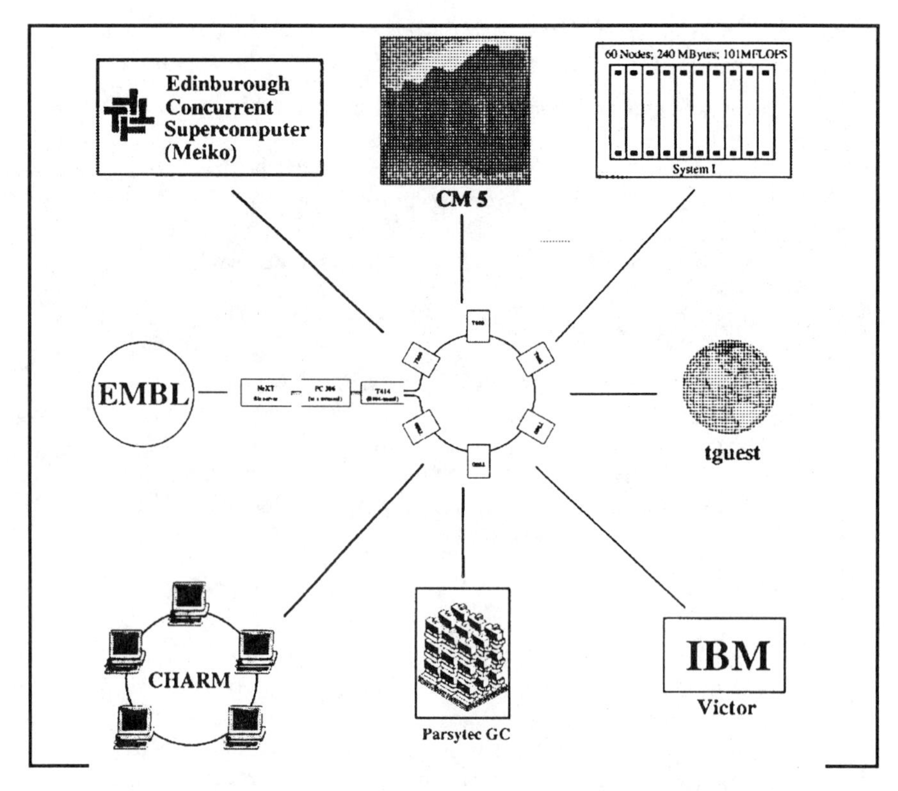

*Figure 5: Dissemination of the parallel molecular dynamics program EGO. Most installations use the **occam II** version, but the CM5 and CHARM [22] installations use the new C-language version of EGO.*

The hardware consists of a systolic double ring [25] of Transputer nodes (see Fig. 3), and is fully described in [14,16,15]. The main parts of each computational node are an INMOS T800 Transputer running on 22.5 MHz, 4 MBytes of parity checked DRAM (100 ns access time), a parity checking logic and a status indicator. The latter signals the internal state of each node to the user by means of three colour-coded LED's (one for a set error flag, one for a parity error, and one for access of the external DRAM). Deadlock situations and an uneven distribution of the work-load can be easily detected by observing the DRAM access indicator lights.

Six computational nodes fit onto one six-layer, double eurocard sized (160 mm by 233 mm) board. Ten of these boards fit into one 19'' chassis, totalling 60 nodes which for molecular dynamics calculations match the computational power of a Cray 2 processor (see Section 5). Such a powerful machine can be placed on a desktop and does not take up more space than a PC (see Fig. 6).

All link connections determining the topology of the network (here a systolic double ring) are provided by the backplane of the chassis, which can be exchanged to allow for different topologies.

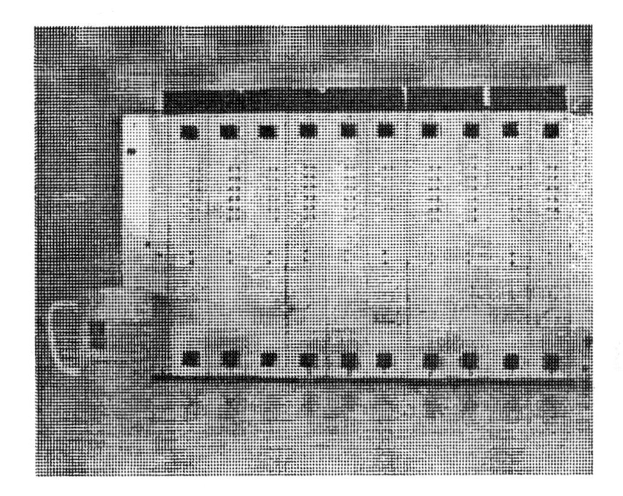

Figure 6 :Left: A 60 node/240 Mbyte DRAM Transputer-based parallel computer with the performance of a Cray 2 processor. This machine was designed and built for $60,000 by members of our group. Nine graduate students and a postdoctorate each soldered a board (note their signatures on the respective boards). The machine, programmed for large-scale molecular dynamics simulations, is shown running a simulation of a 24,000 atom water-membrane system using the X-PLOR compatible program EGO; the 60 lit status lights indicate that all nodes run at full speed demonstrating excellent load balance. The cup on the left indicates the small size of the machine.
Right: Picture of a single board. Dominant component of the footprints of the six nodes is the Transputer chip; the 4 MBytes of parity checked memory for each node are mounted on the other side of the board.

Parity checking is important, since we run a system with 60 nodes and 240 MBytes of solid state memory. With such a large system, a parity error is likely to occur every other month. Assuming a runtime of several months for the molecular dynamics code, this probability is unacceptably high. Together with an automatic restart mechanism in the software, the parity checking allows sufficiently long, unsupervised dynamics calculations without the risk of corrupting data due to memory faults.

As shown in Fig. 3 and Fig. 7 the systolic double ring is interfaced to a PC (or a Silicon Graphics Workstation). Using Sun's Network File System (NFS) large disks with hundreds of megabytes are connected to the Transputer system to store the huge amount of output data, typically 70 Mbytes per day. Since the output is split over several files it is even possible to look at a trajectory and analyse it (using X-PLOR on a conventional machine) *while* later portions of it are still being computed by EGO. Such preliminary analysis allows detection of obvious errors and monitoring of the progress of the computation.

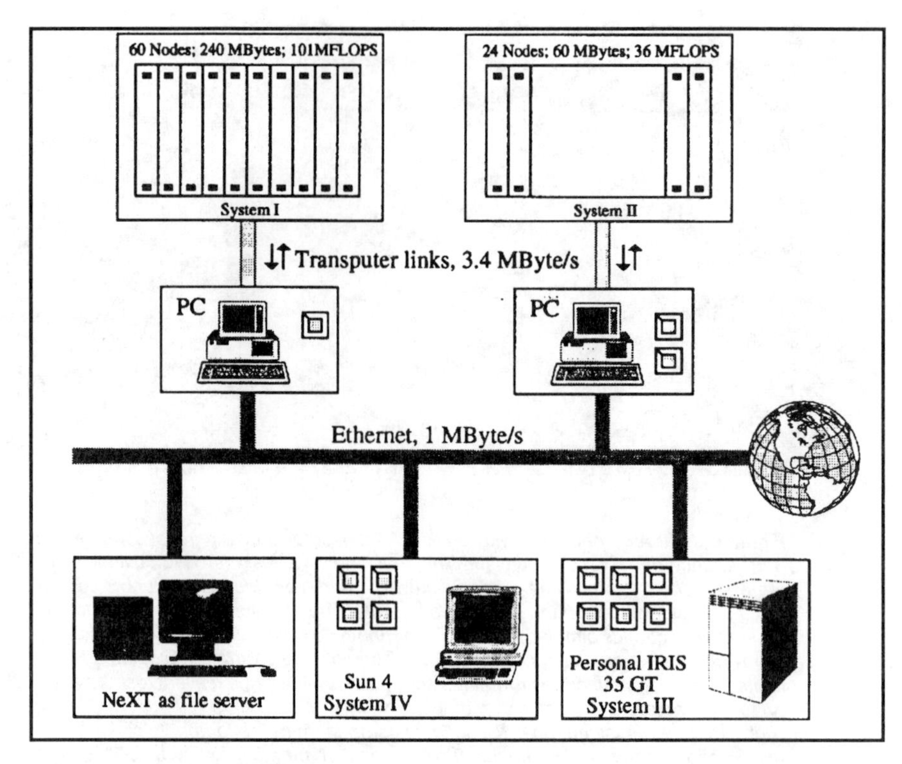

Figure 7: Networking of the Transputer system in the theoretical biophysics group of K. Schulten. In addition to **occam II** *and the TDS3 of INMOS, the C compilers of Par.C by Parsec and ANSI-C by INMOS are available. Although designed for molecular dynamics simulations, we successfully run neural network simulations, magnetic resonance imaging (MRI) simulations, and percolation simulations on our Transputer systems. Using the PC-SERVER of Luigi Rizzo, not only the UNIX based systems but also the IBM-PC based systems are fully accessible remotely over the network from virtually everywhere in the world.*

5 Benchmarks

5.1 Comparison of the transputer system with other systems

Several benchmarks comparing the molecular dynamics performance of our Transputer system with other computers have been carried out. Different programs were used on the individual machines; the results are given in Table 1. To compare the performance of our Transputer-based parallel computer with other machines, we computed on several machines a short (100 steps) molecular dynamics trajectory of *bacteriorhodopsin* with 2,166 atoms, *bacteriorhodopsin* with additional water totalling 4,039 atoms, and the membrane bilayer with water cups and embedded *bacteriorhodopsin* totalling 23,208 atoms.

The distance class algorithm [13] used by EGO provides a speed-up of six to ten (compared to calculations without cut-off) but does not introduce the errors connected with neglecting far range Coulombic and van der Waals interactions. Therefore, it must be considered as being somewhat in between those methods using

Program	Machine	2,166 atoms cut-off	2,166 atoms no cut	4,039 atoms cut-off	4,039 atoms no cut	23,208 atoms
EGO	GC: 128 T805 nodes	0.66		1.33		32.1
EGO	60 T800 nodes	0.97		2.51		83.8
EGO	CM-5 32 nodes	≈ 58		–		–
EGO	CM-5 64 nodes	≈ 90		–		–
MD	CM-2, 32k procs	n / a	–	n / a	2	–
X-PLOR1.5	CRAY2	0.64	3.39	3.52	20.4	–
X-PLOR1.5	Convex C2	3.3	21.3	14.0	88.55	–
X-PLOR1.5	Stardent 3000	8.2	46.5	28.5	196.7	–
X-PLOR1.5	SG1 4D / 220GTX	15.54	121.8	43.88	–	–
X-PLOR1.5	Stardent 1500	18.22	123.16	62.2	–	–

Table 1: The table shows the average CPU time taken for one integration step of three sample systems with 2166 atoms, 4039 atoms, and 23,208 atoms respectively. The machine entry "GC" stands for the Parsytec GCel machine, which we could use for a couple of days. The timings for the CM-5 are very preliminary as that version of EGO is not yet completely debugged and not at all optimized. The implementation effort is still ongoing. If not noted differently, the program is run using only one CPU. The two timings for X-PLOR indicate computations done with cut-off (11Å) and without cut-off. MD [33] does not allow for a cut-off. EGO, the dynamics program on the Transputer machine, uses a distance class algorithm [13].

a cut-off and those computing all N_a $(N_a - 1)/2$ interactions, as indicated in Table 1.

From Table 1 one can easily see that a 60 node Transputer machine running EGO outperforms most of the other tested supercomputers while it costs only a fraction of their price. Our actual costs were about $60,000 for the whole machine, which excludes costs for development and assembly.

A detailed performance analysis of EGO can be found in [30].

5.2 Scaling performance of the transputer system

Aside from the absolute performance, another key property of parallel machines is how the performance scales with the number of nodes working on a given problem. This scaling performance is shown in Fig. 8.

The performance is given by the number of integration steps that can be completed per time unit, i.e. per second. For an ideal machine this performance would increase linearly with the number of nodes; the actual behaviour comes very close to this linear dependence. The deviation from the linear behaviour is smaller for the larger system. Experiments showed that a minimum number of ≈ 50 atoms per node is necessary to achieve a good scaling of the performance.

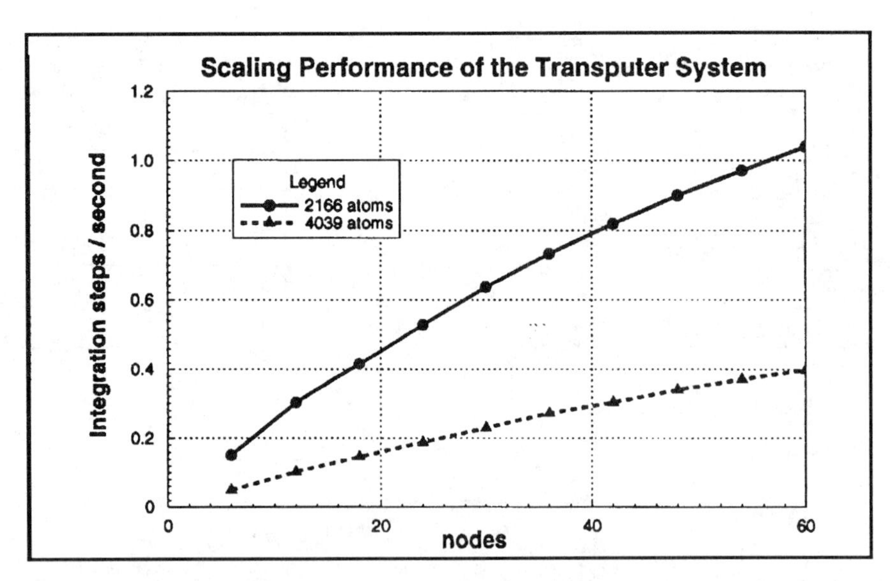

Figure 8: Scaling of the performance with the number of nodes. Since only complete boards with six nodes could be removed, the measured values (dots and triangles) are in increments of six nodes. The scaling behaviour of the performance was measured for two systems with 2,166 atoms and 4,039 atoms, respectively.

6 Simulation results[1]

Fig. 9 shows some results of the membrane simulation. The deuterium order parameter profile, in the upper left, shows the typical decrease of the order parameter with increasing chain position (towards the free end of the chain). The absolute values are higher than those found experimentally (typically between 0.1 and 0.2), probably due to the short simulation time.

The water density in the membrane region, shown in the lower left of Fig. 9, was computed at three different times. The increased penetration of water into the lipid region at later times is clearly visible.

The picture on the right in Fig. 9 shows the trajectories of the centres of mass of each lipid molecule as a projection into a plane perpendicular to the membrane plane. The inward pointing trace of the outer (restrained) lipid molecules is a result of the contraction of the reference coordinate set. Due to these restraints the diffusion of the outer lipids is remarkably smaller than that of the free lipids in the center. The lipids show considerable movement perpendicular to the membrane plane, but the simulation time of 115 ps is too short to show lateral diffusion with lipids exchanging places.

---

[1] A detailed description of this membrane simulation is currently being published with Michael Schaefer elsewhere [17]

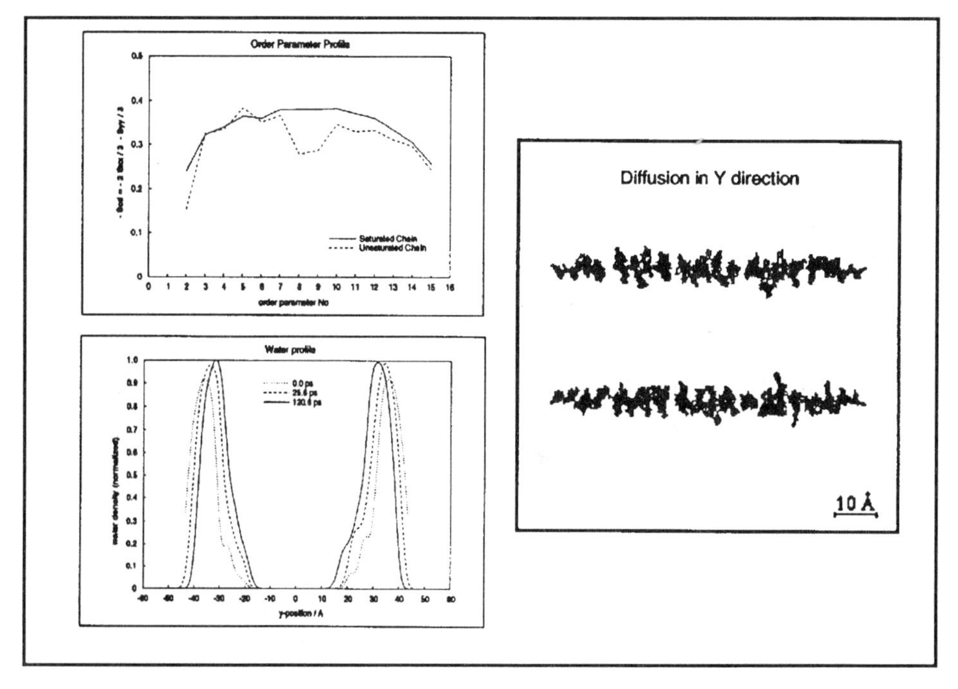

Figure 9: Simulation results from the equilibration phase of the membrane-water system (without bacteriorhodopsin*).*
Upper left: Order parameter profile for the two fatty acid chains of the lipid molecule. The decrease in the order parameter for the carbon atoms at the end of the chain (high order parameter No.) is typical.
Lower left: The water penetration profile, recorded at the start, at an intermediate point, and at the end of the simulation, shows that the penetration of water molecules into the headgroup region increases with time.
Right: Shown is the Y-component of the centres of mass of each lipid molecule over the course of the simulation.

7 The next generation of transputers: T9000

In October 1985 INMOS introduced the Transputer family [8]. Since then, several improved members have been added to the family. In 1987 the T800 was introduced [6], and with only a few modifications it has remained a top-of-the-line product. We currently employ the T800 processor in our parallel computers.

While other manufacturers introduced their new processors (e.g the i860 by Intel) over the past four years, INMOS worked on the design of the T9000 which was officially unveiled in April 1991. Unfortunately INMOS was not able to deliver actual products at that time and by now (May 1992) more than a year has passed without the T9000 being available. The new announcement of INMOS is that T9000 will be available by the end of 1992.

The T9000 provides an increase in speed of about a factor of ten in comparison to its predecessor, the T800, in both communication as well as computation [21]. This improvement can easily be utilized as the new T9000 is binary compatible with the

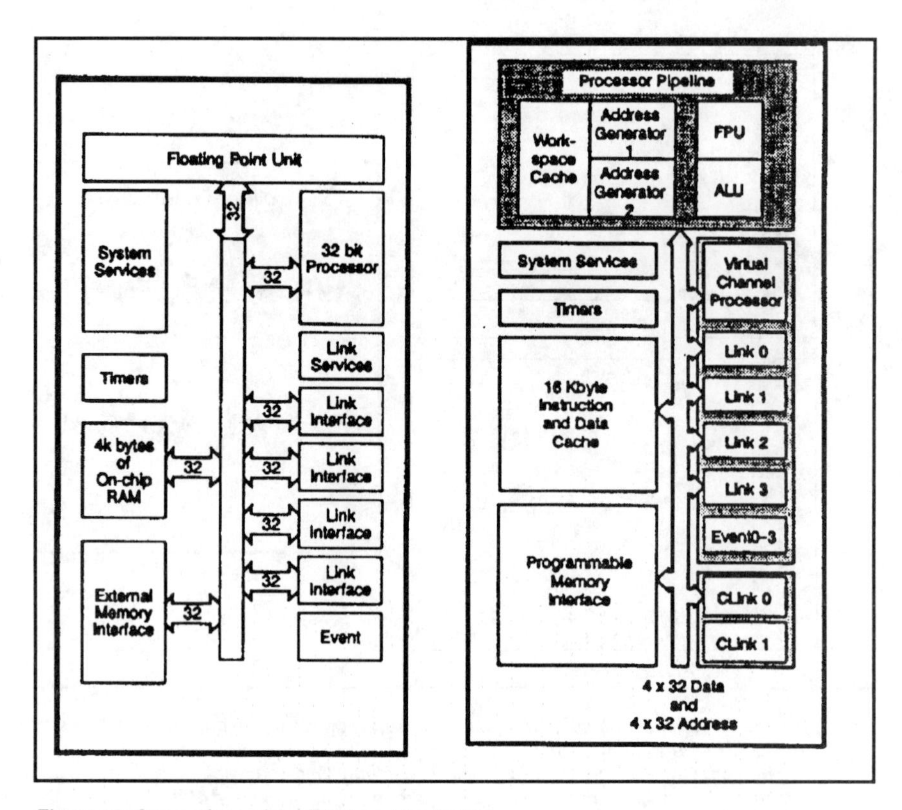

Figure 10: Comparison of the T800 and the T9000
Left: Block diagram of the T800 (20). Integrated on one chip one finds not only ALU, system services and a programmable memory interface, but also a 64-bit FPU, 4 KBytes of on-chip RAM, and four independent link interfaces.
Right: Block diagram of the T9000 (21). All the basic building blocks of the T800 are visible: ALU, system services and programmable memory interface, 64-bit FPU. New is an additional workspace cache. The cache/RAM has been extended to 16 KBytes and, most importantly, the four link interfaces have been equipped with a virtual channel processor.

T800. It is possible to carry over all the programs developed for the T800 Transputer, with only minor modifications necessary to handle the improved link structure that allows transparent communication between any two T9000s.

From our experience with current Transputers (T800), it is very reasonable to expect that the T9000 will achieve the claimed values for sustained performance, which are 60 MIPS[1] (150 MIPS peak) and 10 MFLOPS[2] (20 MFLOPS peak) [21], in our application using our program EGO.

[1] *M*ega *I*nstructions *P*er *S*econd

[2] *M*ega *FL*oating point *O*perations *P*er *S*econd

The T800 (see Fig. 10 left side) is a 32 bit processor with a 64 bit floating point coprocessor and four independent communications agents (links) integrated on a single chip [19, page 5]. The links provide point to point communication between any two Transputers. Having only four links, each Transputer can connect with only four other Transputers. This limits, in general, the portability of software, which must be adapted to different network topologies. Also, the programmer must address the problem of communication explicitly; for instance, routers must be written and included in the source text.

The IMS C004, a 32 x 32 programmable crossbar switch [20] was designed to increase the connectivity in a network of T800 Transputers. It can connect any of its 32 input links with any of its 32 output links, programmed by commands on a control link. Unfortunately, it introduces a 1.75 bit delay in the exchanged messages (reducing the bandwidth from 1.7 to 1.3 MBytes/s per link), and has a very high latency for reconfiguration. Therefore it is not suitable for *dynamic* reconfiguration of Transputer networks.

The communication problems of the T800 and the C004 are remedied by the virtual channel processor of the T9000 (see Fig. 10, right side) in conjunction with the new C104 crossbar switch with submicrosecond latency. The virtual channel processor can multiplex any number of links onto a single physical link, making them "virtual links" [23]. Each message on a channel is split into a sequence of packets; packets from different messages on different virtual links can be interleaved over each physical link, thus preventing deadlock. The C104 will be able to route *individual messages* to their destinations, rather than simply providing a reconfigurable link topology like the C004. A switching latency of about 0.5 μs (low load conditions) and a total bandwidth of about 5 GBytes/s are expected. The following explanation is taken from [26]:

> "Inmos chose interval routing, which is complete (every packet eventually arrives), deadlock-free, inexpensive, fast, scalable (good for any size network), versatile (good for any type topology), and can be near optimal (shortest routing followed). A number assigned to each node is used for its address. At every routing switch each link knows which header values it should recognize. Intervals are nonoverlapping and numbered so every header falls into just one interval. Each interval is a pointer to the subtree containing all these processor numbers. The attraction in this scheme is that routing can be computed with a single comparison. The C104 allows 1 or 2 Byte headers, giving up to 65,536 Transputers in a single network.
>
> Algorithms exist to prevent cyclic paths that are deadlock free. An optimal labelling scheme for an arbitrary network is unknown, but the more common topologies such as rings, trees, and two-dimensional arrays are known. Grid and hypercube schemes can be constructed from these topologies.
>
> When too many messages go through a single node, a hot spot develops, stalling a large number of packets. This can be avoided by evenly

> distributing the flow. The C104 can perform this using universal routing by choosing a random node to initially send the packet to, where it is forwarded to its destination. Latency is increased and throughput is decreased, but the worst-case performance is near optimum . . .
>
> Packet switching removes certain restrictions in **occam II.** The present version is a static language in which resources must be determined at compile time, so new processes cannot be added at run time on remote processors. Also each channel must be explicitly allocated to a physical link at compile time. The T9000 and the C104 allow full **occam II** to be implemented."

Portability to any routed array of Transputers is possible without any change, although efficiency will be determined by the topology of the underlying hardware. We expect an increase in speed by at least a factor of ten between a T800 system and a T9000 of equal size. The improved communications scheme eliminates the need for our software routers, making the EGO program shorter, less complicated, and faster.

Acknowledgements

We would like to thank Lyndon Clarke from the University of Edinburgh who contributed the idea of the *double* ring topology and who helped to implement it on the Edinburgh Concurrent Supercomputer (ECS).

We would further like to thank Michael Schaefer for his contributions to the membrane simulations and Helmut Grubmüller who was one of the chief designers of the whole Transputer system. Barbara Napier designed and rendered the pictures Fig. 1, Fig. 5, and Fig. 7.

This work was supported in part by the Beckman Institute at the University of Illinois in Urbana-Champaign and the National Institutes of Health Resource 'Concurrent Biological Computing' under grant number 1 P41 RR05969*01.

A. How to use the Transputer Guest Account

Contact person:	Helmut Heller
Phone:	(217)244-6914
E-mail:	heller@lisboa.ks.uiuc.edu
Office:	Beckman 3rd floor, room 3143
Address:	University of Illinois at Urbana–Champaign Beckman Institute Attn: Mr. Helmut Heller 405 North Mathews Avenue Urbana, Illinois 61801, U.S.A.
logon:	tguest@oslo.ks.uiuc.edu, IP-number: 128.174.214.4 you will obtain the password if you send a short e-mail stating your name, address, and why you would want to use the guest account to heller@lisboa.ks.uiuc.edu

Appendix A1 What is available

A.1.1 Hardware

- Production sites
  - 60 T800 nodes (22.5 MHz) with 4 MByte each and parity checking,
    host: B004 (T414 (17.5 MHz) with 2 MByte with parity checking) in '386 PC
  - 25 T800 nodes (20 MHz, 22.5 MHz) with 1 MByte
    host: B008 (T800 (25 MHz) with 2 MBytes) in AT compatible PC
- Development sites
  - 5 T800 nodes (20 MHz) with 1 MByte each
    host: B014 (T800 (25 MHz) with 4 MByte) in Silicon Graphics IRIS4D machine, supports remote login (main tguest machine)
  - 4 individual T800 nodes with 1 MByte each on a Parsytec VMTM board supporting four independent users hosted in a SUN4

A.1.2 Software

- *T*ransputer *D*evelopment *S*ystem from INMOS, TDS3 D700E
  - occam II compiler producing very fast code
  - folding editor
  - network debugger
  - very stable environment
- Par.C from Parsec, V 1.3
  - C compiler for parallel programs
  - the executable code produced seems to be slower than that of the occam II compiler
- Helios from Parahelion, V 1.1
  - UNIX-like operating system for transputer networks up to 20 Transputers
  - C compiler for sequential programs
  - V 1.0 proved to be highly unstable (therefore not used by us)

A.1.3 Parallel Applications

- EGO, a molecular dynamics simulation program, I/O-compatible to X-PLOR format (H. Heller)
- software for simulation of Magnetic Resonance Imaging (written in C utilizing a farm concept, B. Puetz)
- software for simulation of neuronal networks (K. Obermayer)
- software for investigations of percolation phenomena (C. Kurrer)

Bibliography

[1] K.Boehncke, H.Heller, H.Grubmüller, and K.Schulten: 'Molecular Dynamics Simulations on a Systolic Ring of Transputers', in A.S. Wagner, editor, *'NATUG 3: Transputer Research and Applications 3'*, pages 83–94, Amsterdam, 1990. North American Transputer Users Group, IOS Press.

[2] B.R. Brooks, R.E. Bruccoleri, B.D. Olafson, D.J. States, S.Swaminathan, and M.Karplus: 'CHARMM: A Program for Macromolecular Energy, Minimization, and Dynamics Calculations', *J.Comp.Chem*, **4***(2)*, 187–217 (1983).

[3] C.L. Brooks III, M.Karplus, and B.M. Pettitt: *'Proteins: A Theoretical Perspective of Dynamics, Structure and Thermodynamics'*. John Wiley & Sons, New York, 1988.

[4] A.T. Brünger: *'X-PLOR'*, The Howard Hughes Medical Institute and Department of Molecular Biophysics and Biochemistry, Yale University, New Haven, CT, May 1988.

[5] W.J. Dally and C.L. Seitz: 'The torus routing chip', *Distributed Computing*, **1**, 187–196 (1986).

[6] P.Eckelmann: 'Transputer der 2. Generation', *Elektronik*, (1987), issue **1**, 2, and 3.

[7] W.Fenton, B.Ramkumar, V.Saletore, A.Sinha, and L.V. Kale: '*Supporting Machine Independent Programming on Diverse Parallel Architectures*', (in press), 1991.

[8] S.Ghee: *'IMS B004; IBM PC add-in board'*, INMOS, February 1987, Technical note11, #72TCH01100.

[9] L.Greengard: *'The Rapid Evaluation of Potential Fields in Particle Systems'*. MIT Press, Cambridge, 1988.

[10] L.Greengard and W.D. Gropp: *'A parallel version of the fast multipole method'*, Technical report, Rept. YALEU/DCS/RR-640, 1988.

[11] L.Greengard and V.Rohklin: 'A Fast Algorithm for Particle Simulation', *J. Comp.Phys.*, **73**, 325–348 (1987).

[12] L.Greengard and V.Rokhlin: '*On the efficient implementation of the fast multipole algorithm*', Technical report, Rept. YALEU/DCS/RR-602, 1988.

[13] H.Grubmüller, H.Heller, A.Windemuth, and K.Schulten: 'Generalized Verlet Algorithm for Efficient Molecular Dynamics Simulations with Long-Range Interactions', *Molecular Simulation*, **6** (1–3), 121–142 (1991).

[14] H.Grubmüller: '*Dynamiksimulation sehr großer Makromoleküle auf einem Parallelrechner*', Master's thesis, Physik-Dept. der Tech. Univ., Munich, Germany, 1989.

[15] H.Heller: '*Dynamiksimulation sehr großer Makromoleküle am Beispiel des photosynthetischen Reaktionszentrums von Rhodopseudomonas viridis*',

Diploma Thesis, Technical University of Munich, Physics Department, T30, James-Franck-Street, D-8046 Garching / Munich, 1988.

[16] H.Heller, H.Grubmüller, and K.Schulten: 'Molecular Simulation on a Parallel Computer', *Molecular Simulation*, **5**, 133–165 (1990).

[17] H.Heller, M.Schaefer, and K.Schulten: 'Construction, Molecular Dynamics Simulation and Analysis of a Lipid Bilayer', *Biochemistry*, (**1992**), in preparation.

[18] W.D. Hillis and J.Barnes: 'Programming a highly parallel computer', *Nature*, **326**, 27 (1987).

[19] INMOS: '*Reference Manual: Transputer Architecture*', July 1987, #72TRN04803.

[20] INMOS: '*The Transputer Databook*', 1989.

[21] INMOS: '*H1 transputer*', 1990, advance information.

[22] L.V. Kale: 'The Chare Kernel Parallel Programming Language and System', *Proc. of the International Conf. on Parallel Processing*, **2**, 17–25 (1990).

[23] D.May and P.Thompson: '*Transputers and Routers: Components for Concurrent Machines*', Technical report, INMOS, April 1990.

[24] J.A. McCammon and S.C. Harvey: '*Dynamics of proteins and nucleic acids*'. Cambridge University Press, Cambridge, 1987.

[35] N.S. Ostlund and R.A. Whiteside: 'A machine architecture for molecular dynamics: the systolic loop', in B.Venkataraghavan and R.J. Feldman, editors, 'Macromolecular Structure and Specificity: Computer-Assisted Modeling and Applications', pages 195–208. *Annals of the N. Y. Acad. of Sciences* **439**, New York, 1985.

[26] D.Pountain: 'Virtual Channels: The Next Generation of Transputers', *Byte*, E&W 3–12 (1990).

[27] A.R.C. Raine, D.Fincham, and W.Smith: 'Systolic Loop Methods for Molecular Dynamics Simualtion Using Multiple Transputers', *Comput.Phys.Commun.*, **55**, 13–30 (1989).

[28] R.Schwyzer: 'Peptide-Membrane Interactions and a New Priciple in Quantitative Structure-Activity Relationships', *Biopolymers*, **31**, 785–792 (1991).

[29] W.W. Shu and L.V. Kale: 'A Dynamic Load Balancing Strategy for the Chare Kernel System', *Proceedings of Supercomputing '89*, 389–398 (November 1998).

[30] A.Sinha, H.Heller, and K.Schulten: 'Performance Evaluation of the Parallel Molecular Dynamics Program EGO', *Comput.Phys.Commun.*, (**1992**), in preparation.

[31] W.F. van Gunsteren and H.J.C. Berendsen: 'Moleküldynamik-Computer-simulationen; Methodik, Anwendungen und Perspektiven in der Chemie', *Angew. Chemie*, **102,** 1020–1055 (1990).

[32] L.Verlet: 'Computer 'Experiments' on Classical Fluids. I.Thermodynamical Properties of Lennard-Jones Molecules', *Physical Review*, **159***(1)*, 98–103 (1967).

[33] A.Windemuth and K.Schulten: 'Molecular Dynamics on the Connection Machine', *Molecular Simulation*, **5**, 353–361 (1991).

[34] A.Windemuth: '*Dynamiksimulation von Makromolekülen*', Master's thesis, Technical University of Munich, Physics Department, T30, James-Franck-Street, D-8046 Garching / Munich, August 1988.

Autonom: a computer program for the generation of IUPAC systematic nomenclature directly from the graphic structure input

Janusz L. Wišniewski

Beilstein Institute, Varrentrappstr. 40-42, 6000 Frankfurt/Main 90, Germany

Abstract

The general subject of this paper is AUTONOM, a fully automatic and practical computerised system for generation of IUPAC compatible systematic nomenclature directly from the graphic structure input. The algorithm developed for AUTONOM analyses the compound's structural diagram, input via a graphic interface, and generates the name purely on the basis of the resulting molecular connection table. Fundamental aspects of the programming approach are presented and discussed along with the overview of the performance, accuracy and reliability of the computer system.

Introduction

Contrary to well-established, standardised, and internationally acknowledged structural diagram conventions, a complete, comprehensive grammar for systematic chemical nomenclature does not exist to date. The system of recommendations [1] developed by the IUPAC Commission On Nomenclature of Organic Chemistry (hereafter referred to as "IUPAC") has not become a truly universal standard, presumably because of the complexity of the recommended rules, but also because of the frequent alternatives allowed in name assignment. It must also be said that a certain reluctance on the part of chemical industry to recognise the need for fully systematic nomenclature can be detected, and this leads in practice to the use of much quasi-systematic and inconsistent nomenclature.

Both of the leading producers of chemical information reference works (Beilstein Institute, and the Chemical Abstracts Service (CAS)) follow IUPAC recommendations, but only to a certain extent. In order to obtain consistency in the selection of preferred index names both CAS and Beilstein have devised their own non-documented *ad hoc* nomenclature rules or practices. These 'dialects' of IUPAC are nowhere clearly defined for the average user. This has not contributed to solving the difficulties in establishing an unambiguous nomenclature standard. As long as such a standard does not exist, the practising chemist will find himself alienated from systematic nomenclature. I believe that a truly unambiguous computer-based 'structure-to-name' translator could help the emergence of such a nomenclature standard.

Except for previous AUTONOM papers [2,3], to date there are no reports of a general purpose computer program which uniquely translates structural diagrams into chemical names. The first naming program for a very limited class of compounds (addition compounds, alloys, co-polymers, and mixtures) was reported by Van der Stouw *et al.* in 1976 [4]. Five years later, Mockus *et al.* reported [5]on CAS plans to introduce a computer program for automatic structure-to-name translation in the generation of CA Index Names. The paper presented in detail the design of the algorithm, however, no subsequent report of the implementation of such an algorithm has so far followed. Recently, Meyer and Gould [6] have confirmed the feasibility of a personal computer-based program that would accept structure input and generate IUPAC-like systematic names, but the level of sophistication of their solutions appears to be still relatively low.

The AUTONOM (AUTOmatic NOMenclature) computer-based system presented in this paper is the latest attempt by the Beilstein Institute to provide the chemical community with a fully automated 'structure-to-name' translator.

The IUPAC rules have been adopted as the basis of rational naming. There are, however, some 'respectable' nomenclature practices being used which, although yielding succinct and intelligible names, have not yet been sanctioned, codified or documented by IUPAC Commission on Nomenclature of Organic Chemistry. These usages are also incorporated in AUTONOM.

Substitutive nomenclature has generally been preferred except where other styles, i.e., subtractive, radicofunctional, or additive, are widely recognised and / or unavoidable.

Although the following discussion will concentrate on the first PC-based commercial version of the program, AUTONOM 1.0, it should be stressed that the concepts and solutions presented here apply generally to any 'structure-to-name' translation software.

The algorithm

The algorithm applied in AUTONOM analyses a structure diagram of an organic compound entered via a graphic package and generates a chemical name purely on the basis of the connection table of the input structure. The communication between the structure-drawing package (user-operated) and the naming part of the algorithm is realised by the interface module of AUTONOM. Once the complete structure has been entered, this interface fills the input buffer with data. The internal construction of the buffer is static and the interface is responsible for handling the differences in coding of the input structure by various drawing packages. The text name generated by AUTONOM can be directly displayed on a monitor, stored as a database query term, or re-routed to another application for further processing. The algorithm operates on a systematic step-by-step basis; while a chemist can often apply nomenclature rules without conscious effort by just 'seeing' the structure, the algorithm of AUTONOM has no such ability and must systematically collect, order and store numerous items of information about the structure during the processing cycle. Since the system has been designed for use as a personal computer application, much attention had to be paid to optimal usage of limited storage resources. The items of information about the structure collected in the process of naming are stored in dynamically allocated computer memory segments and kept, only for as long as they are needed, in a currently executed decision-making task and then immediately discarded. The memory segments are then deallocated and returned for further use.

The general approach adopted for the processing of a structure is based on the concept of localising (and possibly selecting) the smallest structural entities. These entities are nothing more than skeletal units i.e., catenated or cyclically connected atoms forming discrete nameable objects such as functional groups, chains or ring systems. The naming cycle of a structure is conceptually divided into

1. initialisation and ring system perception
2. functional group recognition
3. ring system identification
4. parent structure selection
5. name tree creation
6. name assembly

The selection and identification of objects takes place during the first three phases of the naming cycle. On completion of the third phase all atoms of the input structure are then distributed among the identified objects. So far no hierarchy among the objects has been established. This is now done using nomenclature rules during the fourth and fifth steps (parent structure selection and name tree creation). The descriptive term 'name tree' used here is by no means abstract. The objects and their mutual relations are, on completion of the fifth phase, represented by a data structure having a form of tree with the parent structure as the root and substituents as branches of the tree. During the last phase of the naming cycle hierarchically ordered objects are related with corresponding text name fragments which are then compiled into the complete name of the input structure.

Initialisation and ring system perception

After the input structure has been entered, using the structure editor, AUTONOM fills the input buffer with data. The data are then used by the initialisation routines for constructing a two-dimensional Boolean atom connection matrix and atom vector. Each component of the vector corresponds to a single atom of the input structure and contains a data record with complete information on the atom and its bonding to other atoms in the structure.

The ring system identification currently implemented in AUTONOM is a two-step process initiated by the ring system perception. The reliability of the ring perception routine is the crux of the naming algorithm.

The model adopted for AUTONOM can be considered as representative for the 'smallest set of smallest rings (SSSR)' category of algorithms. From the point of view of nomenclature it was concluded that the perception of such a set of rings is the most suitable one for successfully naming the cyclic entities occurring in the structure. Once the SSSR for the structure is generated the algorithm finds all rings that share at least one atom with one or more other rings and so builds a list of nameable ring systems. On completion of the process all the atoms involved in cyclic closures are marked and distributed among the ring systems.

Once all the ring systems have been localised and their atoms marked, the algorithm generates for each ring system a corresponding hash code, which contains, in a very compact form, information on atom characteristics and atom interconnections of the ring system. The simple non-reversible hashing scheme used for the coding is unique in the sense that the same ring system is always tagged with the same hash string, but it does not mean that each system has a unique code. The main purpose of the hashing, as will be explained in greater detail later in this paper, was to relate all the rings in the input structure with classes of rings of similar structure.

Functional group recognition

Acyclic portions of the structure are scanned to select substructure units bearing functional group characteristics, such as hetero atom compositions with unsaturated bonds. In the subsequent process of searching in a dictionary, each selected unit is compared with entries of pre-defined functional groups. Each entry in the dictionary is associated with a pre-determined priority rank and, in the case of a match, the priority rank is used to generate a sorted list of all functional groups present in the structure. Additionally, each selected and ranked group from the list is characterised by a 'base' atom, i.e the atom through which it is connected to parent candidate skeletal units, such as ring systems or chains. The highest ranking entries from the list are then used in the next phase(parent structure selection).

The term 'highest ranking' derives from the IUPAC recommendations [1] and is supposed to set clear conditions on selection of the principal functional group, which later (in the completed name), is cited as the suffix and is the main criterion for choosing the parent structure skeletal unit. In many cases, particularly in the area of acids and their derivatives, the recommendations formulated by IUPAC might lead to ambiguous names. AUTONOM follows the recommended group ranking, but additionally, (when necessary) intersects the rank principle with the 'maximum multiplicity' rule, mainly in order to simplify and shorten the name.

Functional group recognition, although here presented as the sovereign phase of the naming cycle, is not a single run process which retires after the ranked list of functional groups has been determined. The list must be frequently updated automatically in the course of the naming of a single structure. The functional groups have been identified with preferred localisation of the parent structural skeleton. Since in this phase of the naming cycle no topological hierarchy of structural objects is yet established the preferred 'direction' is imposed by the established natural topology of atoms belonging to a functional group, e.g.

- expected parent skeleton ← R_1-C(=O)-O-R_2

If it later proves that this assignment of 'direction' of the carboxylic acid happens to be correct then no changes to the group are necessary. However, if the parent skeleton is finally determined to be localised in the opposite direction relative to the group, i.e.

- R_1-C(=O)-O-R_2 → to parent skeleton

then the initially identified acid must be split into two other groups, namely (-O) with the prefix descriptor 'oxy' and (=O) with the prefix descriptor 'oxo'. The splitting takes place sometime later, during the name tree creation, after the mutual topological relation between the parent structure skeleton and the skeleton of the group has been determined.

Another example of complex problems approached already in this phase is the nomenclature of hydrocarbon chains whose terminal CH_3 units are replaced by functional groups which contain carbon as the central atom. This is the case for aliphatic acids, acyl halides, aliphatic nitriles, and (thio)(seleno)(telluro)aldehydes. In all such cases the central carbon atom of the corresponding group, depending on the status of the chain (parent or substitute chain) to which it is attached, may stay as a part of the group or must be included into the set of atoms constituting the chain. This problem is mentioned here in order to illustrate that the elegant model of three types of separate objects i.e., chains, rings and groups does not always conform to reality and must be occasionally corrected.

Ring system recognition

The ring perception phase discussed earlier in this paper delivers a complete list of all localised ring systems. Each entry in the list contains a data record of the atoms positions of its sub-rings in the previously generated SSSR, and the hash code. Each such record is then processed by the ring recognition routines.

In the first step, (obligatory for each perceived ring system), a dictionary scan for ring systems with trivial names is carried out via an atom-by-atom search mechanism . The atom-by- atom matching is conducted in a special dictionary containing connection tables together with prescribed numbering and names. Much attention has been paid to optimising the organisation and construction of the dictionary in terms of efficient computer processing demands.

Each ring system chosen for inclusion in the dictionary is coded with exactly the same hashing algorithm used for localising the ring systems of the input structure during the ring perception phase. Thus, the complete set of trivial-name ring systems is grouped into small pots with a common hash code as an address. At the time of writing, the measured average rings/pot ratio was approximately 7. Although the statistical distribution of rings among pots is far from being uniform, the maximum registered number of rings in a single pot is less than 25. This means, in practice, that in the worst case each localised ring system must be matched, atom-by-atom, against a maximum of 25 dictionary entries. To speed up searching even more, only inter-ring atom connections with no information on bonding are stored in addition to atom characteristics. In this way no bond restoration is necessary in this phase of the naming cycle, thus saving the time normally spent on bond denormalisation and unscrambling during searching.

With the growing number of entries in the dictionary, problem of storage will become more explicit. It is planned for next AUTONOM versions, to use an effective data compression algorithm [7] in order to significantly reduce storage requirements.

During the dictionary search each perceived ring system is compared with the entries in a single dictionary ring pot until an exact match is obtained. Then the algorithm, using the array of fixed locants stored together with each dictionary entry, continues with the complex routine of ring-atom numbering.

For nonsymmetrical ring systems, the only possible composition of locants is the one stored in the array of the matched dictionary entry. For symmetrical rings this is not the case. Depending on the number of symmetry axes, a single atom can be numbered with many equivalent locants. The idea of storing all possible permutation of locants together with the dictionary entry was rejected as introducing unacceptable degradation in system performance. Instead, a routine based on look-ahead techniques has been designed. The routine is able to re-generate all possible permutations of locant compositions on the basis of the one composition stored in the dictionary and connection table of the matched ring system. Once all the permutations are available, the final ring-atom enumeration is decided by applying the following criteria, in the order given:

a. principal functional groups
b. 'indicated' hydrogens
c. multiple bonds in compounds whose names indicate partial hydrogenation (cycloalkenes, pyrazolines, and the like)
d. number of substituents
e. lowest locants for substituents named as prefixes
f. lowest locants for substituents named as prefixes in alphabetical order of citation.

For a wide spectrum of ring system classes dictionary access has been replaced with purely algorithmic ring identification. The dictionary does not contain any entry that falls into one of these classes; AUTONOM generates the names of such ring systems exclusively on the basis of the analysis of their internal composition. The policy strictly followed during programming AUTONOM is to cover the greatest possible number of ring system classes using algorithmic identification, and reference to the dictionary is only carried out if the latter proves to be distinctly more efficient from a system performance point of view. The classes of ring systems which are recognised purely by an algorithm include:

- monocyclic hydrocarbons
- bicyclic alkanes
- monospirocyclic alkanes
- dispirocyclic alkanes
- tricyclic alkanes
- heteromonocyclics named by the Hantzsch-Widman method [8]
- replacement ('a' terms) nomenclature heteromonocyclics

- replacement ('a' terms) nomenclature heterobicyclics
- replacement ('a' terms) nomenclature monospiroheterocyclics
- replacement ('a' terms) nomenclature dispiroheterocyclics
- replacement ('a' terms) nomenclature heterotricyclics.

There are also classes of ring systems for which a combined dictionary and algorithmic scheme of identification is desirable. This combined approach is applied for all fused polycyclic hydrocarbons and heterocyclics with monospiro or dispiro connections to the ring systems classified in the above list.

Ring assemblies localised in the input structure are treated in a special way. According to IUPAC a ring assembly is a linear composition of two or more identical ring systems joined by acyclic single or double bonds, not necessarily at equivalent positions. The part of ring assembly identification conducted in assembly connections and the recognition of these systems not as isolated rings but as members of a bigger cyclic configuration. In nomenclature practice, this means inclusion of specific nomenclature criteria in the process of atom enumeration (such as special priority status for the atoms at the point of attachment, or the assignment of non-primed, primed or multi-primed locants to atoms constituting particular ring systems of the assembly).

Parent structure selection

This phase of the naming cycle begins with identification of potential parent chain(s). The chain identification process has been built around a single recursive routine using techniques common for graph transition methods [9].

Once candidate parent chains have been identified, the algorithm ranks the selected candidate structural units (ring(s) and functional parent(s), if any, have already been identified in the previous phases) according to the relevant nomenclature principles. If more than one candidate structural unit of the same rank competes for selection as a parent, the following sequence of principles, in the order given, is applied until a decision is reached:

a. greatest number of the principal functional groups cited as a suffix
b. preferred hetero atom content
c. rings preferred to chains
d. seniority of rings (if ring candidates present)
e. seniority of chains (if chain candidates present)
e. lowest locants for the substituents
f. lowest locants for the substituent cited first as prefixes in alphabetical order.

Seniority of ring systems is decided by applying 12 criteria which are, in general, in agreement with the IUPAC recommendation C.0.14 (p. 101 in [1]). The senior chain, i.e. the chain upon which the nomenclature and enumeration of all atoms in the structure will be based, is chosen by applying successively 10 criteria formulated by the IUPAC recommendation C.0.13 (p. 97 in [1]).

The structural objects which cannot be assigned the status of parent automatically become substituents on the selected parent unit.

Name tree creation

The hierarchic principle underlying the approach to a construction of a chemical name (parent, substituent, substituent-on-substituent, etc.) was the decisive factor in the design of the data format for effective name generation analysis. The data structure finally implemented has been built around the concept of an ordered binary tree [9].

The parent fragment selected in the preceding phase is established as the root of the name tree, while the other structural units become branches of the name tree. Starting at the root, the tree is traversed in an upward direction and mutual relationships among the branches (e.g. type of bonding, locants of connections, indicated hydrogen locants for ring systems, locants of multiple bonds in chains, etc.) are established and data concerning these relationships are added to the branch descriptors. Concurrently, if necessary, new structural objects are generated and corresponding branches added to the tree. The formation of new objects is encountered either in the case of newly generated functional groups resulting from the operation of the group splitting or in the case of newly identified substituent chains.

In the case of multiple candidate substituent chains starting at the same atom, the usual sequence of IUPAC rules is applied to find the senior one.

The structural objects no longer needed are immediately discarded and the memory segments they have occupied are deallocated and returned for further use. This is the case with the original functional groups which need to be split or with ring systems involved in an assembly junction. In the latter case one catenated ring assembly structural unit and corresponding single common assembly data record are generated while the individual assembly members are discarded.

The full description of all the possible objects and their mutual relationships is complete once the tree has been completely traversed in an upward direction. The structure is now fully mapped onto hierarchically ordered computer data records containing complete information on nameable structural fragments present in the structure. The data records are related to one another by a system of pointers organised as an easily manageable ordered binary tree.

Name assembly

Once the complete name tree is created, the algorithm begins to traverse the tree downward starting at most distant branch and moving to the root. The data records indicated by the pointers of visited branches are processed and resulting textual name fragments are generated. Track of the sequence of visited branches is kept in the form of labels organised as a dedicated stack. In the next step, the stack is successively emptied and the name fragments indicated by the current top item of the stack are taken and combined into longer fragments. The job of combining smaller name fragments into longer fragments, and finally into a complete chemical name, is executed by a set of sophisticated routines. These handle such complex operations as proper alphabetisation, multiplication, punctuation, redundant vowel elision, suppression of unnecessary locants, insertion of superscript or italic strings, etc.

The whole name assembly process is constantly monitored by an intelligent 'triviality' controller. The controller is a set of complex routines responsible for tracking and replacing (when necessary), the sort of overdone systematic pedantry which computers are prone to, and replacing this with concise IUPAC-sanctioned traditional nomenclature. The example from Figure 1 well illustrates the pithiness which has been introduced into the name generated by AUTONOM (for an extremely complex and fictional compound!)

General remarks on autonom 1.0 package

Figure 1 demonstrates (original output screens) the use of AUTONOM 1.0 in a PC-based implementation. The structure was drawn using mouse and the program returned the name (in the form shown) in 12 seconds.

The current version 1.0 handles single component, uncharged organic species of less than 100 heavy atoms (atoms other than H), neglecting for the moment salts, mixtures, polymers, alloys, coordination compounds, and inorganic materials.

AUTONOM 1.0 is not intended to produce 'the' unique name for a structure (IUPAC recommendations are so consciously formulated allowing considerable freedom in their application); on the other hand, the names produced by any particular version of the program should be reproducible and acceptable in practice. Analysis of the results produced from the current version (1.0) on the basis defined above shows a general success rate (measured by the production of a name on a non-steric basis) of well over 71 % of the structures in the sample. The sample of 43.000 structures was extracted from Beilstein 4.3 mln structure file on a random basis. In the remaining 20-30% the program refuses to produce a name, and prints a fatal diagnostic. The degree of acceptability of the names so produced is difficult to measure, since this is always a matter of taste and individual convention. In my eyes, naturally, the level of total unacceptability is marginally low.

The program runs on an IBM-AT (or compatible, also 80386/80486 based) normal configuration with a minimum of 512 Kb of RAM and a hard disk with the minimum of 4.0 MB free for storage of the dictionaries. On average, depending on the complexity of the structures, AUTONOM 1.0 names up to 17 structures per minute (measured on Compaq Deskpro 386).

Programming the system to its current state was a substantial task resulting in 34.000 PASCAL program lines and 221 different routines and functions.

Acknowledgement

The development of the AUTONOM 1.0 project was generously supported by the Bundesministerium für Forschung und Technologie (BMFT). The author and the Beilstein Institute gratefully acknowledge this support.

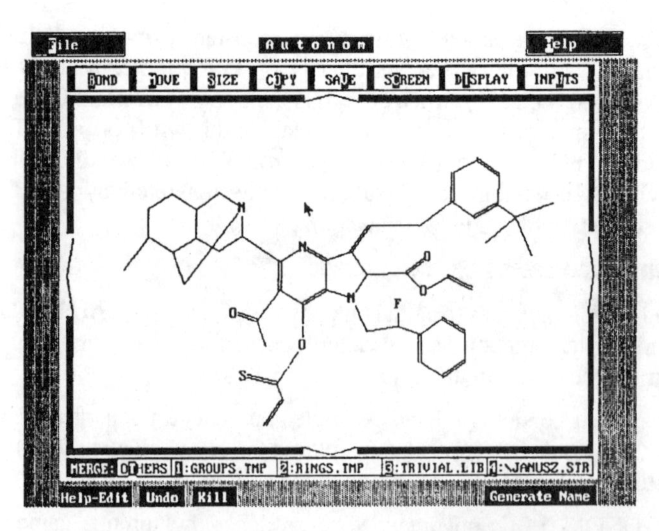

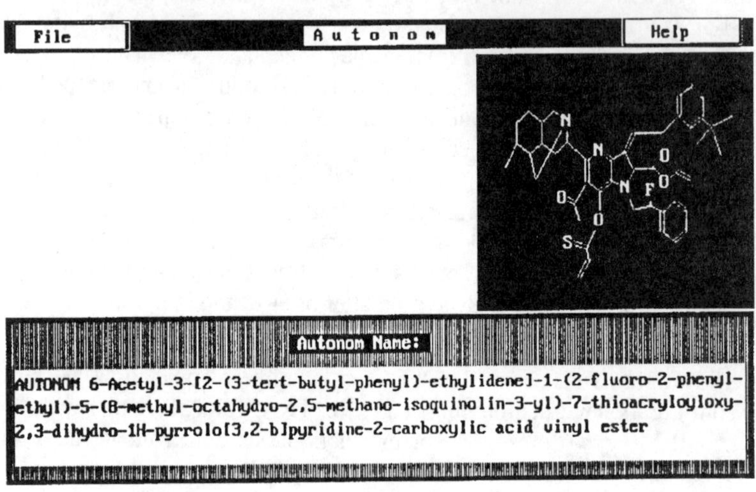

Figure 1: Structure and name output screens from AUTONOM program

References

[1] International Union of Pure and Applied Chemistry. *'Nomenclature of Organic Chemistry, Sections A-F and H'*; Pergamon: Oxford, U.K., 1979.

[2] Wišniewski J. L. 'AUTONOM: System for Computer Translation of Structural Diagrams into IUPAC-Compatible Names. 1. General Design'. *J. Chem. Inf. Comput. Sci.* **1990**, *30*, 324-332.

[3] Goebels L., Lawson A. J., Wisniewski J. L. 'AUTONOM: System for Computer Translation of Structural Diagrams into IUPAC-Compatible Names. 2. Nomenclature of Chains and Rings'. *J. Chem. Inf. Comput. Sci.* **1991**, *31*, 216-225.

[4] Vander Stouw, G. G.; Gustafson, C.; Rule, J. D.; Watson, C. E. 'The Chemical Abstracts Service Chemical Registry System. IV. Use of the Registry System to Support the Preparation of Index Nomenclature'. *J. Chem. Inf. Comput. Sci.* **1976**, *16*, 213-218.

[5] Mockus, J.; Isenberg A. C.; Vander Stouw G. G. 'Algorithmic Generation of Chemical Abstracts Index Names. 1. General Design'. *J. Chem. Inf. Comput. Sci.* **1981**, *21*, 183-195.

[6] Meyer, D. E.; Gould, S. R. 'Microcomputer Generation of Chemical Nomenclature from Graphic Structure Input'. *Am. Lab.* **1988**, *20(11)*, 92-96.

[7] Wišniewski, J. L. 'Effective Text Compression with Simultaneous Digram and Trigram Encoding'. *J. Inf. Sci.* **1987**, *13*, 159-164.

[8] International Union of Pure and Applied Chemistry. 'Revision of the Extended Hantzsch-Widman System of Nomenclature for Heteromonocycles'. *Pure & Appl. Chem.* **1983**, *55(2)*, 409-416.

[9] Tenenbaum, A. M.; Augenstein, M. J. *'Data Structures Using Pascal'*. Prentice-Hall: Englewood Cliffs, N.J., USA, 1981, p 318.

Stereochemistry in the CAS Registry File

Lisa M. Staggenborg

Chemical Abstracts Service, P. O. Box 3012, Columbus, OH 43210, USA

Abstract

Chemists have long been aware of the importance of stereochemistry in reactions and biological activity. In the past few years, state-of-the-art synthesis, separation, and analysis techniques have promoted greater application of stereochemistry. Access to information about specific stereoisomers is becoming increasingly important in areas such as molecular modeling and drug design. CAS has launched a high priority effort to improve stereochemistry handling as part of an overall renovation of the CAS Chemical Registry System.

Stereochemical representation has been an integral part of the Registry System since its inception in 1965. Stereochemical information is currently recorded, using a controlled vocabulary, in the CAS chemical name and the stereochemical descriptor field. Stereoisomers are considered to be different, individual substances, and each is recognised as unique. Approximately 25% of the over 11 million substances in the Registry File contain stereo information.

Although a nomenclature approach to identifying stereoisomers uniquely has served CAS's internal requirements for support of CA indexing and registration, this approach does not completely address the needs of scientists accessing Registry via STN because it does not support stereo structure display or search. To provide these capabilities, CAS is adding specific atom and bond stereochemistry in the connection table to make the stereo information more amenable to computer handling and manipulation.

The general strategy for this undertaking is to upgrade as many as possible of the approximately three million stereospecific substance records using computer algorithms. Approximately 1.2 million substances have been converted. This paper focuses on the algorithmic conversion and stereo display of these substances. Plans for additional stereochemistry work in the CAS Registry File are also outlined.

I. Stereochemistry

Stereochemistry has long been of vital interest to chemists because of its effect on the properties of molecules, especially in biological processes. In many cases a single stereoisomer is responsible for biological activity. Other isomers are often inactive or may have unrelated or innocuous side effects. However, in some cases, the effect of stereochemistry can be dramatic. The most infamous case of stereospecific biological activity is the drug thalidomide. Thalidomide was prescribed in the 1950s as a sedative. Tragically, the drug was also teratogenic, resulting in thousands of deformed infants. D-thalidomide appears to be responsible for the sedative effect while L-thalidomide is the teratogen.[1] Stereoisomers can differ in other properties as well. For example, the *S* enantiomer of carvone has a caraway-like odor, while the *R* enantiomer has a spearmint-like odor.[2] Stereoisomers, particularly geometrical isomers, can also differ in their physical properties. For example, the *Z* isomer of 2-butenoic acid (isocrotonic acid) is a liquid at room temperature while the *E* isomer (crotonic acid) is a solid.[3]

Stereochemistry is increasingly important to chemists. Approximately 50% of the drugs in the marketplace have at least one stereocenter.[4] Improvements in the areas of synthesis, separation and analysis are giving chemists greater insight into and control over stereochemistry.

By definition, stereochemistry is the spatial nature of a molecule. However, the concept of 'chirality' or 'handedness' differs from the 3-D coordinates of a molecule in space. This distinction is clearer if an analogy is made to left and right hands. A three-dimensional picture captures a specific 'conformation' such as a clenched fist or an open hand. However, the 'chirality' of the hand, whether or not the hand is left or right, is totally independent of whether the hand is clenched or open. The CAS Registry system deals with substances on a configurational level (chirality) as opposed to a conformational level. A substance and its enantiomers are recognised as unique and therefore registered, but specific conformations are typically not registered separately. This paper describes CAS's handling of stereochemistry in the sense of chirality, not in the sense of 3-D coordinates such as those used in molecular modeling.

II. Stereochemistry in the CAS Registry System

The Chemical Abstracts Service (CAS) Chemical Registry System is a computer system that identifies substances uniquely on the basis of their molecular structure. Begun originally to support substance indexing for Chemical Abstracts (CA), the Registry System is now perhaps best known as the source of CAS Registry Numbers and the Registry File on the STN International network.

The Registry System has evolved over a period of years. The initial 1965 version, Registry I, was designed to register only fully defined organic compounds. In 1968 the second version of the CAS Registry System, Registry II, was introduced, which extended machine registration to include inorganic substances, coordination compounds, polymers, mixtures, alloys and certain incompletely defined substances. Registry III, the current version of the Registry System, became operational in late 1973. Although no changes were made in the basic algorithmic techniques for registration, Registry III made a major adjustment to the Registry records so that the system would more effectively support CAS index nomenclature generation and computer-based structure output. The design, content, functions, statistics, special features, and input structure conventions of Registry III have been described in detail in previous papers.[5-15]

The boundaries of the Chemical Registry System are continually being extended, both in content and the manner in which information is added. In the past 25 years the CAS Chemical Registry System has changed from a production tool for CAS publications and services to a vital, international service for the scientific community. With this expanded role, CAS intends that the Registry System and related services will become even more responsive to the needs of scientists and engineers.

CAS is currently designing and developing Registry IV. Improvements include enhanced alloy processing, improvements in Registry File display on STN, faster registration through a new online input system, addition of biosequence data, and introduction of display and search of stereochemical information in the Registry File. Additional items being investigated include more flexible searching via a variety of conventions for representing substances and better support for industrial and engineering users of information about metals and alloys, polymers, ceramics and composites.

Stereochemical representation has been an integral part of the CAS Chemical Registry System since its inception in 1965. Stereoisomers are considered to be different, individual substances, and each is assigned its own CAS Registry Number. The unique identification of stereoisomers permits the storage and retrieval of information collected from scientific literature about a specific isomer. Approximately one-fourth of the over 11 million substances in the Registry File contain stereo information.

Substances on CAS Registry File:	11,637,649	
Stereo-containing substances:	3,065,000	26.3%

CAS Registry File Statistics (20 June 1992)

The unique identification of stereoisomers in the Registry system is accomplished via a controlled vocabulary field called the 'stereochemical descriptor' which was introduced with the CA Ninth Collective Index (9CI) in 1972 along with many other nomenclature changes. Prior to 1972, the stereo information was recorded less systematically. Although additional nomenclature changes have been made, CAS still uses nomenclature and stereochemical descriptors based primarily upon rules introduced in 1972.

The stereochemical descriptor describes the stereo in the overall substance. This has the advantage that stereoisomers appear at the same index heading in printed Chemical Abstracts. For example:

Androstan-1-one, (5α)

Androstan-1-one, (5β)-

The CAS stereochemical descriptors use standard stereochemical terms such as alpha, beta, cis, trans, endo, exo, syn, anti, *R*, *S*, *E* and *Z*.[16-17] *R** and *S** are used

RN 2244-16-8
CN 2-Cyclohexen-1-one, 2-methyl-5-(1-methylethenyl)-, (S)- (9CI) (CA INDEX NAME)
CN (S)-Carvone
DES 1:S
Absolute stereochemistry.

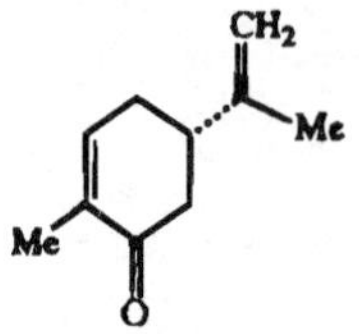

281 REFERENCES IN FILE CA (1967 TO DATE)

Figure 1

RN 6485-40-1
CN 2-Cyclohexen-1-one, 2-methyl-5-(1-methylethenyl)-, (R)- (9CI) (CA INDEX NAME)
CN (R)-Carvone
DES 1:R
Absolute stereochemistry.

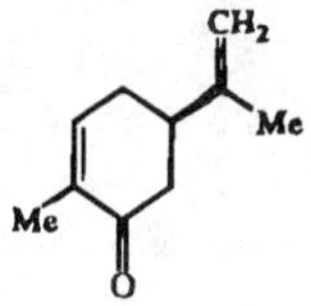

422 REFERENCES IN FILE CA (1967 TO DATE)

Figure 2

for relative stereochemistry. The Cahn, Ingold, Prelog (CIP) sequence method is used extensively.[18-19]

Figures 1 and 2 show two stereoisomers.

Note that each isomer has a unique Registry Number (RN). For the first isomer, RN 2244-16-8, the stereochemical descriptor, shown in the field 'DES', is '1:S'. This indicates that the stereochemistry is absolute and has a CIP value of S. The stereochemical descriptor information is also contained at the end of the CA index name in the 'CN' field, in a slightly different format, as '(S)'.

The stereochemical descriptor has proven itself to be an invaluable tool in the CAS Registry System. Stereochemical descriptors have been used to encode the stereo information for over three million diverse substances reported in the literature.

Stereochemistry is currently expressed in stereochemical descriptors in three ways: systematic stereochemical descriptors (types 1, 2, 3 and *), stereoparents (types 4, 5 and 6), and coordination compound stereochemical descriptors (type 7). Systematic descriptors are used with systematically named compounds. Stereoparent descriptors are provided for natural products. Stereoparent names, such as pregnane and morphinan, use trivial names which concisely imply certain complex stereochemistry.

The eight specific types of stereochemical descriptors used in the CAS Registry System are:

	Stereo-descriptor Type	Example CA Index Name	Example Stereo-chemical descriptor	CAS Registry Number
1	Absolute	Propanoic acid, 2-hydroxy-, (S)-	1:S	79-33-4
2	Relative	Cyclohexane, 1,2-dichloro-, *trans-*	2:TRANS	822-86-6
3	Optical	Propanal, 2,3-dihydroxy-, (+-)-	3:(+-)	56-82-6
4	Stereo-parent	Cholest-5-en-3-ol (3β)–	4:3B,CHOLEST	57-88-5
5	Amino acid / carbohydrate	α-*D*-Glucopyranoside,β-*D*-fructo-furanoyl	5:B-D-ARABINO, A-D-GLUCO	57-50-1
6	Trivial name	Retinol	6:RETIN	68-26-8
7	Coordination compound	Platinate(2-), dibromotetrachloro-, (OC-6-12)-	7:OC-6-12	35026-29-0
*	Known or implied	Bicyclo[2.2.1]heptan-2-ol, 1,3,3-trimethyl-,(1*S*-*exo*)-	*	470-08-6

CAS stereochemical descriptors

Although the stereochemical descriptor has been effective for the CAS Registry System, it is not amenable to the type of computer manipulation necessary to support features such as stereo display and stereo structure search that are of interest to CAS customers. Interpretation of stereochemical descriptors requires knowledge of CAS nomenclature and CIP and the ability to identify stereocenters in a substance. CAS is converting the stereochemical descriptor to an atom/bond-specific representation in the connection table in order to provide improved stereochemical features to users of the CAS Registry File.

Several different techniques are to be used to represent stereochemistry in the Registry connection table record, depending on the type of stereocenter being described.[20] Tetrahedral stereocenters, stereogenic double bonds and allenes are described by using a parity descriptor similar to that described by Petrarca, Rush and Lynch[21] in 1967 and adapted by Wipke and Dyott in their Stereochemically Extended Morgan Algorithm.[22] The relative CIP ranks of the neighboring atoms are recorded along with the parity to allow straightforward calculation of *R*, *S*, *E* or *Z* for the center or bond according to the CIP rules. Parity is described in greater detail later in this paper. Coordination centers will be described by using a geometry descriptor and a configuration descriptor.

A fundamental assumption of the stereo effort at CAS is that as many stereo substances as possible in the CAS Registry File should be enhanced to support features such as stereo display. Thus the project faces an enormous backfile of over three million stereo substances as well as the challenges of developing new stereo features. Users of the CAS Registry File have encouraged CAS not to wait until the effort is complete before beginning to deliver enhanced online capabilities.

CAS began its multi-year stereochemical efforts with organic substances containing only tetrahedral stereo centers and double bonds. Algorithmic conversion of over 1.2 million substances has been completed for substances with systematic stereochemical descriptors (types 1, 2, 3 and *). Work on the conversion of organic substances with stereoparent descriptors (types 4, 5 and 6) is beginning. Coordination compounds (type 7) will be addressed as a third phase. Atropisomers, hindered biphenyls, and other types of planar and axial stereochemistry, which account for only a small percentage of stereo substances currently seen in the literature, are being postponed.

Stereo Conversion

Phase 1 Systematically named compounds

- 1: Absolute
- 2: Relative
- 3: Optical
- *: Known or implied stereochemistry

Phase 2 Stereoparents

- 4: Stereoparent
- 5: Amino acid/carbohydrate
- 6: Trivial name

Phase 3 Coordination compounds

- 7: Coordination compound

The goal of the stereo project is to provide useful services, such as stereo display and stereo search, to users of the Registry File. Stereo display is the standard graphical representation of stereochemistry. Figure 3 shows a display of a substance in the Registry File where the stereo is contained only in the stereochemical descriptor textual field. The same substance is shown via stereo display in Figure 4 and with stereo labels in Figure 5.

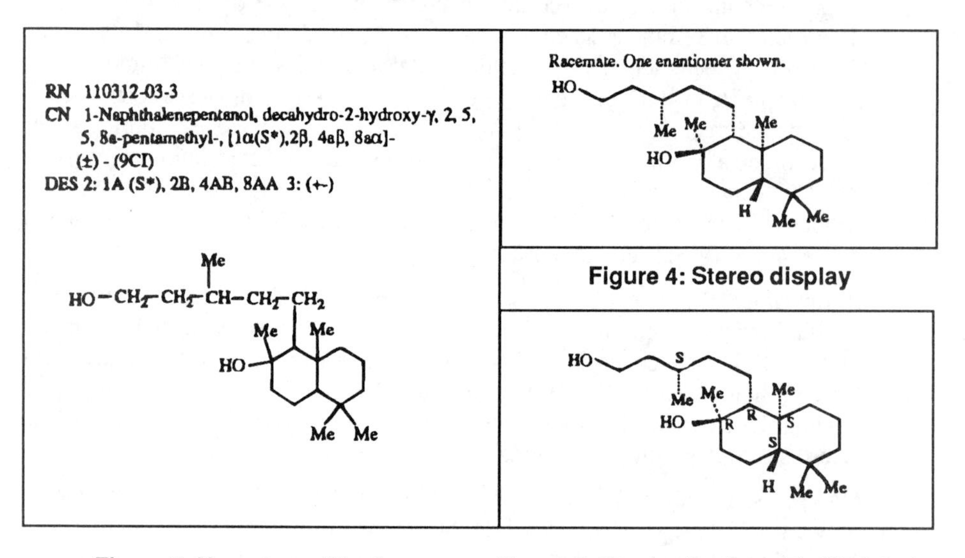

Figure 4: Stereo display

Figure 3: Non-stereo Display

Figure 5: Stereo display with CIP labels

Stereo search is the ability to narrow a search to retrieve only certain stereospecific configurations. Currently, in the Registry File on STN International, all structure searches (exact, family, substructure and closed substructure) retrieve all stereoisomers. Stereo search will be a precision search tool that will allow users to limit search results via stereochemical information.

Features such as stereo display or stereo search require the presence of stereo in the connection table at the node and bond level. Stereo conversion is the process of converting the textual stereochemical descriptor to parity information in the connection table. Figure 6 shows a stylised representation of parity in a connection table. Note that stereo information is indicated specifically at nodes 1, 9 and 10 and bonds 11-13 and 16-18.

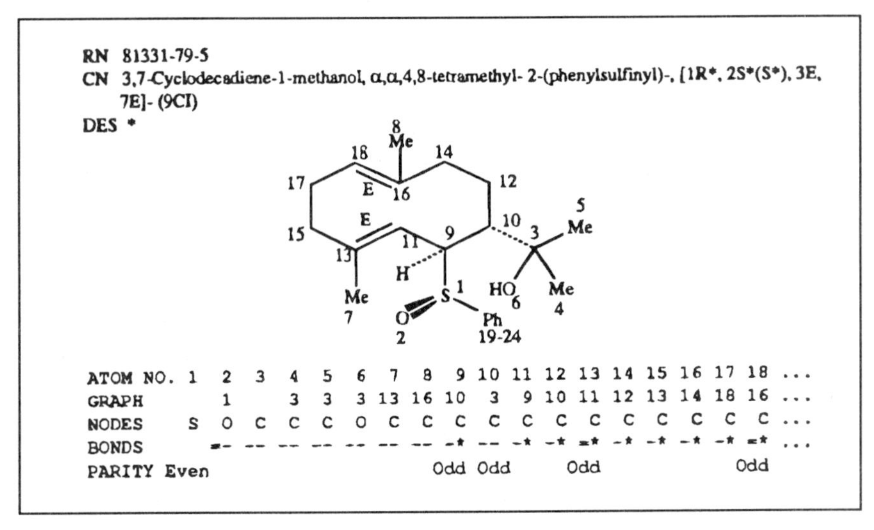

ATOM NO.	1	2	3	4	5	6	7	8	9	10	11	12	13	14	15	16	17	18	...
GRAPH		1		3	3	3	13	16	10	3	9	10	11	12	13	14	18	16	...
NODES	S	O	C	C	C	O	C	C	C	C	C	C	C	C	C	C	C	C	...
BONDS		=-	--	--	--	--	--	--	-*	--	-*	-*	=*	-*	-*	-*	-*	=*	...
PARITY	Even								Odd	Odd			Odd					Odd	

Figure 6

The stereo project at CAS is actually many interrelated stereo projects: stereo conversion, stereo display, stereo search, stereo input and stereo registration. This paper focuses on the stereo conversion and display of substances with systematic stereochemical descriptors.

III. Stereo Conversion

Stereo conversion is the translation of the textual stereochemical descriptor into stereospecific node and bond information in the connection table. There are two potential methods for stereo conversion. The first is manual input of stereo information. The second is algorithmic conversion via computer software. CAS is using a combination of these approaches.

The highest priority for stereo conversion has been placed on the largest group of substances, those with systematic stereochemical descriptors of type 1, 2, 3 or *. Algorithmic conversion of over 1.2 million substances with stereochemical descriptors of type 1, 2, 3 and * is already complete.

Type	Description	Number	Percentage
*	Known or implied stereochemistry	722,744	23.6%
1	Absolute	340,279	11.1%
2	Relative	687,182	22.4%
3	Optical	138,152	4.5%
	Phase 1 Subtotal	*1,888,357*	*61.6%*
4	Stereoparent	297,042	9.7%
5	Amino acid / carbohydrate	553,959	18.1%
6	Trivial name	19,614	0.6%
	Phase 2 Subtotal	*870,615*	*28.4%*
7	Coordination compound	306,028	10.0%
	Phase 3 Subtotal	*306,028*	*10.0%*
	Total Stereo Substances	**3,065,000**	**100.0%**

CAS Registry File statistics (20 June 1992)

Algorithmic conversion consists of a series of computer programs. Conversion requires the connection table, name and stereochemical descriptor for a given substance. This information is translated into parity, much the same way that a chemist would draw a graphical stereo diagram from the same information.

CIP and Parity

In order to appreciate algorithmic conversion, it is first necessary to have a basic understanding of both parity and CIP. Parity is an encoding of stereochemistry similar in many ways to CIP. Both parity and CIP are independent of nomenclature and rely exclusively on the spatial arrangement of atoms and bonds.

The basic CIP method involves priority ranking of the attachments to the stereocenter. Attachments are ranked by descending atomic number. In organic compounds it is generally necessary to compare the ranking of two or more carbon bonds. This is done by proceeding outward one step at a time until a decision is reached. Once priorities are assigned, the stereocenter is viewed with the lowest ranking attachment away from the viewer. *R* is assigned to a clockwise (right-handed) sequence of the highest ranking attachments, while *S* denotes a counterclockwise (anticlockwise) sequence.

A chemist assigning a CIP descriptor to a substance such as 2-chlorobutane, RN 22157-31-9 (see Figure 7), actually completes a number of steps. First the stereo center must be identified, in this case, node 1. The configuration of the attachments must be known, in this case shown by the Cl atom, node 4, above the plane of the paper. Next the chemist must assign priorities, via the CIP sequence rules, to the attachments. Cl, node 4, is the highest priority since it has the highest atomic number. H is the lowest priority. To determine the priority of the two carbons, nodes 2 and 3, the attachments to these nodes must be considered. Node 2 is considered higher priority because it is attached to a carbon, node 5, while node 3 has only hydrogen attachments. The attachments to the stereo center, node 1, can be listed in decreasing priority Cl, C-2, C-3, H. Holding the lowest priority attachment away from the viewer, the attachments are in clockwise order and so have a CIP descriptor of '*R*'. If the attachments had been in a counterclockwise order the CIP descriptor would have been '*S*'.

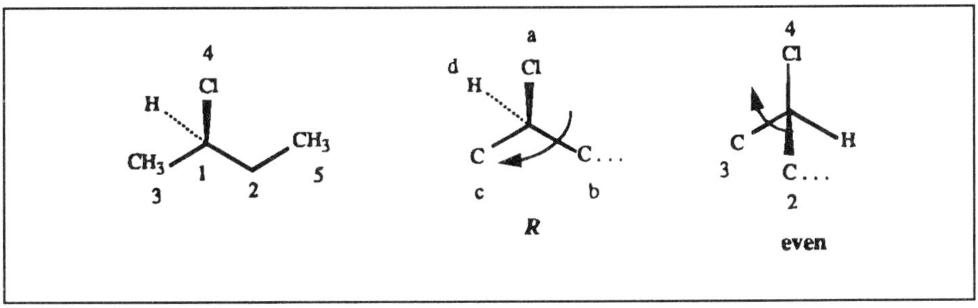

Figure 7

Assignment of parity is very similar to assignment of the CIP descriptor. Again, the stereocenter (node 1) and its configuration must be known. Instead of assigning priorities to ligands, connection table node numbers are used. H is considered to have a higher node number than any node number in the connection table. When the low numbered attachment is held toward the viewer, the remaining attachments are in clockwise order; therefore the parity is even. If the attachments had been in counterclockwise order, the parity would have been odd.

CAS makes use of both CIP and parity. CIP is used as input in the stereochemical descriptor and can be provided on stereospecific displays. Users of the CAS Registry File see only CIP and graphical stereo information. Parity functions behind the scenes in the connection table. CIP descriptors could have been used to store stereo information in the connection table, however, parity has the advantage that it requires minimal computer processing. CIP can require extensive analysis of a substance to determine the priority of attachments. Parity relies on connection table numbering which is already present. Additionally, parity is readily usable for either relative or absolute stereochemistry. Even symmetrical substances which cannot have CIP described, such as 1,4-disubstituted cyclohexanes, can be easily described with parity. Parity can be used with any connection table numbering as long as no number is repeated (i.e., a connection table cannot have two node '1's). Parity can be mapped from one numbering scheme to another. CIP ranking can be translated into other connection table numbering by treating the CIP ranks as node numbers (i.e. a=1, b=2, c=3, d=4). When CIP numbering is used, R is equivalent to even and S to odd because of the inherent properties of a tetrahedron.

The Conversion Process

The first step in the stereo conversion process is the selection of candidates from the Registry File. In Phase I, substances with systematic stereochemical descriptors (type 1, 2, 3, *) are selected. Polymers, mixtures, incompletely described, manually registered substances and coordination compounds are excluded. RN 117585-25-8 is an example of a substance which is selected as illustrated in Figure 8.

After a substance is selected, edits are run to compare the DES field to the stereo information contained in the name. Any inconsistencies found are reviewed by chemists. Because many stereochemical descriptors utilise nomenclature locants it is necessary to determine these locants. A program called 'Nomenclature Translation' (NT), which was developed at CAS in the early 70's,[23] is used to determine the nomenclature locants (Figure 9). NT also 'freezes' tautomers and alternating double and single bonds into specific double and single bonds.

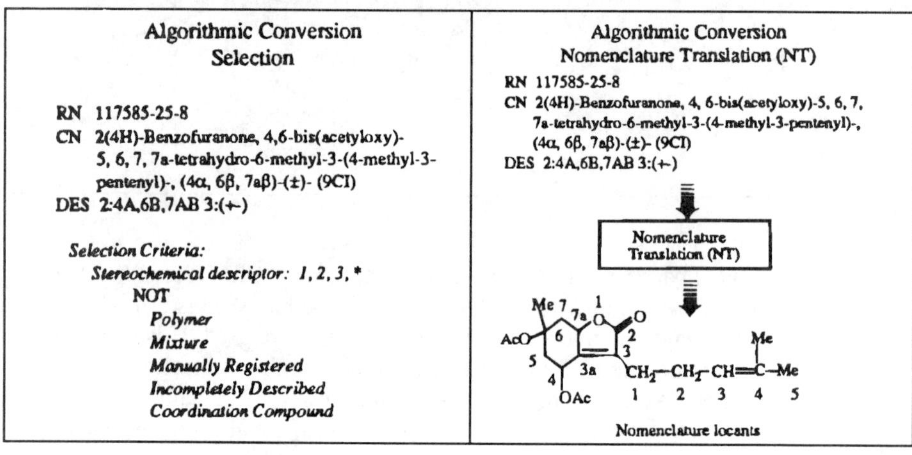

Figure 8 Figure 9

Next, all stereocenters are found (Figure 10), and CIP calculations are done (Figure 11).

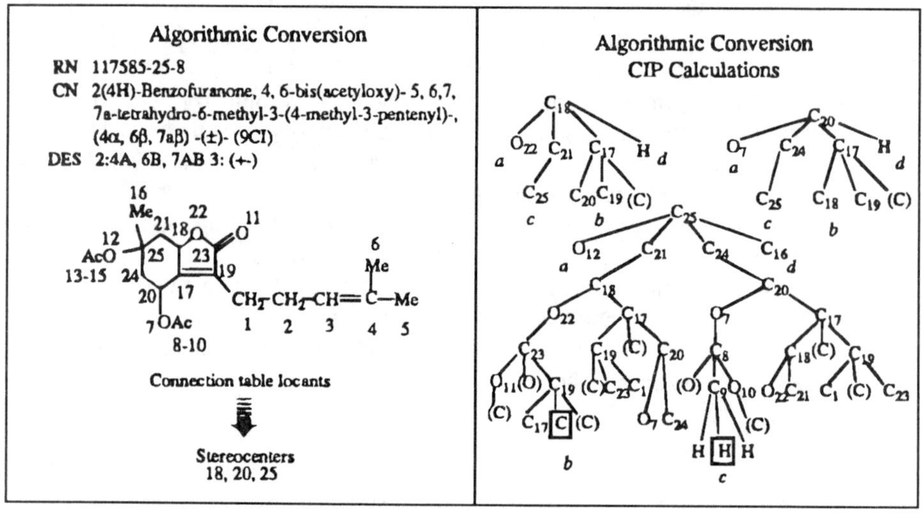

Figure 10 Figure 11

The stereochemical descriptor is parsed to find the specific terms. In the example, the term '4α' is associated with node 20 in the connection table. CIP priority information as well as rules for CAS stereochemical descriptors are used to determine that the 'alpha' refers to node 7 (the highest CIP acyclic attachment) and is arbitrarily positioned 'up'. Each term is interpreted. '6β' is translated to node 25 with attachment 12 'down'. '7aβ' is translated to node 18 with a hydrogen 'down'. The spatial relationships (up and down) are then translated to parity as shown in Figure 12. Finally, the parity and CIP sequence information is written to the connection table. Storing CIP sequence information allows straightforward calculation of *R* and *S* descriptors for the center according to CIP rules. Figure 13 illustrates the use of the conversion results in stereo display.

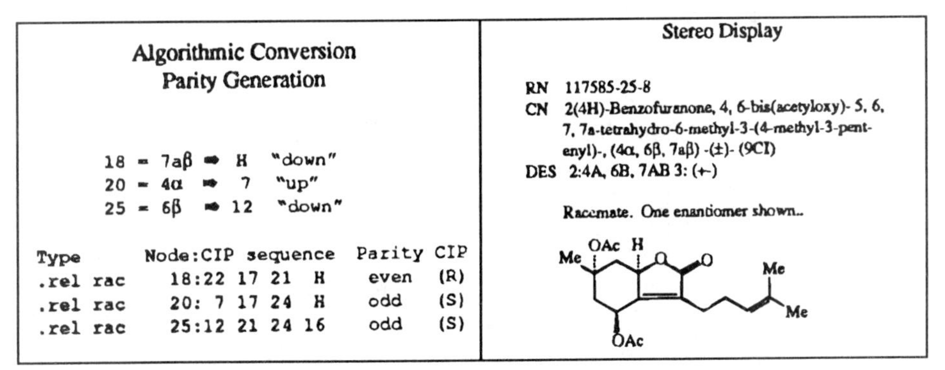

Figure 12 Figure 13

Figures 14-16 provide additional examples of substances which have been processed via algorithmic conversion. A range of descriptors is handled, including terms which require the relating of multiple centers such as 'endo' and 'syn'. The CIP algorithm is capable of resolving complex prioritisations such as that required for RN 117585-25-8 (Figure 11).

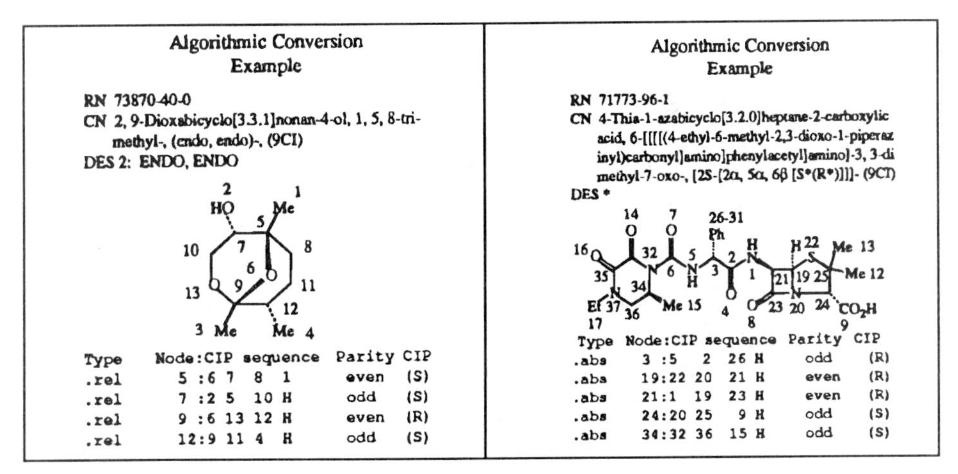

Figure 14 Figure 15

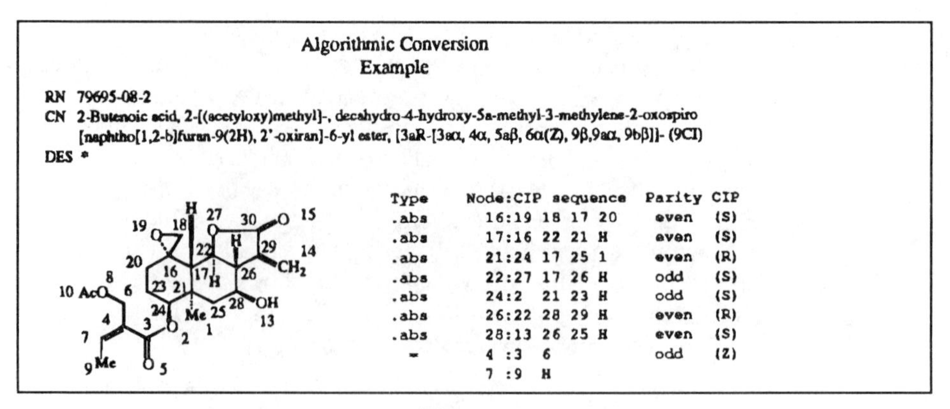

Figure 16

Development of the algorithmic conversion programs was an iterative process. Testing was performed at each stage of development. For example, the CIP algorithm was tested in isolation prior to development of other conversion functions. The Registry File provides a wealth of test cases including many small pockets of specialised substances which require special handling. For example, rings with trivalent nitrogen must be treated specially. CAS typically only cites trivalent nitrogen atoms as stereocenters when they are contained in bridges. Nitrogen atoms at fusion points in rings are not routinely cited. Therefore, the conversion program was required to distinguish between nitrogen atoms in bridges and nitrogen atoms at fusion points. Figure 17 shows a substance with a nitrogen in a bridge at node 10 with parity. Figure 18 shows a substance with a nitrogen at a fusion point at node 6 without parity.

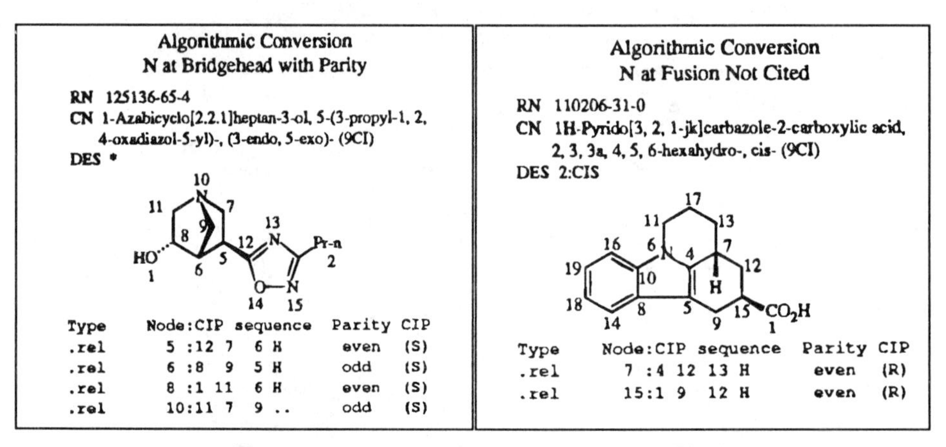

Figure 17 **Figure 18**

Approximately 10,000 substances were tested by intellectually reviewing the parity values. These values were saved in an authority file for regression testing. Additionally, an audit was run on initial conversion results to verify that no two substances in the same stereo nest (identical connection table information, excluding stereo) had the same stereo information in the connection table. The audit also verified that all substances in a nest have the same CIP priority ligands. The audit was valuable for finding latent program bugs as well as for verifying data integrity on the CAS Registry File. Corrections were made to both stereochemical descriptors and the conversion programs as appropriate.

Statistics for Phase I Conversion

The goal of algorithmic conversion is to generate correct parity information. Although it is desirable, it is not a requirement that algorithmic conversion translate stereochemical descriptors for 100% of substances. In some cases it is simply more cost-effective to input the stereo manually.

The overall conversion rate for the algorithmic conversion of substances with systematic descriptors (types 1, 2, 3 or *) is 67% (1.24 million out of a total 1.86 million candidates). This can be broken down by descriptor type as follows:

Stereochemical descriptor type	*Percentage conversion*
1	88%
2	92%
3	96%
*	29%

Conversion statistics by descriptor type

It is not surprising that substances with stereochemical descriptors of type * are least likely to pass algorithmic conversion. The * stereochemical descriptor is used only when it is not appropriate to use one of the other descriptor types (1-7). As a result, many special cases end up with an * stereochemical descriptor.

The goal of the stereo effort at CAS is to convert as many stereo substances as possible — either algorithmically or manually. Additional algorithmic conversion of substances with systematic stereochemical descriptors (type 1, 2, 3 and *), especially of special cases with large numbers of substances, is under investigation. The decision of whether to proceed algorithmically or manually will be based on cost estimates of each approach.

Future Conversion Efforts

Future efforts for algorithmic conversion will involve substances with stereoparent (type 4), amino acid/carbohydrate (type 5), trivial name (type 6), and coordination compound (type 7) stereochemical descriptors.

Development of algorithmic conversion programs for substances with stereoparent (type 4) descriptors is in progress. These substances use a stereoparent name to give the basic stereo. Additional or abnormal stereo, as well as features such as ring modifications and side chain stereo, are specifically cited in the stereochemical descriptor. For example, 4:3B,5A,9B,17B.ANDROST is used for an androstane derivative. Conversion of type 4 substances is based on templates. Known stereo from the templates is transferred to the file substance, and then modifications, as specified in the stereochemical descriptor for the substance, are applied. As with the conversion of substances with systematic stereochemical descriptors (1, 2, 3 or *), it is necessary to map nomenclature locants to connection table numbering to determine stereocenters, calculate CIP priorities, and produce parity values. And, of course, the special cases must be made to either pass conversion correctly or be routed for manual processing.

Conversion of substances with trivial name (type 6) stereochemical descriptors is expected to be similar to conversion of stereoparent (type 4) substances. However, many additional special cases are expected in this area, and due to the low number of substances, it may be more cost-effective to convert many of these substances manually. Work in this area has not yet begun.

Substances with amino acid/carbohydrate (type 5) stereochemical descriptors will require special approaches. This effort may be unified with other CAS efforts in this area, such as the work on biosequences.

The third phase of algorithmic conversion will address coordination compounds (type 7). These substances contain molecular geometry at coordination centers such as square planar (*SP*-4), trigonal bipyramid (*TB*-5), and octahedral (*OC*-6). Substantially different approaches are needed for algorithmic conversion of these substances.

IV. Stereo display

Stereo display is the first visible accomplishment of the stereo project at CAS. Stereo display relies upon the existence of parity in the connection table; therefore, stereo conversion is a prerequisite. The stereo display described in this paper, unless specifically noted otherwise, is to support substances with systematic stereochemical descriptors (type 1, 2, 3 or *).

There are two possible approaches to stereo display. The first is to manually draw all of the stereo substances and save the graphical image for use as a playback image. This approach is effective in many systems. Data entry is often accomplished via graphical input, and it is logical to reuse this input for display purposes. The second approach to display is to algorithmically generate display images from a connection table. CAS has been using algorithmic display since the late 70s.[24] CAS utilises keyboarding staff to rapidly key chemical data with an average input rate of approximately 90 structures per hour. Not surprisingly, aesthetics are not a consideration for input, and so these keyboarded images are not useful for display purposes. Algorithmic display provides high quality images that are typically highly consistent for similar substances.

CAS's algorithmic structure display (ASD) program requires as input a connection table and basic ring shapes. The ASD program uses a computer-readable authority file of basic ring shapes, the Ring Image File, in constructing structure diagrams. The Ring Image File contains basic geometrical data for each unique ring shape identified by the Registry system for all ring systems contained in the Registry File.

Figure 19 contains the ring image for RN 11785-25-8 (shown in Figure 13). Note the absence of heteroatoms and double bonds in the ring image.

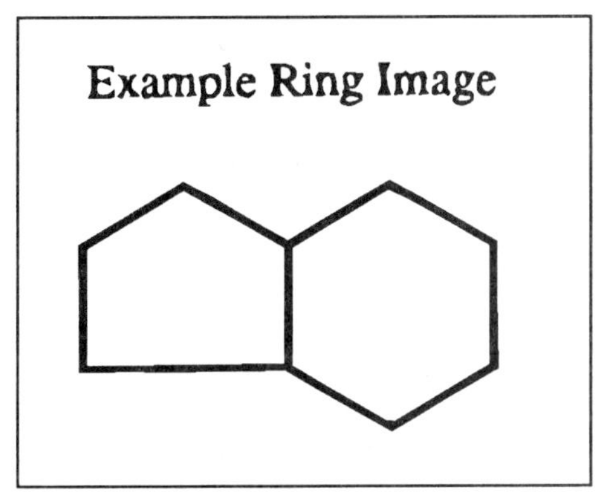

Figure 19

CAS determined that it would be more cost-effective to start with a new ASD program to support stereochemistry rather than to make extensive changes to the original program. The Shanghai Institute of Organic Chemistry (SIOC), which was already doing work in this area, was contracted to provide development support. The first step in the process to develop stereo display was to develop a new non-stereo program. New non-stereo ('flat') images were made public on the CAS Registry File on STN International in December, 1991.[25]

The first priority of CAS stereo display is the creation of clear and unambiguous stereochemical images. The aesthetics of the images, while important, are second priority.

The general style of CAS's stereo display is consistent with chemical community practices, especially those found in the printed literature. Wedged bonds are used for bonds above the plane of the paper. Non-directional dashed bonds are used for bonds below the plane of the paper. Other styles of bonds were considered and eliminated. For example, a directional dashed bond is often used for below the plane. However, there is inconsistency in which end of the bond is pointed toward the stereocenter. Software packages typically have the narrow end of the bond at the stereo center, consistent with solid wedges. Printed images sometimes have the wide end of the dashed bond near the stereocenter, consistent with a diminishing-view graphical perspective. Since the primary objective is a clear and unambiguous image, directional-dashed bonds were not used.

Online images differ little from offline print images. However, online images display an open wedge for above the plane bonds, while offline prints depict a filled wedge. STN Express is supporting filled wedges online. Online displays are provided via Tektronix 4010 graphics. Unfortunately, Tektronics does not directly support the concept of a filled polygon. Thus, the only way to create a filled wedge is via some sort of trick such as drawing a series of small lines. Experimentation with such techniques resulted in an increase in the number of bytes transmitted that had a noticeable impact on response time at 2400 baud speeds while drawing filled wedges.

CAS stereo display images utilise a zig-zag style of diagrams. A hexagonal grid is used whenever possible. This is highly effective for chiral centers, but does pose a challenge for display of double bonds of unknown geometry, since it is virtually impossible to display a double bond in a zig-zag style without appearing to imply stereochemistry.

CAS display images include an explicit statement about the nature of the stereochemistry displayed, i.e. relative, absolute, racemic, optical rotation, known double bond geometry or unknown double bond geometry. Although graphical depictions of this information have been proposed, none of these proposals are used widely enough in the community to be considered unambiguous. Any display of a stereo structure diagram, regardless of the format, results in display of a message.

Figure 20 is an example of a substance with absolute stereochemistry as reported by the author in the original literature. Figure 21 depicts a racemate, again as reported by the author in the original literature. Relative stereochemistry, such as that shown in Figure 22, is used whenever an author does not provide sufficient information for CAS to specify absolute or racemic stereochemistry. CAS considers achiral substances such as 1,4-disubstituted cyclohexanes and meso compounds to have relative stereochemistry. The designation of absolute versus relative versus racemic stereochemistry is made based only on the chiral centers in the substance.

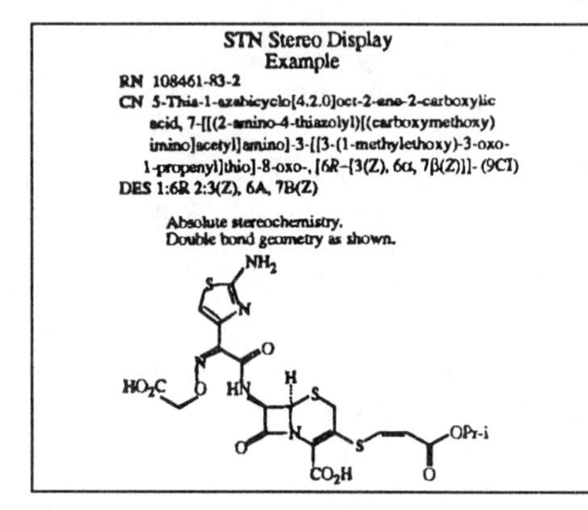

Figure 20

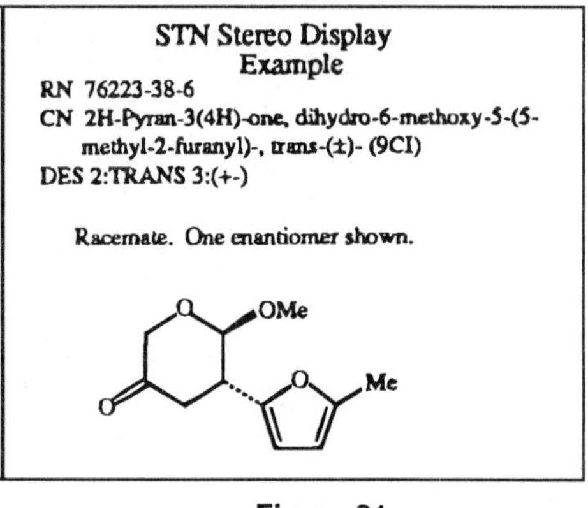

STN Stereo Display
Example

RN 76223-38-6

CN 2H-Pyran-3(4H)-one, dihydro-6-methoxy-5-(5-methyl-2-furanyl)-, trans-(±)- (9CI)

DES 2:TRANS 3:(+-)

Racemate. One enantiomer shown.

O OMe O Me O

Figure 21

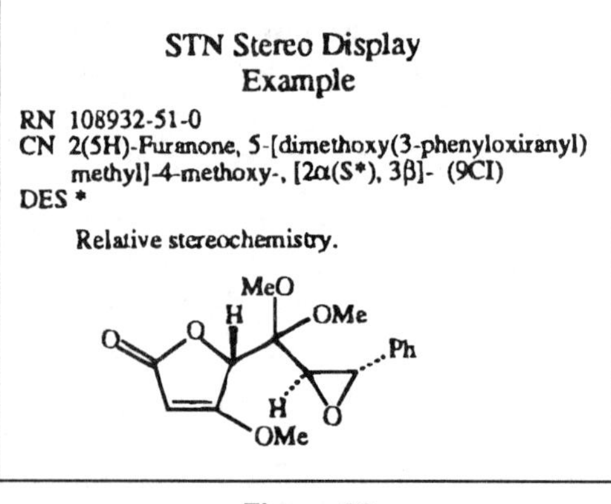

Figure 22

Stereo display in the Registry File on STN International is available on graphics terminals ('type 2'), including STN Express, and offline prints. Stereo display is not supported for text terminals ('type 3'). Three display formats have been defined as illustrated in Figure 23. The default display format, STR, which is included in the standard Registry File default format IDE, provides a stereo display if one is available; otherwise a flat display will be given. The STS format provides a stereo image that includes CIP labels (*R, S, E, Z*) when appropriate. The STF format provides a flat (non-stereo) diagram, whether or not a stereo diagram is available.

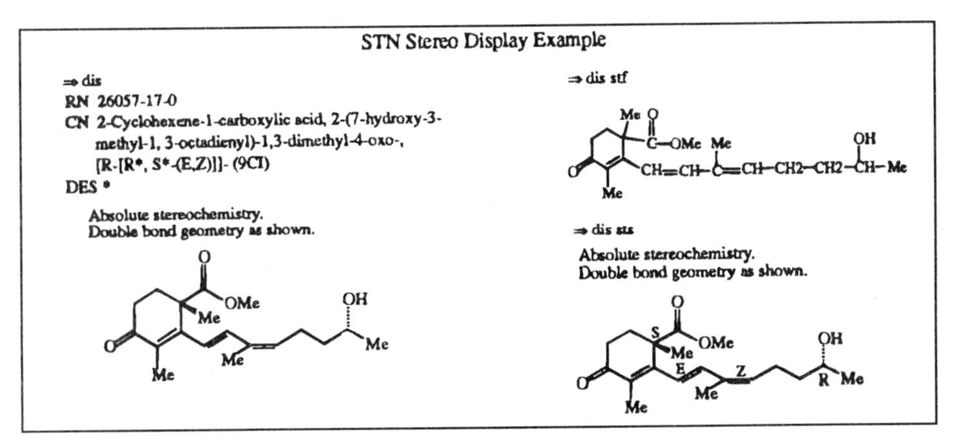

Figure 23

The most challenging aspect of stereo display is to represent spatial relationships unambiguously on paper or a computer screen. A critical rule to avoid ambiguity in stereo display is to prohibit use of a stereo bond between two chiral stereo centers. Figure 24 depicts a substance correctly displayed following this rule. The second structure for the same substance violates this rule and would not be used.

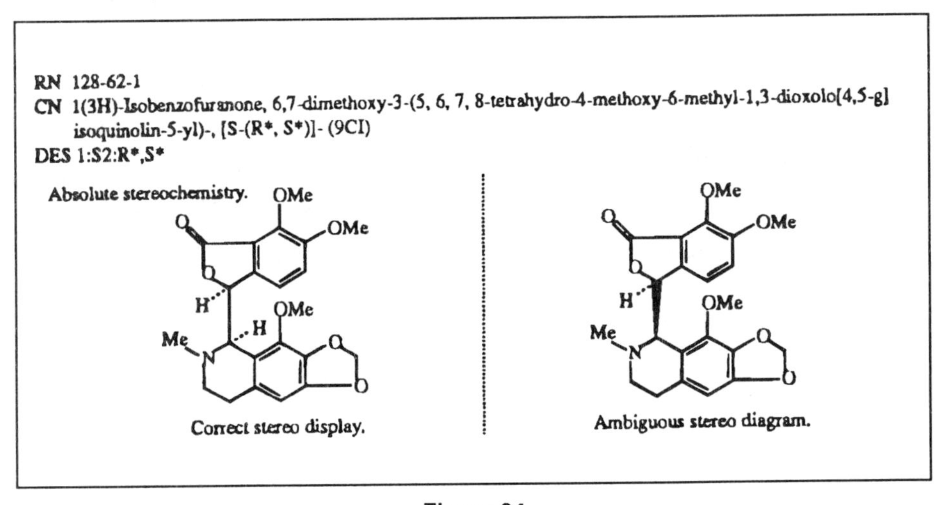

Figure 24

The accepted conventions for displaying stereo bonds at chiral centers are sensitive to the angles and relationships of the bonds. For example, Figure 25 depicts the same stereo center in a variety of ways. Some of the depictions are reasonable; others are nonsense. Nonsense diagrams result in cases where the person reviewing the diagram cannot reconcile the image with a tetrahedral shape. Figure 26 illustrates the special case of bridged rings. It is acceptable to have three nonstereo bonds on one side of a plane and a stereo bond on the other in a small bridged ring even though this same display style is ambiguous when all of the bonds are acyclic. In general, ASD distinguishes cases of acceptable displays of chiral centers by classifying stereo centers by the number of cyclic and acyclic attachments.

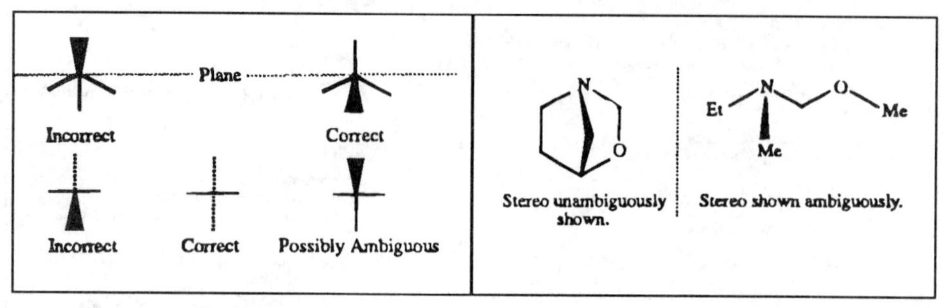

Figure 25

Figure 26

Figures 27-32 provide additional examples of stereo display. These displays reflect a number of stereo and non-stereo considerations such as the addition of hydrogen to stereocenters, the style of *E* and *Z* bonds, and the use of shortcuts.

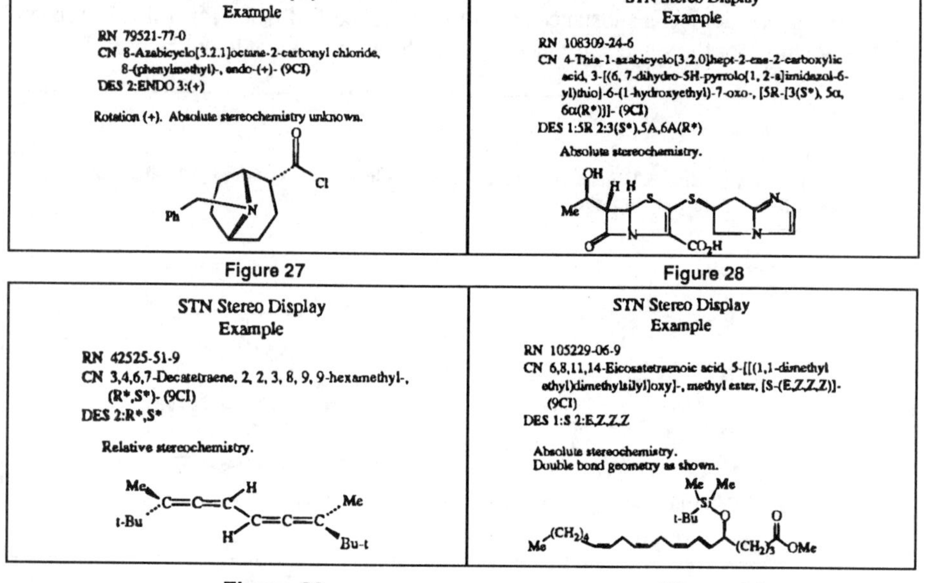

Figure 27

Figure 28

Figure 29

Figure 30

STN Stereo Display Example

RN 54142-58-4
CN Bicyclo[2.2.1]hept-5-en-2-one, 7-(dimethoxy methyl)-, (1R-anti)- (9CI)
DES *

Absolute stereochemistry.

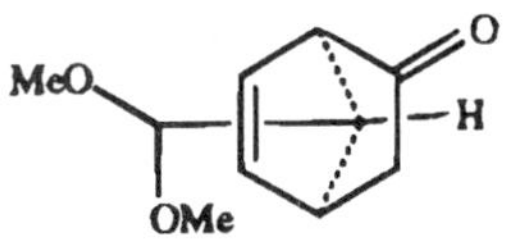

Figure 31

STN Stereo Display Example

RN 120308-91-0
CN 3-Oxabicyclo[4.1.0]heptane-7-carbonitrile, 5-hydroxy-4-(phenylmethoxy)-, [1S-(1α,4β,5α,6α,7α)]- (9CI)
DES *

Absolute stereochemistry.

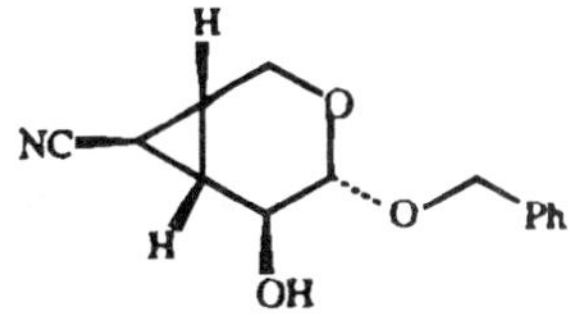

Figure 32

As with conversion, development of stereo display was an iterative process. Cycles of development and testing began with mockups of display examples in the user requirements document. Extensive testing of approximately 10,000 substances was performed within CAS via offline prints. Additional testing was performed online. As always, the Registry File provided interesting and diverse test cases.

Future enhancements will be made to stereo display as parity data becomes available for new classes of stereo substances. Stereoparents will require special techniques to ensure that a consistent orientation is rigorously maintained. Carbohydrates have specialised drawing conventions such as Haworth and Fisher projections that must be investigated. Coordination compounds are likely to require specialised stereospecific templates.

V. Stereo search

STN users currently retrieve all related stereoisomers via any type of structure search (exact, family, substructure or closed substructure) in the Registry File. Stereo searching is a precision searching tool that will allow users to retrieve specified stereoisomers.

The principle features of the planned stereo search procedures are:

- Stereochemistry can be specified in a search query via commonly used structure drawing conventions such as wedge and dashed bonds at chiral centers and geometry about double bonds.
- Search queries can specify stereochemistry in absolute or relative terms.
- Eventually, all stereo searches will automatically produce both a 'flat' answer set that ignores the stereo aspects of the query and a subset 'stereo' answer set that is limited to stereospecific answers. This will safeguard users from overprecise stereo searches.

This approach is illustrated in Figure 33.

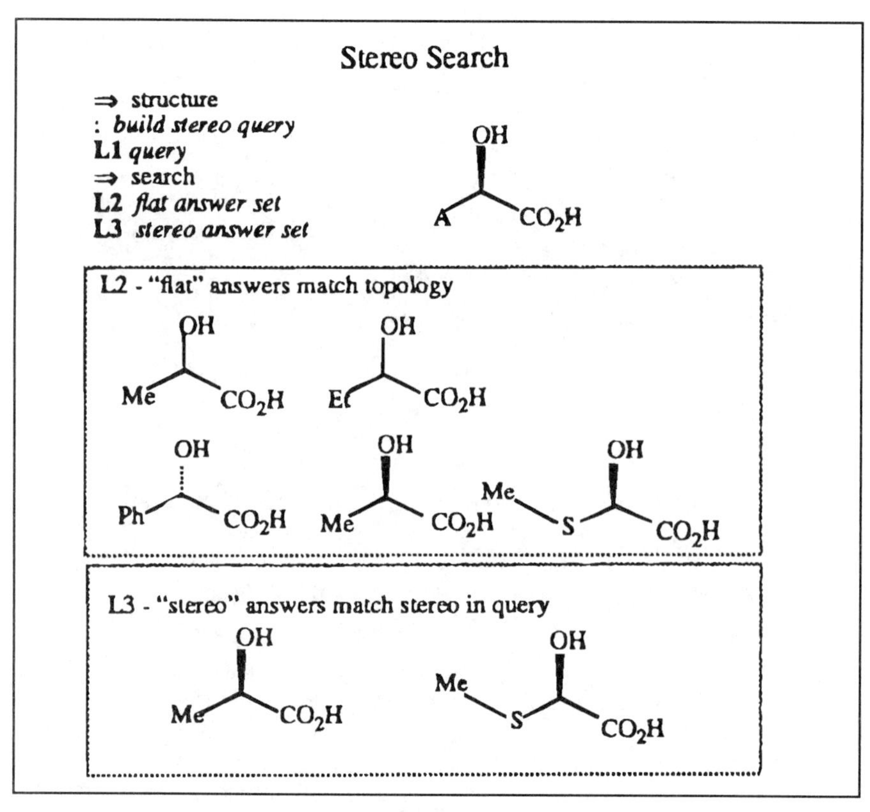

Figure 33

In addition to stereo structure searching capabilities, CAS is also planning to introduce a new field to provide the CAS Registry Numbers of the stereoisomers of a substance, giving searchers an easy way to retrieve the stereoisomers for a substance or for a number of substances in an answer set. A Registry File display on STN might look like:

RN 497-09-6
CN Propanal, 2,3-dihydroxy-, (S)- (9CI) (CA INDEX NAME)
CN Glyceraldehyde, L- (8CI)
MF C3 H6 O3
LC BEILSTEIN, BIOSIS, CAOLD, CASREACT, CJACS, CSCHEM, EINECS, HODOC
RSI 56-82-6, 367-47-5, 453-17-8
DES 1:S

Absolute stereochemistry.

HO CHO OH

REFERENCES IN FILE CAOLD (PRIOR TO 1967)
61 REFERENCES IN FILE CA (1967 TO DATE)

Figure 34

In this example (Figure 34), the proposed 'RSI' (Related Stereoisomers) field lists the Registry Numbers for glyceraldehyde with no stereo specified, DL-glyceraldehyde, and D-glyceraldehyde.

Stereo Input

Manual input of stereochemical information requires stereospecific graphical diagrams. Today most standard, nonstereo input to the Registry File is handled via graphical diagrams, but many changes in the system are necessary to capture stereospecificity. Eventually, most new stereo input to the CAS Registry File will be handled via graphical input, once software is in place to unambiguously interpret stereo information and translate it into parity in the connection table. (See Figure 35).

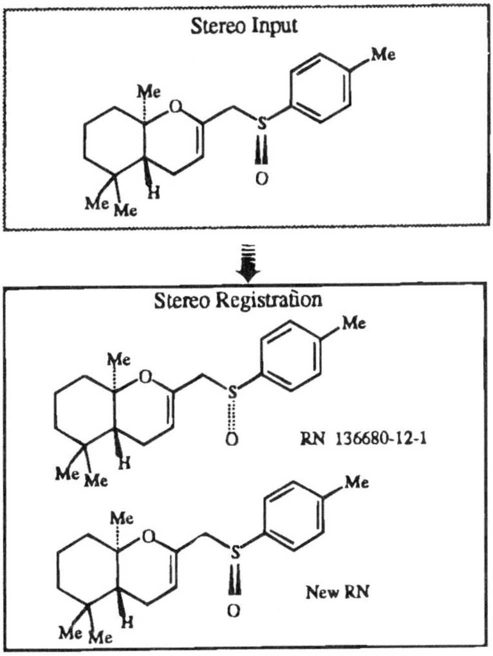

Stereo Input

Stereo Registration

RN 136680-12-1

New RN

Figure 35

Graphical input of stereochemical information will be required for stereo substances on the backfile that fail algorithmic conversion as well as for ongoing input.

VI. Stereo Registration

CAS plans to enhance our CAS Registry File database building by replacing the textual stereochemical descriptor with parity in the connection table as the basis for registration. The parity node/bond specific designation of stereo information is more flexible than the controlled vocabulary textual stereochemical descriptor.

Introduction of parity as the authority for stereochemical information presents new opportunities in the area of stereochemistry. CAS is investigating enhanced computer support for stereochemical registration and stereo nomenclature. Additionally, CAS is reevaluating policies dealing with stereochemical information. For example, substances are frequently reported with incomplete stereochemistry in the literature. CAS is reevaluating the possibility of retaining incomplete stereo information for registration purposes.

VII. Conclusion

CAS is in the midst of a high-priority effort to enhance our stereochemical representation in the CAS Registry File. Users of the Registry File on STN International will see these enhancements in the form of stereo features such as stereo display and stereo search. CAS internal production will benefit from graphical stereo input and stereo registration.

The stereo project is possible at CAS only because of the hard work and dedication of many CAS staff members, including:

Mark Anson, Rich Ball, Bill Bigard, Jim Blackwood, Jim Blake, Paul Blower, Karen D'Angelo, Kathy Deck, Pat Eiben, Bobbi Fiete, Les Goodson, Eva Hedrick, Cindy Huddle, Dave Huddle, Steve Layten, Dwight Lillie, Alan Lipkus, Kent Margraf, Mary McHughes, Joe Mockus, John Peer, Chen Qian, Lisa Staggenborg, Daniel Stossel, Ming Tsao and Chuck Watson.

References

[1] Gaffield, W., Chirality as Manifested in the Biological Activity of Natural Products, *Studies in Natural Products Chemistry,* Vol. 7, (Ed. Atta-ur-Rahman), Elsevier Science Publishers B. V., Amsterdam, 1990, 3-28.

[2] *Loc. Cit.*

[3] *The Merck Index. 11th edition.*, Rahway, N. J., Merck & Co., 1989, pp. 407, 812-813.

[4] *C&E News,* July 9, 1990, 9.

[5] Dittmar, P. G.; Stobaugh, R. E.; Watson, C. E., Chemical Abstracts Service Chemical Registry System. 1. General Design., *J. Chem. Inf. Comput. Sci.*, **1976,** *16,* 111-21.

[6] Freeland, R. G.; Funk, S. A.; O'Korn, L. J.; Wilson, G. A., Chemical Abstracts Service Chemical Registry System. 2. Augmented Connectivity Molecular Formula., *J. Chem. Inf. Comput. Sci.*, **1979,** *19,* 94-8.

[7] Blackwood, J. E.; Elliot, P. M.; Stobaugh, R. E.; Watson, C. E., Chemical Abstracts Service Chemical Registry System. 3. Stereochemistry., *J. Chem. Inf. Comput. Sci.*, **1977,** ***17,*** 3-8.

[8] Vander Stouw, G. G.; Gustafson, C.; Rule, J. D.; Watson, C. E., Chemical Abstracts Service Chemical Registry System. 4. Use of the Registry System To Support the Preparation of Index Nomenclature., *J. Chem. Inf. Comput. Sci.*, **1976,** *16,* 213-8.

[9] Zamora, A.; Dayton, D. L., Chemical Abstracts Service Chemical Registry System. 5. Structure Input and Editing., *J. Chem. Inf. Comput. Sci.*, **1976**, *16*, 219-22.

[10] Stobaugh, R. E., Chemical Abstracts Service Chemical Registry System. 6. Substance Related Statistics., *J. Chem. Inf. Comput. Sci.*, **1980**, *20*, 76-82.

[11] Mockus, J.; Stobaugh, R. E., Chemical Abstracts Service Chemical Registry System. 7. Tautomerism and Alternating Bonds., *J. Chem. Inf. Comput. Sci.*, **1980**, *20*, 18-22.

[12] Moosemiller, J. P.; Ryan, A. W.; Stobaugh, R. E., Chemical Abstracts Service Chemical Registry System. 8. Manual Registration., *J. Chem. Inf. Comput. Sci.*, **1980**, *20*, 83-8.

[13] Ryan, A. W.; Stobaugh, R. E., Chemical Abstracts Service Chemical Registry System. 9. Input Structure Conventions., *J. Chem. Inf. Comput. Sci.*, **1982**, *22*, 22-8.

[14] Hamill, K. A.; Nelson, R. D.; Vander Stouw, G. G.; Stobaugh, R. E., Chemical Abstracts Service Chemical Registry System. 10. Registration of Substances from Pre-1965 Indexes of Chemical Abstracts., *J. Chem. Inf. Comput. Sci.*, **1988**, *28*, 175-9.

[15] Stobaugh, R. E., Chemical Abstracts Service Chemical Registry System. 11. Substance-Related Statistics: Update and Additions., *J. Chem. Inf. Comput. Sci.*, **1988**, *28*, 180-7.

[16] Blackwood, J. E.; Giles, P. M., Jr., Chemical Abstracts Stereochemical Nomenclature of Organic Substances in the Ninth Collective Period (1972-1976)., *J. Chem. Inf. Comput. Sci.*, **1975**, *15*, 67-72.

[17] *Chemical Abstracts Index Guide*, Appendix IV, 1989, pp 180I-99I.

[18] Cahn, R. S.; Ingold, C.; Prelog, V., *Angew. Chem., Int. Ed.*, **1966**, *5*, 385-551.

[19] Prelog, V.; Helmchen, H., Basic principles of the CIP-system and proposals for a revision., *Angew. Chem., Int. Ed.*, **1982**, *21*, 567-83.

[20] Blackwood, J. E.; Blower, P. E.; Layten, S. W.; Lillie, D. H.; Lipkus, A. H.; Peer, J. P.; Qian, C.; Staggenborg, L. M.; Watson, C. E., Chemical Abstracts Service Chemical Registry System. 13. Enhanced Handling of Stereochemistry., *J. Chem. Inf. Comput. Sci.*, **1991**, *31*, 204-212.

[21] Petrarca, A. E.; Lynch, M. F.; Rush, J. E., A method for generating unique computer structural representations of stereoisomers., *J. Chem. Doc.*, **1967**, *7*, 154-65.

[22] Wipke, W. T.; Dyott, T. M., Stereochemically unique naming algorithm., *J. Am. Chem. Soc.*, **1974**, *96*, 4834-42.

[23] Vander Stouw, G. G.; Elliott, P. M.; Isenberg, A. C., Automated conversion of chemical substance names to atom-bond connection tables., *J. Chem. Doc.*, **1974**, *14*, 185-93.

[24] Dittmar, P. G.; Mockus, J.; Couvreur, K. M., An Algorithmic Computer Graphics Program for Generating Chemical Structure Diagrams., *J. Chem. Inf. Comput. Sci.*, **1977**, *17*, 186-92.

[25] Better structure diagrams in REGISTRY, *STNews*, Nov/Dec, 1991, Vol. 7, No. 6, p 10.

Chemical literature data extraction. Bond crossing in single and multiple structures

F. Kam, R. W. Simpson, C. Tonnelier, T. Venczel, and A. P. Johnson*

School of Chemistry, The University of Leeds, Leeds, LS2 9JT, United Kingdom

Abstract

The procedure to convert a scanned image of a page of chemical structure diagrams (with accompanying text) into a set of connection tables is one of the primary aims of the CLiDE project. These connection tables can then be used in a variety of computer-based applications such as building and maintaining databases. The image is decomposed into component graphics and text which are further analysed to find the lines, wedges, and chemical text strings. In an interpretation phase the connection tables for the molecules are built from these items. The correct interpretation of chemical bonding in the image is often hampered by the constraints of representing a three-dimensional molecule in two dimensions where one bond may be drawn over another. A method of identifying and successfully dealing with these situations is described. A related situation where a bond is drawn crossing a ring implying an undetermined point of attachment is also solved. Examples are presented to illustrate these situations and the rules implemented to handle these structures within the CLiDE program discussed.

General description

Online and in-house chemical databases allow chemists to make both rapid and complex searches for information extracted from the literature. This might include molecular structures, chemical reactions [1] or reaction schemes. Textual information such as an abstract, references or keywords usually accompanies the purely structural information.

Abstracting information from the literature and manually entering it into these databases is a laborious and expensive task requiring a trained chemist. CLiDE (Chemical Literature Data Extraction) is a new software system under development that attempts to solve this problem by computerising the procedure [2]. Programs of this type have been developed extensively in other subject areas [3,4]. Within chemistry, McDaniel and Balmuth [5] have described a commercial product for the interpretation of single chemical structures (Kekulé: Optical Chemical (Structure) Recognition). Contreras *et al.* [6] have presented details of a similar program that has been applied to simple molecular images already isolated from a journal page.

The aim of CLiDE is to process whole page(s) of chemical information from journals and books to extract the chemical structures, reaction schemes and other relevant chemical information. Another important requirement is to perform this task as far as possible without human intervention. In those cases where some ambiguity exists, the program is designed to prompt the user for additional information.

The different features which might be found on a page, such as structures, reaction schemes, diagrams and tables, cannot all be processed in the same way. Their characteristics are different, so a specific process is required for each of these object families. However, some processes will be able to use subprocesses already implemented for dealing with simpler objects. The interpretation of a single chemical structure is a fundamental process that is also required for interpreting reaction schemes and for the understanding of reaction tables containing chemical structure diagrams.

The interpretation of (multiple) chemical structures is, at present, the main achievement of the CLiDE project and opens the way for the processing of more complex objects. The CLiDE program is able to produce a computer-readable file after scanning most organic chemical structures. As will be shown, this includes complicated structures involving crossing bonds, molecular groups and generic groups.

The tasks performed to generate the connection table corresponding to a drawing of a chemical structure are detailed below. Both isolated structures and structures embedded in text or graphics can be interpreted by the current program. All the structures present on a page are converted into their corresponding connection tables and relevant text is associated with the correct structures.

System overview

The CLiDE program [7] determines the connection table of a scanned chemical structure in three steps: a recognition phase, a text grouping phase, and an interpretation phase. The general process is shown in Figure 1. The Primitive Recognition Phase takes the bitmap image of a structure and extracts the primary information concerning the graphic and character components that make up the page. This includes line detection and segmentation as well as Optical Character Recognition (OCR) routines. These graphic components and characters are called primitives. Text grouping is necessary to produce chemically meaningful words and other text items (such as lines and blocks of text) from the set of recognised characters. It is essential that this is done before the interpretation phase because the chemical information is crucial to the correct construction of the connection table. The interpretation phase determines the correct chemical context and meaning of the graphic primitives and text that form a chemical structure. These primitives are converted into higher level data types termed items (atoms, bonds,

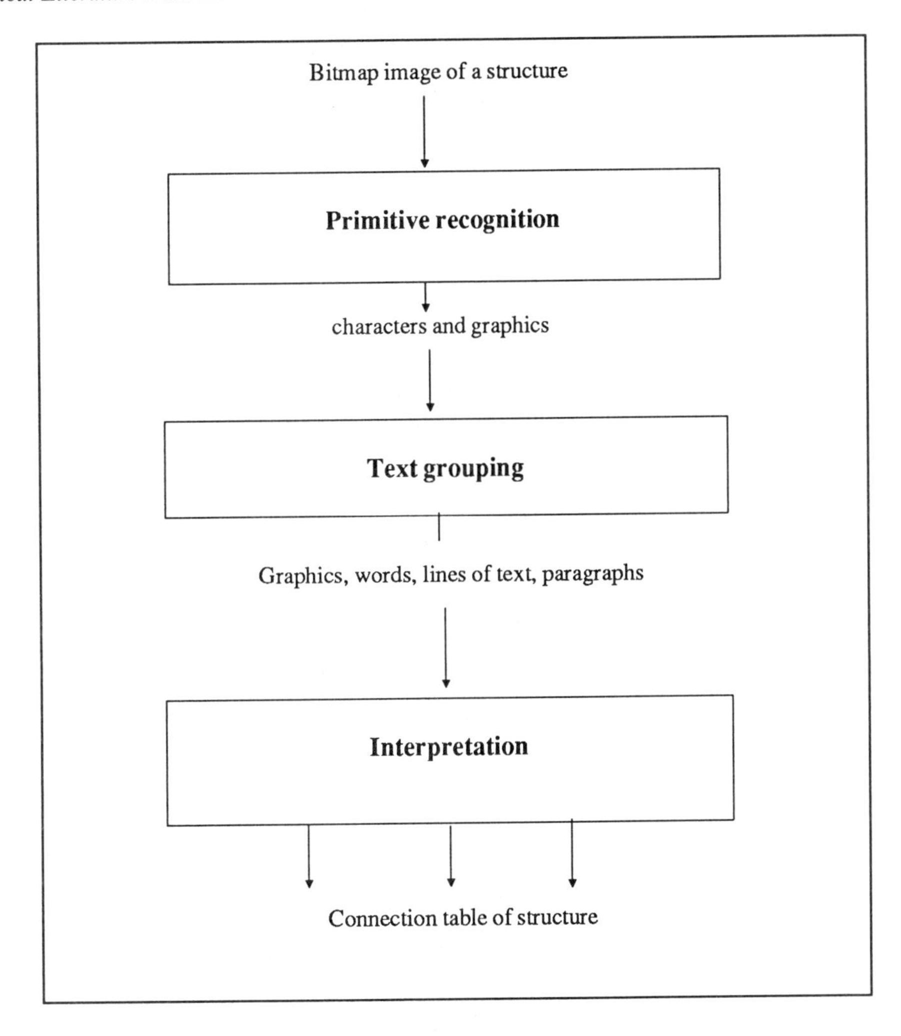

Figure 1: The CLiDE system for chemical structure interpretation proceeds in three phases. These phases combine to transform the original bitmap image of a structure into a chemically useful format.

wedge bonds, etc.) For example, two parallel, adjacent lines become a chemical double bond at this stage. The result of this processing is a connection table that can be stored in an appropriate format.

The processing of a single chemical structure is illustrated in Figure 2. The details of the individual steps are discussed in the following sections.

Primitive recognition

The binary scan of a journal page produces a fine grid of black or white dots, called pixels (Figure 2(a)); the grid itself is referred to as a bitmap. Recognition of primitives in the bitmap is the most time consuming task of the CLiDE process due

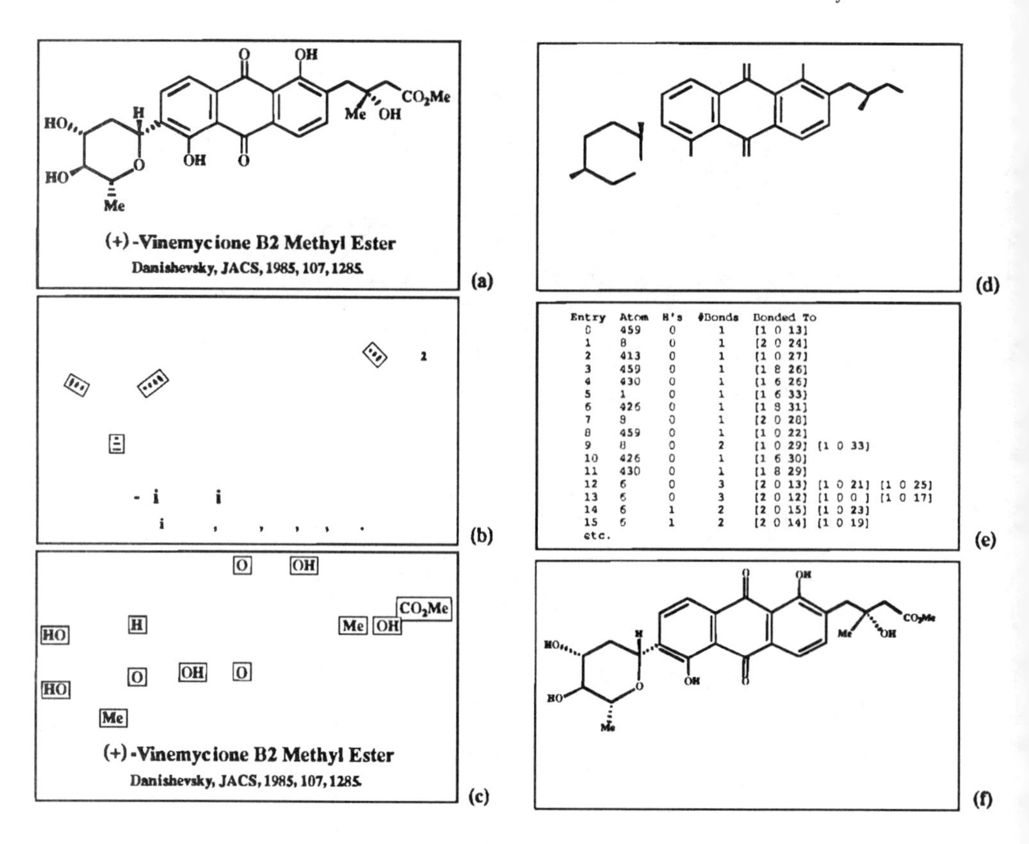

Entry	Atom	H's	#Bonds	Bonded To		
0	459	0	1	[1 0 13]		
1	8	0	1	[2 0 24]		
2	413	0	1	[1 0 27]		
3	459	0	1	[1 8 26]		
4	430	0	1	[1 6 26]		
5	1	0	1	[1 6 33]		
6	426	0	1	[1 8 31]		
7	9	0	1	[2 0 28]		
8	459	0	1	[1 0 22]		
9	8	0	2	[1 0 29]	[1 0 33]	
10	426	0	1	[1 6 30]		
11	430	0	1	[1 8 29]		
12	6	0	3	[2 0 13]	[1 0 21]	[1 0 25]
13	6	0	3	[2 0 12]	[1 0 0]	[1 0 17]
14	6	1	2	[2 0 15]	[1 0 23]	
15	6	1	2	[2 0 14]	[1 0 19]	
etc.						

Figure 2: The CLiDE processing of a single chemical structure. (a) Original scanned image. (b) Dash-like connected components. Those dashes which form dashed lines are found and are shown here boxed. (c) Characters. The full set of characters is examined to find text words that may be part of a chemical structure. The resulting superatom strings are shown in boxes. (d) Lines and larger graphics are processed to find the elementary wedged and straight line segments. (e) Representation of the chemical structure as a CLiDE connection table. (f) The chemical structure redrawn from a saved ChemDraw file.

to the amount of data to process and the local view the computer has of the data. Using a scanner resolution of 300 dpi produces approximately 1/4 Mb of data to be processed for a typical journal page.

During the first step of primitive recognition, the bitmap image is segmented into connected black regions which we term connected components. The connected components are represented by their interpixel contours which are defined as the discontinuous lines surrounding the black regions at the edges of the outermost black pixels. This representation significantly reduces the size of memory needed to store an image. The contours are defined as the coordinates of a starting point and a sequence of four directions (N, S, E, W). The method of Ahronovitz-Bertier-Habib [8] is used to create the contour code and to find the connected components.

Each connected component is composed of one or more primitives. The recognition of graphic primitives can be done on the connected components separately because each primitive is, or belongs to, only one connected component. The primitive recognition is performed using the contours without the reconstruction of the original bitmap. The connected components are divided into three subclasses: characters, graphics and dashes. Due to the embedding of isolated characters in graphic regions, text/graphic separation methods based on string detection [9] could not be used. The separation of the characters, dashes and graphics is based on the relative size of connected components, on their surface area and on the ratio of the sides of their surrounding box. This method results in a reasonably accurate separation. Any errors are corrected during the recognition step by moving those connected components which cannot be interpreted from one group into another.

The small connected components which form the dashes group are analysed using a Hough transform [10,11] to detect those sub-groups of dashes that fall on straight lines and are also close together, as shown in Figure 2(b); these become dashed-line primitives. The characters group (which now includes dashes that were not part of dashed lines) is sent to the artificial neural-network-based OCR module for classification (Figure 2(c)). The contours of the remaining group, the graphics, are analysed to find the component line segments and wedges (Figure 2(d)). Details of this procedure have been described previously. [7]

At this stage of the CLiDE process all the elementary (chemical) graphics and characters have been identified. The next phase assembles these primitives into chemically meaningful groups and finally complete chemical structures. For a single chemical structure this information is stored internally as a connection table (Figure 2(e)). Externally, this information can be saved in a number of different formats. For example, a ChemDraw file can be produced, as shown in Figure 2(f).

The grouping phase

The grouping of characters creates small text regions, such as paragraphs, chemical strings and structure labels. The text grouping begins by grouping into words those characters which are adjacent and on a line. Words are then grouped into lines, and finally lines are grouped into blocks of text. Each character is associated with one block of text.

Once the text blocks are identified, we are particularly interested in those which consist of only one word, because they may be text in, or associated with, structures. These words are shown Figure 2(c). Small blocks of text are examined to see if they contain information relating to generic groups in a nearby structure. This information is highly structured and can be extracted and used in building the connection table. Those blocks which are clearly paragraphs of text are not used by the program at present.

Superatom	'O-'	'O'	'HO'	'OMe'	'COCO'
Code	8	498	426	462	499
Connections	2	1	1	1	2
Equivalent	'O'	'O'	'OH'	'OMe'	'COCO'
Hydrogens	0	0	1	0	0
Charge	–1	0	0	0	0
Stereochemistry	0	0	0	0	0
Negative charges allowed	0	1	0	0	0
Positive charges allowed	0	1	0	0	0
Letter bonded list	1	1	2	1	1 3
Sub-connection table				8 [1,1,2]	6 [1,1,2]
					[2,1,3]
				6 [1,1,1]	6 [1,1,1]
					2,1,4]
					8 [2,1,1]
					8 [2,1,2]

Table 1: Superatom information, for several superatoms, stored in the database. The sub-connection table for a superatom is encoded as follows: For each atom in the superatom the subconnection table specifies its atomic number and for each bond to that atom there is a code. This code is three numbers indicating the bond order, the bond style (normal, wedged, etc.) and the number of the atom that the bond is made to.

Interpretation phase

A chemical structure is composed of two types of objects; graphics and text. As has been discussed previously, the information from the page has been processed to produce text items and graphic primitives that may form part of a chemical structure or structures. It is the function of the interpretation phase to correctly identify the chemical context of these primitives and to build a connection table from them. The text occurring in a structure may correspond to an individual atom or a molecular group (e.g. Cl or Me respectively) and is referred to as a superatom. All of the relevant information concerning superatoms is contained in a database. The identification of superatoms in the CLiDE process is done by referring to a look-up table. In fact the task is more than identification; the database not only allows one to check the validity of the superatom items but also contains information required to build the connection table. Each superatom is represented by a card (Table 1) containing several items of information listed as follows:

- A unique code, e.g. atomic number for atoms.
- The number and position of external bonds required by the superatom (valency for the atoms).
- The description of the charge and stereochemistry.
- The number of negative and positive charges allowed. This information will be used for further checking of the validity of the connection table.
- The number of explicit hydrogens bonded to the superatom.

- An equivalent item which will replace the current superatom in the reformatted connection tables in order to obtain a consistent representation. For example 'C', together with the number of hydrogens and the charge, is equivalent to 'CH', 'CH_2', 'CH_3', 'C^-'; 'C' in the equivalent item of these superatoms.

When required, the connection table of the superatom will be used to produce the expanded connection table for the structure, i.e. the connection table in which all the non-hydrogen atoms are explicitly described. The totally expanded connection table contains only superatoms which have no sub-connection table, i.e. which are atoms.

The graphic primitives are, or are part of, chemical bonds in a structure. It is relatively straightforward to interpret these graphic primitives as chemical bonds. The main complication arises when one bond is crossed by another as will be discussed later. The information defining each bond (order, style, coordinates) and the superatom information is combined to form the connection table. The bond-atom connections are determined before the bond-bond connections.

Other studies [5] base the recognition of a join between two bonds or between a bond and an atom on the distance separating the components; gaps of less than a certain threshold result in the components being joined. However, this is not accurate enough for reliable automatic operation. In Figure 3 the bonds (i), (ii) and (iii) are all close to the superatom OAc but bonds (ii) and (iii) are not pointing towards the bonding atom (oxygen) of OAc. The perpendicular displacement of the bonding atom from the direction of a bond is combined with the actual Euclidean separation to determine whether they should be joined as shown in Figure 4(a). Bond ends are joined if the separation is less than a threshold value (Figure 4(b)).

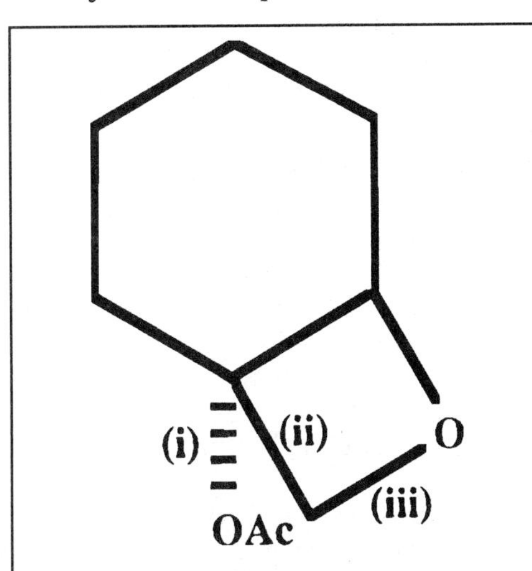

Figure 3: An example of a structure in which the superatom is closer to two (solid) bonds (ii and iii) than to the dashed bond (i) to which it is actually joined.

After perception, the information about a structure is stored as a connection table. Many types of bonds are perceived including single, double, triple, thick, wedged, dotted, dashed and dashed wedged. For most of these bonds, the sense of the direction of the bonds is not important, but for the wedged bonds and dashed wedged bonds, the connection table preserves the direction of the wedge. Other chemical characteristics, e.g. stereochemistry and charge, are also stored in the connection table. The two-dimensional coordinates of the structures obtained from the original scan are retained so that the structure may be re-drawn.

Thus the two main tasks during the interpretation phase are:

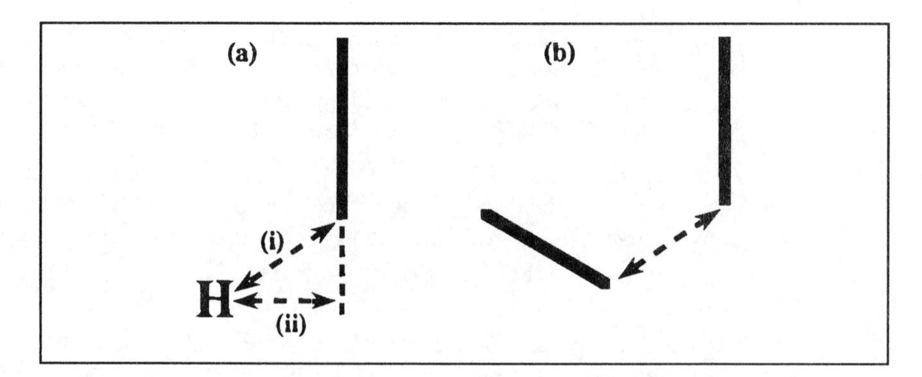

Figure 4: Rules are used during the construction of the connection table: (a) The separation (i) and perpendicular distance (ii) of an atom from a bond end are used to determine if they should be connected. (b) Bond ends are joined together if they are sufficiently close to each other.

(i) Identification of the text present in the structure and extracting the relevant information about this text from the superatom database.

(ii) Building the connection table for each chemical structure by constructing the atom-bond and bond-bond connections.

The internal representation of the information concerning chemical structures identified by CLiDE is a connection table which describes each atom and its environment. The connection table, once reformatted, can be stored for searching and external use [12].

Finally, the remaining small pieces of text are examined. If any text items are close to the completed structure, the information is retained with the connection table as potential structure labels or information.

The last step of the process consists in converting our internal connection table into a suitable output format. The current version of CLiDE outputs chemical structure information in MOLfile [13], ChemDraw [14], and CLiDE formats and can create a Postscript file of the redrawn structure.

Examples

An example set of structures which have been successfully processed using CLiDE is presented in Figure 5. These structures have been chosen to illustrate the scope of the current program. Figure 5(a) is a simple structure including single, double and triple bonds. The only text is the two hydrogen atom labels. In the second structure, Figure 5(b), the two text strings NH_2 and H_2N are determined to be equivalent using information contained in the superatom database. The stereo-chemical information contained in dashed and solid wedged bonds shown in Figure 5(c) is correctly identified. The last example presented, Figure 5(d), is a structure containing a superatom (SO_2) which requires two connections to the structure through the sulphur atom. Figure 5(d) also contains a generic atom (X) and its associated generic text block. The generic text is parsed and the appropriate

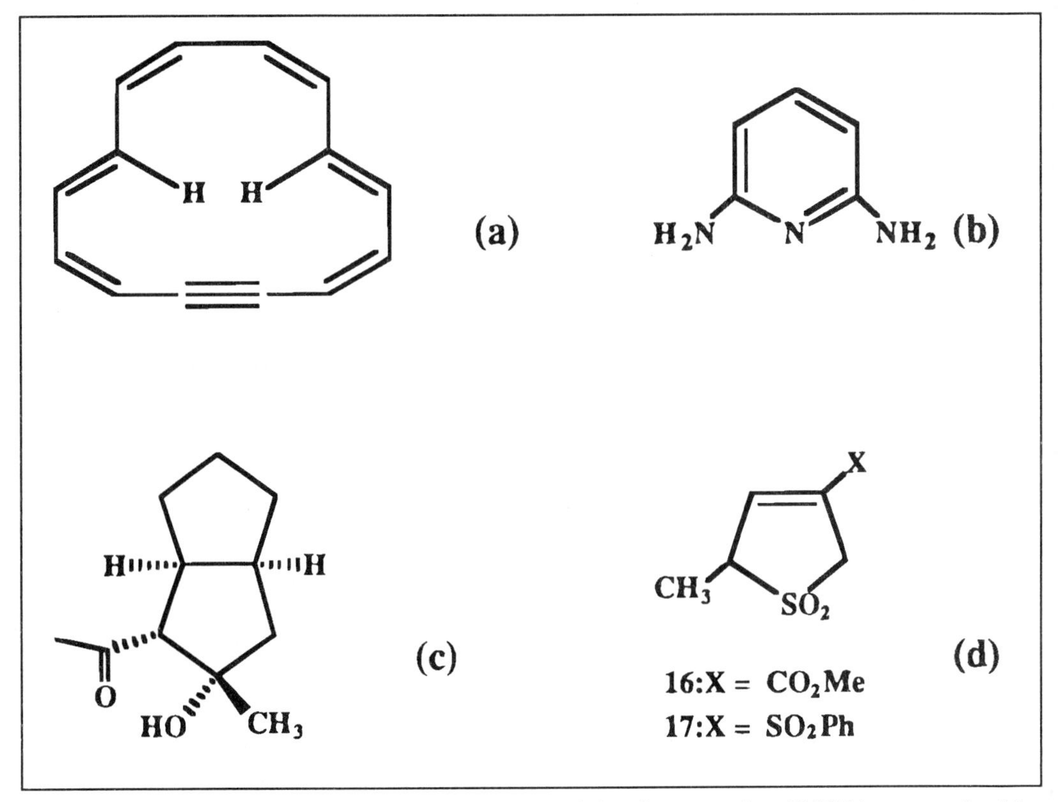

Figure 5: A set of example structures containing features that CLiDE is currently able to process successfully.

replacement groups identified and stored. Substitutions of the two superatom groups for X are made into the structure, resulting in two connection tables. As previously mentioned, the current version of CLiDE can successfully parse most of the common generic text blocks and make the appropriate substitutions.

Crossing bonds in chemical structures

Often chemical structures contain bonds that are partially obscured, or overwritten, by other bonds. Usually this occurs in bridged structures when the drawing attempts to preserve some sense of the three-dimensional shape of the molecule. Topological concerns are of lesser importance since there are only a few isolated cases of molecules which cannot be represented by planar graphs [15,16]. The other reason for drawing a structure in which a bond is drawn crossing another is where the bond being crossed is a ring bond and the crossing bond indicates that the position of attachment of this bond to the ring is undetermined. In structures in which there is a bond crossing situation, usually one bond 'cuts' another although there are cases where there is no apparent gap in the bond that is crossed. A number of structures containing crossing bonds that can be successfully interpreted by the current version of CLiDE are shown in Figure 6.

The correct interpretation of the graphic primitives to produce a chemical connection table requires that these situations are detected and dealt with appropriately. In Figure 7(a) and 7(b) are shown the two cases of normal crossing bonds. Without specific rules to detect that these structures contain crossing bonds, an incorrect connection table is built; these are illustrated in Figures 7(c)-(e).

To correctly identify those structures containing a bond that is cut (e.g. Figure 7(a)) the following rules are used; for all the bonds in the structure find those pairs of bonds that are collinear, separated by a small distance and where a third bond passes through this gap. Each pair of bonds is replaced by a new bond that spans the original distance. The new bond is marked as being behind. The construction of the

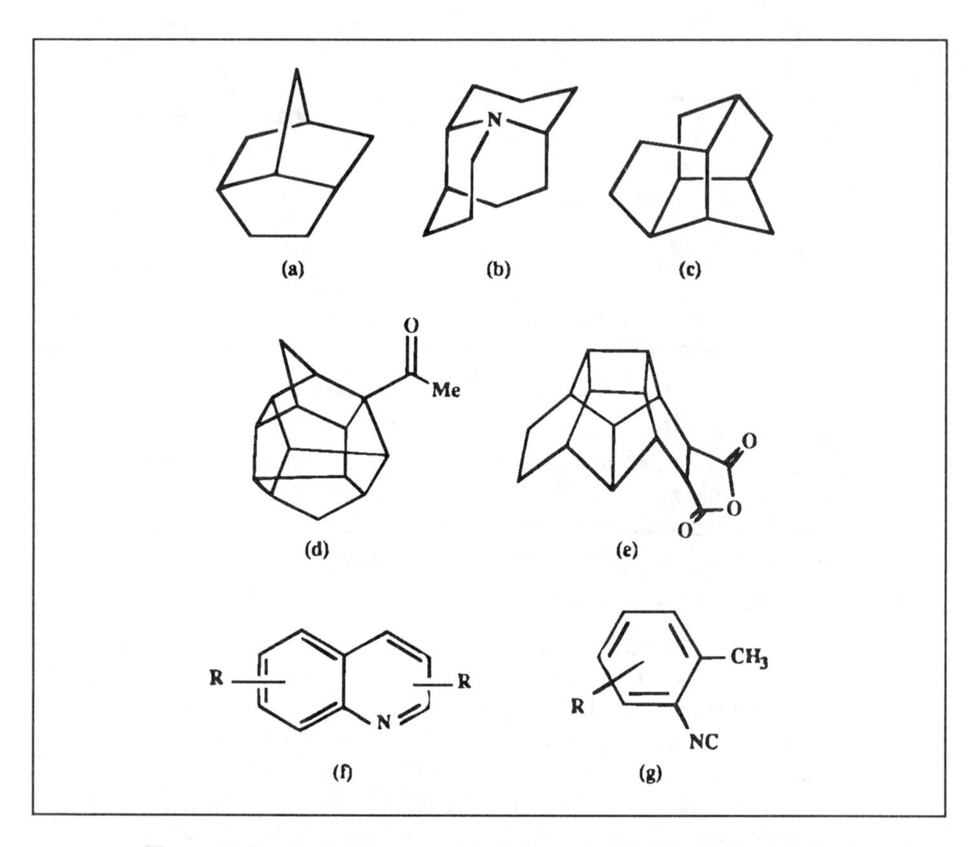

Figure 6: A set of structures taken from the literature illustrating bond crossing situations: (a)-(c) show the bond that is crossed being cut by the other bond, (d) and (e) show bonds which are overdrawn, and (f) and (g) are undetermined attachment bonds depicted by one or more isolated bonds cutting or drawn over a ring bond.

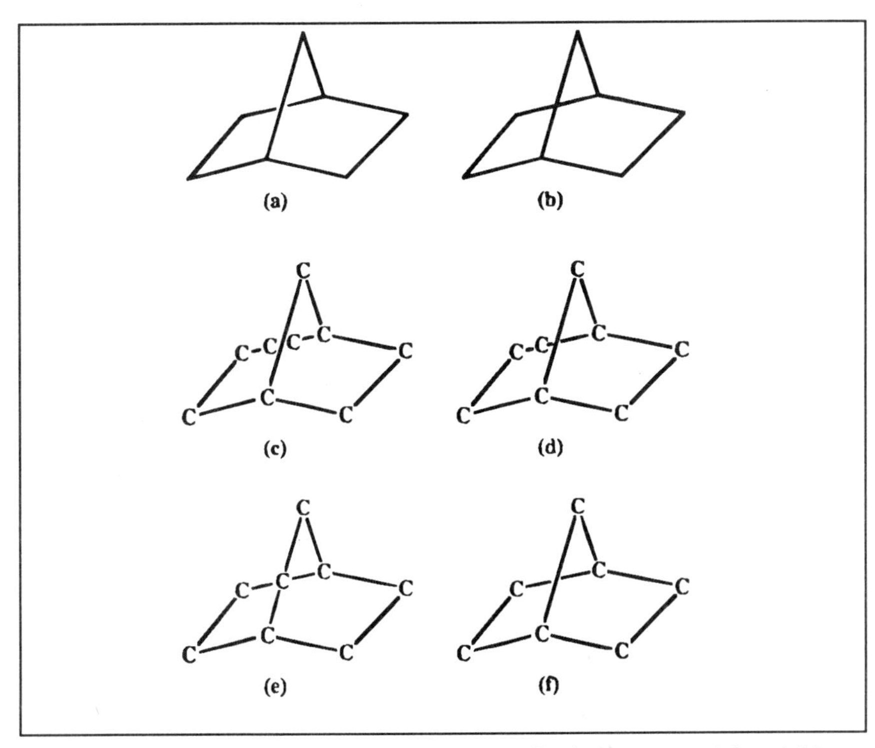

Figure 7: Bond crossing situations. Both structures (a) and (b) represent the same structure. Incorrect construction of the connection table for (a) can lead to either (c) or (d) being formed. If (b) is processed without detecting the crossing then (e) is formed. In both cases (f) is the required structure.

connection table is continued in the normal way. The behind mark on a bond could be used to determine stereochemistry in some cases, although at present it is only used for redrawing purposes.

For structures containing bonds drawn over each other (e.g. Figure 7(b)) the situation is more difficult. Initially, the incorrect connection table is formed (Figure 7(e)). Then all carbon atoms making four explicit bonds are examined. If any of these atoms has two pairs of collinear bonds it is tested further. The atom is removed by replacing each collinear pair of bonds by a new bond. One of these new bonds must belong to a ring. The removal of this atom must still leave a connected structure, avoiding the case shown in Figure 8 except in one important case discussed below. If either of these tests fails the atom is reinstated and the original connection table reformed.

The other main crossing bond situation, the undetermined attachment bond (Figure 6(f)-(g)), is dealt with using a similar set of rules. If a potential bond crossing situation is detected and the removal of the central carbon atom leaves a structure with two new crossing bonds where one bond is a member of a regular ring and the

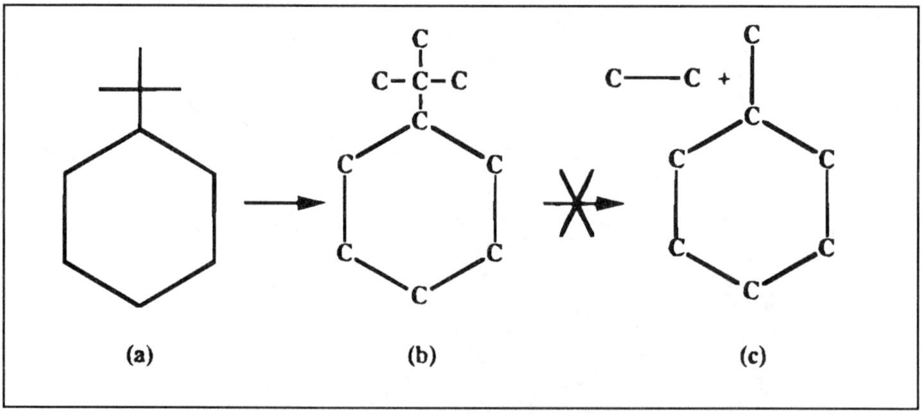

(a) (b) (c)

Figure 8: The structure (a) contains a potential bond crossing since after the first phase of connection table creation it contains a carbon atom at the center of two collinear pairs of bonds, (b). Removal of this atom results in a disconnected structure where neither new bond is a member of a ring, (c), so the central atom is reinstated.

other is an isolated single bond, this indicates a ring crossing situation. Clearly the test for a connected structure will fail for this case. To deal with an isolated bond crossing a regular ring, the interior carbon atom is replaced by a virtual atom. A list of possible attachment points is made and stored with this virtual atom. This ensures we obtain the correct molecular formula and retain the information about the undetermined nature of the attachment. All the original coordinates are also retained for redrawing purposes.

Conclusion

CLiDE is able to produce a list of connection tables from a scanned image of chemical structures embedded in a page of text. The CLiDE prototype has been widely tested and our algorithms have been shown to work in a number of difficult cases.

Some improvements are being made and will be detailed in future publications. As the initial aim of the work was to go beyond single, isolated chemical structure recognition, we have been working towards identifying those structures that form part of a reaction. This reaction may be part of a larger reaction scheme. Given the position of reaction arrows and chemical structures on a page, we have produced an algorithm to extract the chemical reactions from a reaction scheme although this has not yet been included in CLiDE. Also a YACC program has been written to interpret reaction conditions and will be implemented at the same time.

References

1. Borkent, J. H.; Oukes, F.; Noordik, J. H. Chemical Searching Compared in REACCS, SYNLIB and ORAC. *J. Chem. Inf. Comput. Sci.* **1988**, *28*, 145-50.

2. CLiDE is writtenas C++ and currently implemented on a SUN SPARC work station. The images are scanned by an *Agfa Focus S 800GS* scanner at 300 dpi (dots per inch) resolution.

3. Fahn, C. S.; Wang, J. F.; Lee, J. Y. A Topology-Based Component Extractor for Understanding Electronic Circuit Diagrams. *Computer Vision, Graphics and Image Processing* **1988**, *44*, 119-38.

4. Fukada, Y. A Primary Algorithm for the Understanding of Logic Circuit Diagrams. *Pattern Recognition* **1984**, *17*, 125-34.

5. McDaniel, J. R.; Balmuth, J. R. Kekulé: OCR — Optical Chemical (Structure) Recognition *J. Chem. Inf. Comput. Sci.* **1992**, XX, YYYY.

6. Contreras, M. L.; Allendes, G.; Thomas-Alvarez, L.; Rosas, R. Computational Perception and Recognition of Digitized Molecular Structures. *J. Chem. Inf. Comput. Sci.* **1990**, *30*, 302-7.

7. Ibison, P.; Jacquot, M.; Kam, F.; Neville, A. G.; Simpson, R. W.; Tonnelier, C.; Venczel, T.; Johnson, A. P. Chemical Literature Data Extraction. 'The CLiDE Project'. *J. Chem. Inf. Comput. Sci.* **1992** (in press).

8. Ahronovitz, E.; Bertier, M.; Habib, M. Contour Coding for Image Manipulation and Compression. *IAPR 86 IEEE 742*, **1986**, *2*, 1033-1035.

9. Fletcher, L. A.; Kasturi, R. A.Robust Method for Text String Separation from Mixed Text/Graphics Images. *IEEE Trans on Pattern Anal. and Machine Intelligence* **1988**, *10*, 6.

10. Duda, R. O.; Hart, P. E. Use of the Hough Transform to Detect Lines and Curves in Pictures. *Graphics and Image Processing* **1972**, *1*.

11. Illingworth, J.; Kitter, J. The Adaptive Hough Transform. Computer Vision, *Graphics and Image Processing* **1988**, *44*, 87-116.

12. Ash, J. E. 'Connection Tables And Their Role In A System'. In *Chemical Information Systems*; Ash, J. E. and Hyde, E., Eds.; Ellis Horwood Publishers: Chichester **1975**, 157-176.

13. Dalby, A.; Nourse, J. G.; Hounshell, W. D.; Gushurst, A. K. I.; Grier, D. L.; Leland, B. A.; Laufer, J. Description of Several Chemical File Formats Used by Computer Programs Developed at Molecular Design Limited. *J.Chem. Inf. Comput. Sci.* **1992**, *32*, 244-255.

14. ChemDraw, Cambridge Scientific Computing, Inc.

15. Elk, S. B. Interpretation of Kuratowski's Theorem in Graph Theory as Both Topological Abstraction and a Chemical Reality. *J. Chem. Inf. Comput. Sci.* **1990**, *30*, 69-72.

16. Benner, S. A.; Maggio, J. E.; Simmons, H. E. Rearrangement of a Geometrically Restricted Triepoxide to the First Topologically Nonplanar Molecule: A Reaction Path Elucidated by Using Oxygen Isotope Effects on Carbon-13 Chemical Shifts. *J. Am. Chem. Soc.* **1981**, *103*, 1581-1582.

A Markush story

Everett Brenner[1] and Edlyn S. Simmons[2]

[1] 43 Sandy Hollow Road, Port Washington, NY 11050, USA. [2] Marion Merrell Dow, Inc., 21100 East Galbraith Road, Cincinnati, OH 45215-6300, USA

Introduction

[For those unknowledgeable or unclear about what a Markush structure is or how Markush structures differ from other chemical structures, Appendix I includes a brief description including a sample structure and a brief explanation of fragmentation coding.]

Briefly, Markush refers to a class of chemical substances identified by lists of alternative substructures, and it has become the most popular format for generic chemical structures in patents.

In September of last year at this very meeting here at Annecy, Jim Sibley, then of Shell International Petroleum Co Ltd, now retired, presented a paper which included comments on Markush databases which can be searched by structure, i.e. topologically. He maintained that three of these databases were available but have not attracted very much attention in Europe. He also mentioned a fourth service in development. (Figure 1) He questioned whether the database providers had had a clear understanding of the potential customers. Bench chemists in corporations, he further stated, would not likely be given access to a Markush file; academics would not be concerned with using such a database; and the potential number of regular users would in general be very restricted.

Markush System	Host	Database Producer	Database Content
Markush DARC	Questel	WPIM, Derwent	Derwent major countries from 1987. Specific & generic structures except polymers & organometallics
		MPHARM, INPI	EP, US & Fr. from 1985; Fr medicinal 1961–69. Pharmaceutical patents. Specific & generic structures
MARPAT	STN	MARPAT, CAS	Most countries by CA, from 1988. Generic structures only
GENSAL / GREMAS	?	GENSAL / GREMAS, IDC	?

Figure 1: Markush databases, 1992

What I questioned when I heard this paper was how a user-driven project of this kind could be so unsuccessful. I expressed this in an article I wrote for *Monitor* last October and promised an investigation. Edlyn Simmons wrote to me outlining her ideas on the subject (Appendix II) and I in turn sent her letter to individuals I felt had been very involved in the Markush development. Most of them responded and/or were interviewed by me. They were open and willing to present their side of the story.

Edlyn Simmons as co-author of this paper has worked with me to keep the facts in the Markush story as accurate as possible. In addition she is responsible for the appended explanations of Markush and fragmentation codes, the selected bibliography, the content and make-up of the slides and editing. At the close of my presentation, she will in addition deliver her own comments which are essentially an expansion of her remarks in her letter to me (Appendix II) and her reaction to the analysis in this paper.

I admit that when I decided to investigate this matter, I thought I would discover something scandalous about the waste of money for an under-used system. However, when all the facts were in, there was no one to blame but considerable matter to learn from what did go on. The Markush story involves database producers, database vendors, not-for-profit and for-profit organisations, universities, and governments, all in the US, Britain, France and Germany. This being the decade in which cooperative efforts will be on the increase, this report may help to sort out some of the complexities involved in developing new products and systems, and thus help in formulating and implementing future ones.

The organisational players

Organisation	Contribution
University of Sheffield (GB)	Basic research in topological systems
Université de Paris (France)	Basic research in topological systems
BASF (Germany)	In-house development of topological system
Hoechst (Germany)	In-house development of topological system that became the GREMAS system
IDC (Germany)	Consortium of companies that used the Gremas system and subsequent Gremas / Gensal system
Derwent (GB)	Commercial development of Markush DARC
Questel (France)	Government-backed commercial development of DARC and Generic DARC based on Université de Paris research and later of Markush DARC
INPI, France	Government development of Markush DARC based on Generic DARC
Chemical Abstracts Service (USA)	Non-profit database producer and for-profit host that developed CAS Online and subsequent MARPAT system

Figure 2: Organisations active in the development of Markush databases

- The University of Sheffield and the University of Paris represent academia involved in basic research on topological systems.
- BASF AG (Germany) and Hoechst AG (Germany) are large chemical companies with some very early development in the topological area.
- Derwent (England) is a for-profit database producer with the most comprehensive file of scientific patent abstracts and their indexes in existence.
- Chemical Abstracts Service (US) is both a not-for-profit database producer of chemical patent abstracts and their indexes, and a not-for-profit vendor, known as STN.
- Questel (France) is a for-profit-oriented online vendor of databases operating as a department of Télésystèmes which is a French government-supported telecommunications company also for-profit oriented.
- INPI is the French patent office.
- IDC was a database producer and vendor governed by a consortium of eight large German companies. As of 1992, IDC has discontinued its operations.
- Barnard Chemical Information Ltd is a for-profit information company founded by John Barnard, one of the principles who worked on topological systems at the University of Sheffield.

Important events

What were some of the important activities that led to the present state of the Markush/topological search databases?

As early as 1957, Robert Fugmann of Hoechst AG had developed GREMAS, a non-topological fragmentation code system. Hoechst began to build a GREMAS file in 1959. In 1962, BASF and Bayer joined Hoechst in adding to the file, and in 1967, IDC, established as a cooperative input organisation, took over what Michael Lynch of the University of Sheffield calls a "restricted topological" system, and Claus Suhr, recently retired from BASF, agrees that the inherent problems of generic structure indexing had been known for a long time and that early systems could handle at least very simple cases.

Some work under the direction of Jacques-Emile Dubois at the University of Paris indicated a start on topological research in the 1970s. This work was the basis for topological system development in France, resulting in Questel's structural search system, DARC.

In 1978 Michael Lynch at the University of Sheffield decided to take on the generics problem. He claims he was unaware of the Dubois work. He states, "I knew that generics was one of the last frontiers as far as 2-D chemical molecules were concerned, and that it might suit the kind of experience I had built up in unravelling complex data structures and providing retrieval mechanisms for them." The project Lynch eventually developed revolved around "a surface representation called GENSAL, the purpose of which was to formalise the relations in generics which are usually defined informally in patents or patent abstracts."

Kathie Shenton, formerly of Derwent, now of Shennor Associates, traced Derwent's interest in commercialising topological development to 1981. Markush searching of the Derwent files was at that time — and still is — carried out with a fragmentation code, not topological in any way. GREMAS was the equivalent of the IDC fragmentation code system.

In 1982, the University of Sheffield presented a paper at a meeting at which IDC members and BASF staff were present. Their work was well received by Ernst Meyer whose experience and early work was well respected, as well as, and I now quote Michael Lynch, "by Claus Suhr and Arthur Kolb [of IDC] and other representatives of German industry." Future Sheffield developments reflected the early work of BASF via Winfried Dethlefsen, whose contributions to the work are cited by Michael Lynch.

After the 1982 presentation, Michael Lynch sought funding support for further research from CAS and Derwent. CAS contributed with the understanding that the results would be available in the public domain. Derwent shortly thereafter added its support. When CAS funding ended IDC stepped into the breach and Derwent continued for a brief time.

In 1983 at a conference held at the University of Sheffield, Questel proposed a Markush topological solution which was built on the earlier Dubois work at the University of Paris which ended up as the basis for Questel's DARC system for structure searching. The solution involved what Michael Lynch describes as supercharging the already implemented DARC system on Questel. Supercharging is in the form of superatoms which are descriptors representing types of structural elements. Thus Markush DARC is essentially a coded structural system, not totally topological.

Also in 1983 at a Derwent subscribers meeting in the US, Monty Hyams, still Director of Derwent at the time, met with representatives of seven companies. Jim Seals and Nick Farmer of CAS proposed an approach to the Markush problem they intended to pursue. Derwent was ripe to move towards implementing a system within the 1984–85 period. Cooperation with CAS was a possibility, particularly if other cooperative efforts related to the CAS ONLINE structure system could be worked out. Also discussed at the meeting was the DARC approach, and there was some support for the University of Sheffield work. Stu Kaback reports on this meeting saying, "The feelings of the group were quite unanimous: we respected the quality of the work that CAS did, but we felt that the DARC system was much farther advanced, and seemed likely to solve the Markush problem before CAS."

Derwent, Questel and INPI, France's patent office, collaborated on software development to get a system up which they demonstrated in The Netherlands in 1987 and thus began the commercial implementation of Markush DARC with the Derwent files resulting in WPIM (World Patents Index Markush) and INPI's files resulting in Mpharm.

In the meantime, the work of the University of Sheffield was further developed by IDC working with Barnard Chemical Information Ltd, the company founded by John Barnard, a Michael Lynch co-worker in this area at Sheffield. In Michael

Lynch's words, this work was aimed at implementing "a system based on the GENSAL language as a means of creating a database to their (IDC's) high standards of accuracy and for the purpose of generating the GREMAS codes automatically. This involved the development of MicroGensip, an interactive front-end for database creation purposes." The GENSAL/GREMAS development which included Markush input was also a mixed topological/fragmentation system.

In the meantime CAS continued to work on the problem and implemented MARPAT. The MARPAT representation bears a strong resemblance to GENSAL. The MARPAT search algorithm, however, is a CAS development. MARPAT uses a small vocabulary of generic group symbols and thus, is, as are the other two systems, a mixed topological-fragmentation system.

To summarise then, there are in essence three approaches to the Markush searching problem:

1. One can be traced back to the early experience in Germany at BASF and the subsequent basic research by Michael Lynch at the University of Sheffield aided by BASF experience. That work was the basis upon which IDC began development towards topological solutions to Markush in conjunction with Barnard Chemical Information Ltd. With the demise of IDC there is no ongoing commercial development continuing the Sheffield work.
2. Although the CAS development of MARPAT was based on GENSAL, Lynch's 'surface representation,' subsequent development of the system has proceeded at CAS.
3. The third approach can be traced to the Dubois work at the University of Paris further developed into the DARC system of Questel and then further developed into Markush DARC and taken on by Derwent to become the WPIM database and by INPI to become Mpharm.

Commentary

At first glance it seems wasteful that we end up with two divergent systems, Markush Darc and Marpat, for so complex a problem which has been so costly in development. It is also troublesome to think that some superior developments at the now defunct IDC may be neglected. Jim Sibley and Edlyn Simmons have questioned whether even one system is necessary.

I. The users

Edlyn Simmons believes that users who encouraged the development of Markush search systems hoped the topological approach would simplify searching, and instead have found it requires even more sophistication. The users are further confounded by the need to use both the new topological system and also the old fragmentation code system.

But my interviewees and responders are almost unanimously in agreement that a topological approach to the Markush problem was a natural development, a progression from structure searching to solve a focused problem which was becoming particularly acute in the pharmaceutical area where the so-called Markush 'nasties' have been on a frightening increase.

Were the developments toward Markush topological searching user-driven? Yes, of course. Kathie Shenton speaks of user interest at Derwent as early as 1981. "In 1985," she says, "when over 50 companies were visited and interviewed on the topic, the concept continued to be greeted with enthusiasm." BASF's early experience indicated great future benefits in developing topological solutions. IDC, a consortium of users, supported development work. CAS, in the early 1980s, decided it would attempt to increase the use of its patent files and satisfy the high-user industrial corporations who are in general willing to pay a fair price for information, unlike the academic area where cost of the use of online retrieval is highly discounted. A survey sent out by CAS asking about Markush development elicited response that every compound in a Markush structure needed to be retrievable. Both Derwent, as a for-profiter, and CAS, as a non-profiter, were driven by comparatively few, highly sophisticated online users in industry, but why not?

Edlyn Simmons believes that the 'user-driven' development was largely based on mis-communication. But it is difficult to find out who mis-communicated. It is not unreasonable for users to want high quality, high performance products. A company such as BASF is willing to pay a high price for high performance. Petroleum companies, much like the few German companies that cooperated to create IDC, established the American Petroleum Institute's Central Abstracting and Indexing Service because they saw the benefits of paying high prices for high performance, customised products.

As for mis-communication and naivete by the database producers and vendors, Ron Dunn, formerly of CAS, now at Macmillan Publishing Co, states wisely that, "Though few in this or any other industry would argue that user input is not essential to new product development, it is not a simple task to define the user requirements, assess the price/benefit ratio, and forecast from the input of a few, albeit core, users what the ultimate demand for a proposed new service will be. Information providers need to be aware that 'listening to the users' is an art rather than a science."

The development of a Markush topological retrieval solution seems logical based on the needs of the users and the inevitable progress we have witnessed over the years both technically and technologically, i.e., progress we have made in structure searching and in computer capabilities.

II. The 1980s

If we are to place blame anywhere for the development of under-used multiple systems, we must look to the mood or economic environment of the 1980s which I have in some of my recent writing characterised as non-progressive and shallow. In general, the decade represented competitiveness and little cooperation when required. Natalie Dusoulier of CNRS in France has characterised it as a time when competitiveness led to confrontation. For example, it was a decade in the US characterised by hostile company takeovers. It was also a decade in the information science and industry area when it has been especially difficult to breach academic research with practical corporate needs.

It is important to note historically that in the 1960s BASF and Hoechst were developing information systems for their own pragmatic use, just as US companies such as DuPont and Exxon were developing innovative files in the 1950s and 1960s. There was little activity in the universities. Library schools had not even entered into the brave new world. It would be a while before they renamed themselves schools of 'Library and Information Science.' Note also that in the 1970s although Michael Lynch pursued support for his work from the entrepreneurial world, he was primarily intent on laying the scientific basis for a good software engineering solution to the problem. "In some senses," Lynch states, ". . . this was a luxury; we did not have deadlines to meet, nor a system to put into operation." Thus the approach of the 50s and 60s which led to building developmental thesauri and systems within industrial corporations, without much help from information science theory, prevailed in the 1980s. Derwent and CAS wished to put a system into operation and would not wait for further basic research, nor would most of the users.

Thus it seems that Derwent and CAS put systems into operation which were not fully capable of solving all the problems, but hoped to improve their systems on the basis of operational experience just the way online and other services improved over the years, at least technologically. In fact because of the short-term approach so characteristic of the 1980s, it is not difficult to understand that compromised systems would prevail. In the 1950s and 1960s there was little theoretical work. In the 1970s and 1980s there has been theory but little patience.

Lost in the shuffle perhaps is the experience emanating from BASF. BASF personnel to this day maintain they understood retrievals which can emanate from a topological Markush system that many other companies cannot even conceive because of their early hands-on experience in searching. There is indication that the use of the Markush systems will increase when advantages are better understood by the search community, even those already sophisticated. It is a fact that structure searching and other innovative systems have all been under-used in the early years of their commercialisation.

The BASF influence is most evident in the University of Sheffield work. Michael Lynch sought out their help via close work with Winfried Dethlefsen. The Sheffield work is now represented in operating systems only by the early MARPAT work. However, though BASF has made itself available for advice to all the developers, there is no clear evidence how much BASF has been listened to in Markush DARC or MARPAT. Neither system credits them, although CAS acknowledges they have conferred with BASF as well as others over the years.

Did the database producers and vendors proceed too quickly? Perhaps. But there remains a big gap between basic research and practical implementation. The 'long term/short term' dilemma is part of the 1980s. It was not out of line with the 1980s for CAS and Derwent to make short term rather than long term moves. Perhaps here is truly a lesson we have to learn for the future. Perhaps the attitudes of the 'long term' academics and the 'short term' entrepreneurs have to change so that long term is not so long, and short term is not so short.

III. Cooperation in the past

I have pointed out that the 1980s were very competitive years and not very friendly ones. In the early 1980s there was a chance that CAS and Derwent might cooperate, particularly since CAS was not only a database producer but now a vendor through its STN network. The efforts failed, but the failure had nothing to do with Derwent's choice to move to the Markush DARC solution. After all, even an American advisory group unanimously recommended Markush DARC. In fact, the combination of a telecommunication company, a patent office and an online vendor with the accompanying support of government, had to be quite attractive in ensuring a long term commitment in developing and implementing a Markush search system. Derwent was in the 'short term' mode and the DARC approach was further along in development. CAS had another problem. They were and are now a vendor as well as database producer and wished to build a system compatible with their STN MESSENGER software. For them to buy into DARC and its developments perhaps did not even occur to them, or was at least given short shrift.

Is national chauvinism also involved? Of course. One may recall in Europe many information leaders felt they should go their own way and redevelop systems and files because of their fear that one day US information systems might not be made available. Early on in the 1960s and 1970s, Germany certainly went its own way. France has always been accused of playing its cards close to its chest, whether it has been true or not. Perhaps many feel the Télésystèmes (Questel), INPI development represents a chauvinistic approach. Was the lack of promulgation of Dubois' work a result of this? Is it peculiar that Lynch did not know of this work when he began his? Was there a language problem?

Persons and personalities certainly played a part in adding to the problems. Most of the human players have been mentioned in this paper. However there were no good guys and bad guys. There was however a bad decade in which the competitive spirit won over depth in thinking. Though the decade can be criticised as being a short-term decade, there were short-term people involved who had vision and long-term people who tended to be impractical. What was needed was a 'golden mean' approach which in this case did not equate the word 'gold' with 'money.'

Competition has its advantages, and therefore having two systems is not today considered a bad thing. It is my opinion, for whatever it is worth, that none of the players in the Markush developments ever thought twice about any real disadvantage in having competing systems. There was an inevitability in the development of two systems under the prevailing winds in the 1980s in the Western World.

IV. Cooperation in the future

Will the economic environment in the 1990s force us to change our ways? Can we learn from our mistakes and still take advantage of a free society and competition? I have already suggested that the academic world and the entrepreneurial one need to close an existing destructive gap. The Markush problem could easily have been characterised as much needed, complex, and too expensive for all concerned. The problem should have been considered by all the parties as a candidate for a cooperative effort which might speed up the solution to a long term problem. There still exist many, tough, customised, sophisticated problems which are really needed by users. Solutions to these problems require cooperation by competitors, if any progress is to be made. Each player I have mentioned has expertise which, if combined, would have made all of them winners.

I am not sure we can correct the past in terms of the Markush problem. There has already been attrition due to the demise of IDC. Perhaps one of the existing systems will fail due to economic pressure. Each may be forced to continue an operating system which loses money. However, how long can Derwent continue its fragmentation code while essentially duplicating it in its topological files? Both Derwent and CAS feel they are in development stages building towards the future and that improvements plus a build-up of files will lead to increases in use in the future. I would opt for talks by all parties in the near future, Derwent and CAS as database producers, CAS and Questel as database vendors, and the University of Sheffield/ Barnard Chemical Information Ltd and Questel as developers, to see if there is any room for negotiation which can benefit all.

Can friendly competition prevail? Our economic environment in the 1990s is quite different from the 1980s. This new environment will probably force us to find a basis for cooperation. Hopefully we will identify projects similar to Markush searching to further enhance our needs and will handle them more responsibly. Unlike the 1980s, real progress may be possible in the 1990s if we move in this cooperative direction.

Afterthoughts: a user's view, by Edlyn Simmons

The patent literature has changed a lot during the last half of the 20th century. Multinational companies have multiplied in Europe, North America and Asia. These companies spent more and more on research and development, filed more and more patent applications, and filed corresponding applications in more and more countries. Patent offices were unable to keep up with the increased number of patent applications, so most countries changed their patent law so that all the patent applications that are filed are published 18 months after their first international filing, and examination of the patent applications is deferred. In the past, inventions that didn't measure up never appeared in published patents. Now every application that is filed is published, sometimes dozens of times, and no Markush structure is too broad or too ambiguous to join the patent literature.

Patentability law has changed very little. Any compound that has been described in print, specifically or generically, is unpatentable. So it is no less important to be able to retrieve references to compounds embedded in Markush structures today when there are millions of chemical patents with broadly defined Markush structures than it was when there were only a few hundred thousand chemical patents and most of them were narrow in scope.

During the 1950s and '60s, major companies invested large amounts of money and manpower to develop ways to access the information in chemical patents. Fragmentation codes and indexing systems were developed, and when private retrieval systems became too expensive for a single company to support, consortia like IDC and commercial services like Derwent and IFI/Plenum took over the job of database building. In those days nobody expected patent searching to be simple or automatic; they were satisfied that they could do it at all.

As computers became more powerful, topological input and searching for specific chemical structures was developed and systems like CAS ONLINE and DARC were operating by the early 1980s. Users could input a picture of a chemical structure, and the computer returned pictures of chemical structures from the database. By comparison with the pictures you could retrieve from a search of the CAS Registry file, the jumbles of disconnected jigsaw puzzle pieces you could retrieve with a fragmentation code search seemed like too much trouble. Users of the cumbersome fragmentation codes began to ask for a topological system that would work with the Markush structures in patents. And commercial databases tried to give them what they wanted.

Who were the users who demanded topological retrieval systems for patents? The most vocal group were the heavy users of patent information systems employed by corporations that subscribed to the existing retrieval systems, especially users of Derwent's Chemical Patents Index. Information chemists complained that they had trouble learning and applying the Derwent fragmentation code, and they were impatient with the number of false drops they retrieved when they searched with it. They figured that if they had a simple topological indexing system they could put all those troubles behind them. There were also searchers whose companies were not buying a large enough Derwent subscription to provide access to the fragmentation code. Those searchers had no way to retrieve information about generically disclosed compounds from patents, and they were interested in getting the information without investing $50,000, or more, before they could learn to do their first search. The searchers at Patent offices were also interested in better and more convenient ways to do their searches, and patent offices were investing a lot of money in automation during the early 1980s.

In France, the Patent Office itself was able to decide what it wanted and begin work. Derwent subscribers took their demands to Derwent, of course, and their voices were added to those of non-subscribers who took their demands for better patent indexing to the Chemical Abstracts Service. CAS cautiously prepared to expand its coverage of patents by surveying chemical information users of all types. I strongly suspect that many respondents who had never been involved in patent searching expressed support for a topological Markush system even though they hadn't thought about it much before it was offered as an option.

That is where some of the mis-communication took place. And perhaps the mis-communication was simply a case of conflicting expectations. The parties who thought they were communicating began with different, and unstated, basic assumptions. I believe that most of the searchers under-estimated the complexity of the data that had to be encoded to create a database with the appearance of an album of pictures of chemical structures and over-estimated the power of the computers that would do the searches. They imagined that the simple-looking user interface reflected a relatively simple search system and that the cost of searching would be about the same as it had always been. Searchers were accustomed to a certain level of expenditure, both in time and money, and these were already high enough that significantly higher costs were not acceptable. Although users warned about the importance of keeping costs in line, no one knew what it would cost to search the new databases until the new databases were ready for release. I believe that most of the computer program designers under-estimated the complexity of the Markush format. Working with simple prototypes of Markush structures, they were able to design storage and retrieval protocols that could handle generic structures and perform generic searches, and they were rightfully proud of the results of their research.

There was another area of mis-communication. Three groups of professionals were involved with research into Markush structure handling, but each group had its own definition of 'research.' At the universities, research was an end in itself. To be able to work out techniques to graph generic structures and to define a theoretical relationship between specific and generic chemical groups was enough. Successful academic research does not need a commercially profitable product to validate its success. In industry, the background of most of the users generating the demand for a topological Markush system, research is considered as the first step in the development of a product. If industrial research cannot produce a new product that satisfies customer needs as well as existing products, no new product reaches the marketplace. But the publishing business doesn't have much experience with the kind of research we do in industry. In the competitive climate of the time, the goal was to create a marketable search system faster than the competition. As soon as there were operable Markush systems, they were introduced to the marketplace. The fact that they were not fully responsive to market demands or that they could not be sold at a price the customer would pay, was not taken into account. Discarding the results of all that expensive research was unthinkable.

The roll-out of Derwent's Markush DARC database was met with less enthusiasm than anyone had anticipated. The most striking problem was its cost. With subscription costs as high as they were, nobody was prepared to pay an extra $75.00 to search a couple of years of patents. And the system turned out not to be suitable for true Markush searches after all: the software could not translate between generic and specific groups, and the number of variables it could handle was so limited that it was necessary to use free sites and leave most substituents undefined, opening the door to the false drop that had been one of the problems that made the fragmentation code unpopular. More fundamentally, Markush DARC turned out not to be a solution to any of the problems users of the code had complained about. There was still a 25-year backfile of patents indexed only with the fragmentation code, and Derwent had no plans to re-index all those patents for Markush DARC retrieval. Searchers had asked for a different retrieval system, but they had been given an additional retrieval system to learn and apply along with the system they had hoped to escape from. If and when the fragmentation code was phased out, users would have to apply both the troublesome fragmentation code and the new Markush system, but Derwent's staff would be free to forget all about fragmentation coding. Where would subscribers go for training then? Would the fragmentation-coded patents become unretrievable because the retrieval system had been forgotten?

INPI's Pharmsearch database can probably be counted as a success. To the public this was a new database. Pharmaceutical companies, especially companies without full access to the Derwent files, were given an entirely new place to search. But it is another place to search and another fee to pay, and it duplicated information in the Derwent and CAS databases. If the file was not greeted with enthusiasm by the public, it could still meet the needs of French patent examiners.

Like Pharmsearch, MARPAT represented an entirely new database. There is no backfile of Markush structures in the CA Registry database. The software can handle more complex Markush structures, and it can translate between generic and specific chemical groups. But the ability to translate goes further than the Markush structures require: MARPAT produces lots of false drop. And it is by no means inexpensive to use. If CAS and the survey respondents expected that MARPAT would encourage a flood of end-users to pay additional fees to search an additional database to retrieve generic structures from patents issued during the last few years, they were mistaken. The search system is too complicated for casual users, the cost is too high for searchers on a tight budget, and the output is too dirty for a quick search.

The problems that were experienced in producing an efficient and inexpensive topological search system for Markush structures have not been caused by the quality of the research at CAS or Questel, nor by the shortcuts they took to get the new search systems on to the market. The problem is caused by the Markush format itself. Markush structures are not fully topological. Topological search software is based on connection tables that record the bonds that connect each atom in a molecule to the other atoms. Both the atoms and the bonds in a Markush structure are variable. It would take a great many connection tables to index every embodiment of a Markush structure — there are many patents that would generate an infinite number of individual records. Under the circumstances, it is not surprising that both MARPAT and Markush DARC have had to integrate fragment terms to make their systems work, and it is not surprising that the goal of simple input and clean output has eluded them. As long as we are working with Markush structures, a true topological search system will remain out of reach.

Appendix I

What is a Markush structure and how are Markush structures different from other chemical structures? The name 'Markush' comes from an inventor whose patent application was the subject of a landmark legal decision nearly 70 years ago. The term Markush came to refer to a class of chemical substances identified by lists of alternative substructures, and it has become the most popular format for generic chemical structures in patents. A Markush structure represents each of the specific compounds that can be constructed by combining the chemical substructures in its definition, as few as two specific embodiments or an infinite number.

A Markush structure contains at least one substructure that is required in each specific compound in the class. There are one or more variable substructures that are represented by a list of specific chemical groups (things like 'chloro' or 'isopropyl') and generic chemical groups (things like 'halogen' and 'branched alkyl'). These are referred to as Markush groups, and they are usually represented by alphabetic or alphanumeric symbols. In Markush DARC and MARPAT, Markush groups are designated as 'G groups.' In many Markush structures the location of some of the chemical bonds is variable. Any chemical substructure that isn't required or listed as optional in the definition of the Markush structure is forbidden to be present in any of its specific embodiments.

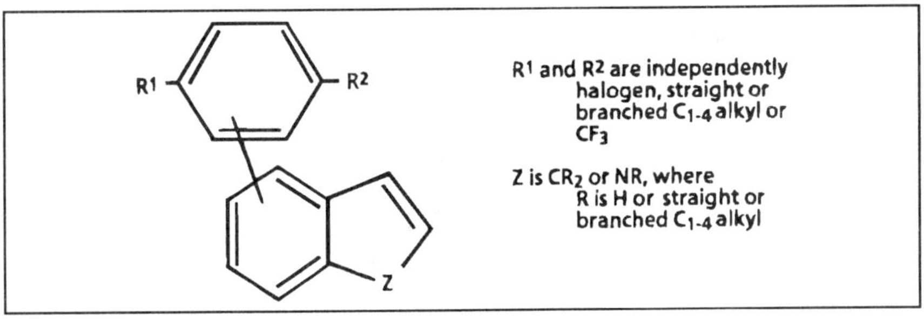

R1 R2 Z

R^1 and R^2 are independently halogen, straight or branched C_{1-4} alkyl or CF_3

Z is CR_2 or NR, where R is H or straight or branched C_{1-4} alkyl

Markush structures can be indexed by means of fragmentation codes, which were developed for use in the earliest computers. A fragmentation code contains code terms that designate chemical substructures. Each database has its own system for defining the fragments that can be searched. To create records for a fragmentation code database, an indexer analyses a Markush structure and identifies each of the substructures with a term in the coding system that can be present in an embodiment of the Markush structure. Some fragmentation codes include special terms that are applied only to substructures that are required in every specific compound in the class defined by the Markush structure. The process is something like separating the pieces of a picture in a jigsaw puzzle, and the result is a record that resembles a box of separated jigsaw puzzle pieces.

A searcher analyses a query structure in the same way and combines the terms for the fragments in a Boolean search strategy. Since it is irrelevant to the searcher whether a group is required or optional in the original Markush structure, records containing code terms that refer to required groups are eliminated from the search results with the Boolean NOT. The result of the search is a set of records that all contain the fragments that are permitted or required in the query structure without any records for Markush structures that are required to contain the fragments that are not permitted in the query structure. The output is another box of jigsaw pieces, and the pieces may or may not fit together to give a picture of the query structure.

Appendix II

December 9, 1991

Everett Brenner, 43 Sandy Hollow Road, Port Washington, New York 11050

Dear Ev:

I've just seen your review of our Annecy paper, and Nancy and I thank you for your kind words. But we feel we have to disagree with your description of it as reviewing every conceivable pitfall a searcher will face in using patent statistics. If we'd had a couple of extra months to work on it and a couple of extra hours to speak we still couldn't have found every conceivable pitfall in patent statistics.

I'd also like to take this opportunity, since you're investigating, to add my comments on Markush databases to Jim Sibley's. Americans — who don't pay much attention to the three Markush databases — are told that the Japanese and Europeans are making greater use of them. Jim is certainly right about the limitations of such a retrieval system. Bench chemists aren't interested in searching generic structure databases and never were. We have information specialists for the very reason that research scientists were too busy doing research to spend much time retrieving information. As soon as there were complex indexing systems for the chemical literature, some chemists became retrieval specialists serving the rest. Although online databases are easier to use than card sorters, bench chemists still only have time for straightforward searches. If any of the Markush files was developed with the expectation of an end-user market, it must have been because the right people weren't asked the right questions about what they would search if it were developed.

In fact, I believe that the 'user driven' development of the Markush systems was largely based on mis-communication. Markush structure retrieval systems were first proposed in the early 1980s, at a time when topological search systems had revolutionised searching for specific compounds. Searchers, who knew very little about the complexities of indexing for topological retrieval and (in most cases) very little about the potential scope of Markush structures in patents, said, "It sure would be wonderful if we could retrieve generic structures from patents as easily as we search the CAS Registry file." Database producers, who hadn't yet learned about the complexity of indexing generic structures for topological retrieval, heard this as a market demand and announced that they were starting development.

But the database producers were trying to satisfy the demand literally instead of trying to find a simpler way to solve the problems that generated the demand. The only methods available to most searchers for retrieving Markush structures were fragmentation codes (usually Derwent's) that required the searcher to learn a complex retrieval system and to analyse the structure for codable fragments. After creating a complex retrieval strategy, the searcher had to obtain patent copies or abstracts and to plough through the descriptions of the Markush structures in those documents to find out if the searched structure was embedded there or if the patent was a false drop. Many customers were asking for something to replace the complex retrieval systems and eliminate the drudgery of weeding out false drops. I believe that they envisioned drawing in a structure and retrieving only hits without imagining that the simple CAS ONLINE retrieval system would have to be modified radically to accommodate complex Markush structures or that the computer would retrieve substantial numbers of records that didn't match the query structure.

Since Chemical Abstracts Service's customers had to go to other database producers and other hosts to retrieve Markush structures from patents, it wasn't illogical to develop Markush structure indexing and retrieval for STN. In developing software that could handle Markush structures, the staff at CAS learned just how complex chemical patents really were. The resulting software is much more complex than the original version, and not as easy to learn to use. It can accomplish most of the tasks a patent searcher would want, including the marvellous trick of recognising the equivalence of a specific chemical group and the genus it belongs to. But the output from a MARPAT search is not what the original requesters wanted: the specific-to-generic translation capability generates false drops, and the chemical structures that display in the record are at least as complicated and hard to interpret as those in the original patent. Drawing a Markush structure for MARPAT retrieval is not particularly straightforward. And the cost of searching such a small file is relatively high.

Of course, even if the MARPAT system were perfect in every way, it can only be used for recent patents. To do a retrospective search you have to go somewhere else, and if you're a Derwent subscriber who can do your whole search in Derwent's files, why invest in another expensive search? MARPAT is available to searchers without a Derwent subscription, but a first-time patent searcher faced with a multi-page Markush representation could easily be intimidated into abandoning MARPAT.

Markush DARC shares the shortcomings of MARPAT with one exception: because it cannot translate between specific and generic groups, it does not generate many false drops. It also requires much longer strings of alternative groups to compensate for the lack of translation, forcing users to choose between narrowing search scopes to avoid system limits and using undefined substitutions that generate false drops. Usage of INPI's MPHARM is limited by the pharmaceutical focus of the database. Usage of Derwent's WPIM is limited by the difficulties Derwent has experienced in inputting the excessively broad generic structures they have named "nasties, supernasties and hypernasties," and by competition from Derwent's own fragmentation code.

I believe that users who asked for topological retrieval were really asking for simplicity. They wanted a replacement for the fragmentation code (i.e., they didn't want to have to use it anymore). What they've gotten from Derwent is a second retrieval system to use in addition to the fragmentation code (it's Derwent staff that won't have to use the fragmentation code anymore). Fragmentation coding has continued during the first years of the WPIM file, and since it is necessary to use the fragmentation code to retrieve older patents, why invest an extra $100 or so to search the newest part of the file topologically, especially when false retrievals were already minimised in the latest revision of the fragmentation code system?

As I hinted at the ACS meeting in August, 1990 (a reprint of the printed version of my paper is enclosed), topological indexing is probably not an appropriate way to handle Markush structures. It's certainly not an economical way. Now that searchers have seen that searching Markush structures topologically isn't as easy as we expected, the results aren't as clean as we'd hoped and the cost is higher than we anticipated, we are reluctant to spend the time and money they cost.

That's why I think customers haven't welcomed the new databases, but it doesn't explain the reasons so much development money was spent. One reason may be that the developers didn't anticipate the magnitude of the cost and didn't know how to stop short of a marketable system and abandon the money they'd already spent. I'm told that the publishing industry doesn't see failed projects as acceptable outcomes for research and development. Mostly, though, I suspect that politics and psychology had a lot to do with it. MARPAT development began around the time negotiations to mount the Derwent WPI file on STN fell through. Work on the two retrieval systems developed the characteristics of an arms race, and you know how much money those cost! I can't wait to find out what your investigations uncover.

Best regards,

Edlyn S. Simmons

Bibliography

1. Sibley, J. F. 'If only . . . a sideways look at some patent databases'. *Recent Advances in Chemical Information*, H. Collier, Ed., Royal Society of Chemistry: Cambridge, 1992, pp. 21-32.

2. Barnard, John M. 'Online graphical searching of Markush structures in patents'. *Database*, **1987**, *10(3)*, 27-34.

3. Barnard, John M. 'A comparison of different approaches to Markush structure handling'. *J. Chem. Inf. Comp. Sci.* **1991**, *31(1)*, 64-68.

4. Schoch-Grübler, Ursula. '(Sub)structure searches in databases containing generic chemical structure representations'. *Online Review* **1990**, *14(2)*, 95-108.

5. Shenton, Kathleen E. 'Graphic retrieval of patent information'. *Proceedings of the 9th International Online Information Meeting*, London 3-5, December 1985, pp. 43-59.

6. Shenton, Kathleen E., and Norton, P. 'Patent Information — Toward Simplicity of Complexity'. Presented at the 1989 Montreux International Chemical Information Conference & Exhibition (unpublished).

7. Roesch, C., Pagis, C. 'Design and production of a pharmaceutical patent database: a patent office experience'. *Proceedings of the 1989 Montreux International Chemical Information Conference*, H. R. Collier, Ed., Springer Verlag: Heidelberg, 1989, pp. 175-185.

8. O'Hara, M. P., and Pagis, Catherine,. 'The PHARMSEARCH database'. *J. Chem. Inf. Comp. Sci.* **1991,** *31(1),* 59-63.

9. Cloutier, Kathleen 'A. Comparison of three online Markush databases'. *J. Chem. Inf. Comp. Sci.* **1991,** *31(1),* 40-44.

10. Schmuff, Norman R. 'A comparison of MARPAT and Markush DARC software'. *J. Chem. Inf. Comp. Sci.* **1991,** *31(1),* 53-59.

11. Simmons, Edlyn S. 'The grammar of Markush structure searching: vocabulary vs syntax'. *J. Chem. Inf. Comp. Sci.* **1991,** *31(1),* 45-53.

12. Wilke, R. N. 'Searching for generic chemical structures'. *J. Chem. Inf. Comp. Sci.* **1991,** *31(1),* 36-40

13. Fisanick, William. 'Requirements for a system for storage and search of Markush structures'. In *Computer Handling of Generic Chemical Structures* (Proceedings of a Conference organised by the Chemical Structure Association at the University of Sheffield, England, March 26-29, 1984). Barnard, John M., (Ed.) Gower: Aldershot U. K., 1984, pp 106-129.

13. Welford, S. M., Ash, S., Barnard, J. M., Carruthers, L., Lynch, M. F., von Scholley, A. 'The Sheffield University Generic Chemical Structures Research project'. In *Computer Handling of Generic Chemical Structures* (Proceedings of a Conference organised by the Chemical Structure Association at the University of Sheffield, England, March 26-29, 1984). Barnard, John M., (Ed.) Gower: Aldershot U. K., 1984, 130-158.

15. Welford, Stephen M., Lynch, Michael F., Barnard, John M. 'Towards simplified access to chemical structure information in the patent literature'. *Journal of Information Science* **1983,** *6 (1),* 3-10.

16. Lynch, Michael F., Barnard, John M., Welford, Stephen M. 'Generic structure storage and retrieval'. *J. Chem. Inf. Comp. Sci.* **1985,** *25(3),* 264-270.

17. Lynch, M.F. 'Generic chemical structures in patents (Markush Structures): the research project at the University of Sheffield'. *World Patent Information, 8 (2)* **1986,** 85-91.

18. Lynch, Michael, Downs, Geoffrey, Gillet, Valerie, Holliday, John and Dethlefsen, Winfried. 'Generic chemical structures in patents — an evaluation of the Sheffield University research work'. In *Proceedings of the Montreux 1989 International Chemical Information Conference,* H. Collier, Ed, Springer Verlag: Heidelberg, 1989, pp 161-174.

19. Lynch, Michael F., Barnard, John M., Welford, Stephen M. 'Computer storage and retrieval of generic chemical structures in patents. I. Introduction and general strategy'. *J. Chem. Inf. Comp. Sci.* **1981,** *21(3),* 148-150.

20. Barnard, John M., Lynch, Michael F., Welford, Stephen M. 'Computer storage and retrieval of generic chemical structures in patents. 2. GENSAL,

a formal language for the description of generic chemical structures'. *J. Chem. Inf. Comp. Sci.* **1981,** *21(3),* 151-161.

21. Welford, Stephen M., Lynch, Michael F., Barnard, John M. 'Computer storage and retrieval of generic chemical structures in patents. 3. Chemical grammars and their role in the manipulation of chemical structures'. *J. Chem. Inf. Comp. Sci.* **1981,** *21(3),* 161-168.
22. Barnard, John M., Lynch, Michael F., Welford, Stephen M. 'Computer storage and retrieval of generic structures in chemical patents. 4. An extended connection table representation for generic structures'. *J. Chem. Inf. Comp. Sci.* **1982,** *22 (3),* 160-164.
23. Welford, Stephen M., Lynch, Michael F., Barnard, John M. 'Computer storage and retrieval and generic chemical structures in patents. 5. Algorithmic generation of fragment descriptors for generic structure screening'. *J. Chem. Inf. Comp. Sci.* **1984,** *24 (2),* 57-66.
24. Barnard, John M., Lynch, Michael F., Welford, Stephen M. 'Computer storage and retrieval of generic chemical structures in patents. 6. An interpreter program for the generic structure description language GENSAL'. *J. Chem. Inf. Comp. Sci.* **1984,** *24 (2),* 66-71.
25. Gillet, Valerie J., Welford, Stephen M., Lynch, Michael F., Willett, Peter, Barnard, John M., Downs, Geoff M., Manson, Gordon, Thompson, Jon. 'Computer storage and retrieval of generic chemical structures in patents. 7. Parallel simulation of a relaxation algorithm for chemical substructure search'. *J. Chem. Inf. Comp. Sci.* **1986,** *26 (3),* 118-126.
26. Gillet, Valerie J., Downs, Geoffrey M., Ling, Ai, Lynch, Michael F., Venkataram, Pallapa, Wood, Jennifer V. 'Computer storage and retrieval of generic chemical structures in patents. 8. Reduced chemical graphs and their applications in generic chemical structure retrieval'. *J. Chem. Inf. Comp. Sci.,* **1987,** *27 (3),* 126-137.
27. Lynch, Michael F. 'Computer storage and retrieval of generic chemical structures in patents: 9; an algorithm to find the extended set of smallest rings in structurally explicit generics'. *J. Chem. Inf. Comp. Sci.* **1989,** *29 (3),* 207-214.
28. Lynch, Michael F. 'Computer storage and retrieval of generic chemical structures in patents: 10; assignments and logical bubble-up of ring screens for structurally explicit generics'. *J. Chem. Inf. Comp. Sci.* **1989,** *29 (3),* 215-224.
29. Dethlefsen, Winfried, Lynch, Michael F., Gillet, Valerie J., Downs, Geoffrey M., Holliday, John D., Barnard, John M. 'Computer storage and retrieval of generic chemical structures in patents. 11. Theoretical aspects of the use of structure languages in a retrieval system'. *J. Chem. Inf. Comp. Sci.* **1991,** *31(2),* 233-253.

30. Dethlefsen, Winfried, Lynch, Michael F., Gillet, Valerie J., Downs, Geoffrey M.; Holliday, John D.; Barnard, John M. 'Computer storage and retrieval of generic chemical structures in patents. 12. Principles of search operations involving parameter lists: matching-relations, user-defined match levels, and transition from the reduced graph search to the refined search'. *J. Chem. Inf. Comp. Sci.* **1991**, *31(2)*, 253-260.

31. Gillet, Valerie J., Downs, Geoffrey M., Holliday, John D., Dethlefsen, Winfried, Lynch, Michael F. 'Computer storage and retrieval of generic chemical structures in patents. 13. Reduced graph generation'. *J. Chem. Inf. Comp. Sci.* **1991**, *31(2)*, 260-270.

32. Fisanick, William. 'The Chemical Abstracts Service generic chemical (Markush) structure storage and retrieval capability. 1. Basic concepts'. *J. Chem. Inf. Comp. Sci.* **1990**, *30(2)*, 145-154.

33. Fisanick, William. U. S. 4,642,762. Assigned to American Chemical Society. February 10, 1987.

34. Ebe, Tommy, Sanderson, Karen A. and Wilson, Patricia S. 'The Chemical Abstracts Service generic chemical (Markush) structure storage and retrieval capability. 2. The MARPAT file'. *J. Chem. Inf. Comp. Sci.* **1991**, *31(1)*, 31-36

35. Fugmann, R., Nickelsen, H., Nickelsen, I., Winter, J.H. 'TOSAR-A system for the structural formula-like representation of concept connections in chemical publications'. *J. Chem. Inf. Comp. Sci.* **1975**, *15(1)*, 52-55.

36. Meyer, Ernst. 'Computer representation and handling of structures: retrospect and prospects'. *J. Chem. Inf. Comp. Sci.* **1991**, *31(1)*, 68-75.

37. Meyer, E., Schilling and Sens, E. 'Experiences with input, translation and search in files containing Markush formulae'. In *Computer Handling of Generic Chemical Structures* (Proceedings of a Conference organised by the Chemical Structure Association at the University of Sheffield, England, March 26-29, 1984). Barnard, John M., (Ed.) Gower: Aldershot U. K., 1984, 83-95.

38. Bois, Roger, Chaumier, Jacques. 'A comparative analysis of the DARC system and the information and documentation system of the IDC'. *World Patent Information* **1980**, *2(2)*, 61-66.

39. Kolb, A. G. 'Topological coding as a basis for the GREMAS file'. *200th National Meeting of the American Chemical Society Washington, DC*, American Chemical Society: Washington, DC, 1990; CINF 23.

Quality and risk assessment in patent searching and analysis

Julia M Fletcher

18 Hillview Road, Hucclecote, Gloucester GL3 3LG, England

Introduction

> "If a little knowledge is dangerous, where is the man who has so much as to be out of danger?" [1]

This quotation from T H Huxley, a contemporary of Charles Darwin's and proponent of Darwin's theory of evolution, embodies the concepts of information (or knowledge) and risk (or danger). Huxley sees knowledge as an active rather than a passive entity, incomplete knowledge being associated with danger.

This paper proposes a model for the relationship between patent information and the risks associated with decisions that are based on such information. The term 'patent information' includes information derived from a study of patents, patent applications and other technical disclosures. Patent information is obtained by a process of searching the literature and analysing the results. This searching and analysis process has the potential to influence critically the risks of subsequent decisions. In order to be totally trustworthy, the patent information must be of the highest possible quality. If the quality is less than perfect, can our decisions ever be free from danger?

The concept of quality applies equally well to the provision of services as to manufacturing functions [2]. In any investigation based on scientific information, poor quality searching and analysis is clearly undesirable and can be positively misleading. The magnitude of the effect this has depends upon the purpose of the investigation.

In the patents area, the results of searching and analysis are often employed strategically in making major commercial decisions. Furthermore, the consequences of decisions based on poor quality information are potentially very costly: lawsuits, abandoned product launches and misdirected research for instance. In the extreme case, a company whose business is founded on a small number of patents could collapse altogether if those patents are established as invalid. It is well known that Glaxo's profitability has, in recent years, been closely related to the sales of its anti-ulcer drug Zantac. Clearly, if its Zantac patents were invalidated, Glaxo's commercial monopoly would disappear and it could become extremely vulnerable to competition.

Context and definitions

It is first necessary to define what we mean by 'quality' and 'risk' in a decision-making context (Figure 1).

Quality

An appropriate definition of quality in this context might be conformance to the requirements of the decision-maker who relies on patent information. This contrasts with the rather nebulous idea, acceptable in common parlance, of quality as goodness. What are the requirements of decision-makers in the patent area? How can the analyst meet those requirements, what standards of performance are acceptable and what methods are there for measuring performance? [3] In order to answer these questions, it is helpful to look more closely at the steps comprising the process of searching and analysis.

Quality	=	Conformance to Requirements
Risk	=	Penalty x Probability of Error

Figure 1: Definitions

The patent analyst receives a search request, translates it into a problem which can be solved using the methods at his or her disposal, analyses the results and presents them in a form that is most comprehensible and useful to the recipient. Clearly, where time allows, the process should be an iterative one; each stage of the search is followed by evaluation of the findings so far and identification of further avenues for investigation. In theory, this process can continue indefinitely until all possible avenues are exhausted and the results are comprehensive but, in practice, it is subject to the constraints of time, money and so on.

There are two principal methods of measuring performance: predicting the likely outcome of an event (such as the consequence of a decision) and assessing the search process in terms of conformance to certain standard procedures. The assumption here is that by conforming to such procedures, the highest quality outcome will be achieved. The outcome of a decision based on a poor quality patent search could be litigation or a patent that is weaker than it might have been. It may be some years before the outcome can be determined.

However, conformance can be assessed immediately, for example in terms of the characteristics listed in Figure 2 [4]; together, these define the quality of the search results. The searching and analysis process must make efficient use of resources such as time, money, information sources, people and so on. Because patent decisions involve both legal and commercial considerations, confidentiality and accuracy are of paramount importance. Timeliness implies adherence to the agreed deadlines which are often related to time periods set by statute or imposed for commercial reasons. The resulting report should be comprehensive in its coverage

Good quality patent information is:

- efficient
- confidential
- accurate
- timely
- comprehensive
- appropriate
- interpreted

Figure 2: Performance Standard

of the relevant subject matter, countries and time periods. It should contain explanation and interpretation which is appropriate to the recipient; good communication and feedback are helpful in understanding what is required. In real life, perfect searches exist infrequently and good quality patent analysis can only reduce uncertainty.

Risk

Risk can be defined empirically as the product of penalty and the probability of error. Here, error could be, for example, the omission from a search report of a pertinent piece of prior art or granted patent, a missed deadline or an inaccurately directed search. Penalty in this context has two facets. Either it relates to the costs of a disaster which can result from a missed reference (for example, an infringement suit initiated by the owner of a patent that should have been found during the search) or it concerns the loss of a profit-making opportunity. A perfect search corresponds to zero probability of error and, therefore, to zero risk. Perfection is, in practice, rarely attainable so there is nearly always a finite probability of error which is in inverse proportion to the quality of the search: high quality, lower probability of error and vice versa.

Figure 3 shows hypothetical risks for patentability and infringement searches. We can begin by setting a notional penalty for misses prior art in a patentability search at, say, 100 units. The probability of error could be 0.01 for a good quality search and 0.1 for a poor quality one. The corresponding risk of a decision to apply for a patent is thus lower in a good quality search. In the case of an infringement search, for example before a product launch, the penalty may be substantially higher: perhaps 10,000 units because of the potential for litigation. By performing a good quality search, we should be able to reduce the risk of litigation based on unlocated prior claims. A high quality search means reduced risk but it cannot remove the risk entirely.

	Penalty	Probability of Error	Risk
PATENTABILITY			
a) Good quality search	100	0.01	1
b) Poor quality search	100	0.1	10
INFRINGEMENT			
a) Good quality search	10,000	0.01	100
b) Poor quality search	10,000	0.1	1,000

Figure 3: Risk Assessment

Decision-making

Most management decisions are based on imperfect information and consequently involve a degree of risk. Figure 4 shows the three major components of a decision made by a patenting company whose competitors have valid patents in the market it is proposing to enter. Although the results of patent analysis are not the only factor involved, they obviously play a major part in decision-making. Other uncertainties are introduced from the scientific and commercial aspects of the decision. As patent analysts, we have to reduce the risk of the patent input to the decisions by taking what steps we can to improve the quality of the information we provide. Herein lies the challenge which underpins all our work.

Patent analysis in practice

From the many areas where patent analysis can be applied, four selected ones will be examined:

1. Patentability searches

According to most major countries' patent laws, novelty is destroyed if information about the invention is made available to the public (orally, in writing or by some other method) anywhere in the world. A patentability search seeks to establish

1	Scientific Optimum	Research output
	+	
2	Patent Situation	Results of infringement and patentability searches
	+	
3	Commercial knowledge	What will sell

Figure 4: Decision-making Process

whether a supposed invention is in fact novel and non-obvious before patent applications are filed to protect its exploitation. An application based on inadequate information could be rejected by the patent issuing authorities or it could be granted and subsequently found invalid. (The Patent Office Examiners' searches are not perfect either).

Clearly, it is an impossibility to search everything and usually one focuses on a combination of what are, from previous experience, known to be the most productive avenues of enquiry. It can be difficult to know when to stop. However, as time and money are never in limitless supply, this problem is often solved by expediency. This approach may be pragmatic but it does not always allow good quality searches to be conducted.

Edlyn Simmonds [5] has described 'the paradox of patentability searching' in terms of the dilemmas which face the searcher. The problem of nomenclature is one aspect: how do you pin down a subject which is so new that its vocabulary has not yet been standardised through usage? Because of this, the probability of error is likely to be higher for an innovation than for an incremental improvement. In addition the penalty will probably be higher for the innovation: in the right commercial climate, a highly innovative product can sometimes create a new market and consequently generate significant revenue for the company whose product it is. In the case of an incremental improvement the technical fields will probably be well known to the requester if not to the searcher and the probability of error is likely to be lower. Thus, the risks associated with patentability searches are likely to be greater for a startling innovation than they are for an incremental improvement.

2. Infringement searches

Prior to launching a new product, many companies would commission an infringement search to check on patent claims which could be infringed by the sale of the product. The search results serve as a map of the minefield the company is about to enter: companies which venture into the market in ignorance of where the mines (claims) are,risk being blown up (sued). Failure to locate a patent whose claims are infringed by the proposed product can result in a costly infringement action in the courts. If more than one country is involved, the costs escalate accordingly. The penalty for missing pertinent claims in a search is high so decisions based on infringement searches are inherently very risky. Even a non-perfect search is better than none at all as it reduces the probability of error and therefore the risk associated with the decision. This assumes that finding relevant prior claims is a positive benefit: in reality, it may lead on to invalidity searches (see later) and/or licensing negotiations and there is no guarantee that these will succeed; litigation can still follow. It is usually preferable, however, to be prepared for the dangers than to remain unaware of them.

3. Invalidation of competitors' patents

This is one of the most challenging types of search and offers considerable scope for the analyst's creativity. It is commissioned when a company identifies a competitor's patent as an obstacle to its present or future commercial activities. The aim of the search is to locate prior art to invalidate the patent claims by proving that the invention was not novel or that it was obvious to a person skilled in the art at the priority date of the patent. Commercially, the potential gains for the company can be enormous: the chance to recoup R & D and production costs as well as to make a profit, provided the company has the freedom to market its products. If the company is already in the market when the blocking patent comes to light, there is potential for all the penalties associated with an infringement action. Attendant risks are thus potentially high also.

4. Background surveys of technology

These are carried out, for example, when a company is planning a new strategy for Research and Development or when a patent agent needs a quick introduction to a new technical area with its associated terminology.

There was one potentially disastrous situation in which I was involved when a blocking patent was identified after an expensive R & D programme and immediately before a product launch. The product was almost certain to infringe the patent's claims. Fortunately, on this occasion, the necessary prior art was found and the patent was successfully invalidated but another time the patent may be valid and all the R & D expenditure wasted. A background search conducted at the early stages of planning of the R & D would have given two options: to attempt to challenge the patent (if indeed it had been granted or its corresponding application had been published) or to plan the research around it.

Here, the possible penalty for a company is its commitment to a futile and costly R & D programme with no marketable product at the end of it. A good quality background survey with low probability of error leads to minimum risk of this occurring.

Application

This section looks at the investment in resources required to achieve various levels of quality and value in patent searching and analysis.

As we have seen earlier, risk is dependent both on the penalty for a given situation (fixed) and the probability of error (inversely proportional to search quality). The risk associated with a decision can be lessened by reducing the probability of errors which impinge directly on the making of that decision, that is, by improving the quality of the search for the information. However, high quality searching implies higher costs. It would be desirable to have some means for determining when the expected improvement in quality justifies increased expenditure on searching.

It is feasible to devise a model for decision-making based on the cost, quality and value of patent information. One possible model is outlined below, derived from my own observations and experience. No doubt different models can be proposed which reflect more accurately the experience of others in the field. I hope that useful discussions will result from a consideration of the example.

Figure 5 shows in graphical and greatly simplified form the relationship between the quality of information and the costs (expressed in some measurable way, e.g.

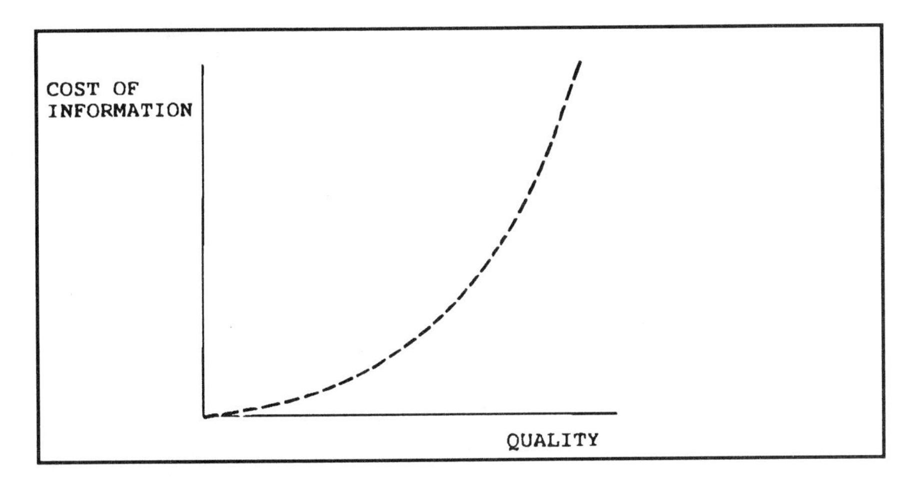

Figure 5: Cost and quality of information

in pounds, man-hours) of obtaining it. Quality embodies many factors and this graph could be seen as a two-dimensional representation of the multi-dimensional diagram. For the sake of simplicity, the aspect of quality which concerns us here is comprehensiveness: perfect searches are 100% comprehensive. The cost/quality curve is shown as asymptotic to the perfect quality measurement and this is where very high costs must be incurred for only an infinitesimal improvement in quality. It is assumed that perfection is unattainable' in some, perhaps rather simple searches, it may in fact be attainable.

In Figure 6, there is shown graphically the possible relationship between value and quality, for a search based on patent information. The value of the information can be expressed in terms of financial gain or prevention of financial loss. It is linked to the penalty incurred in the absence of the information and hence to the risk of a decision based on it. An increase in the quality of the information leads to a reduction in risk and thus to an increase in its value. A search of low quality is of

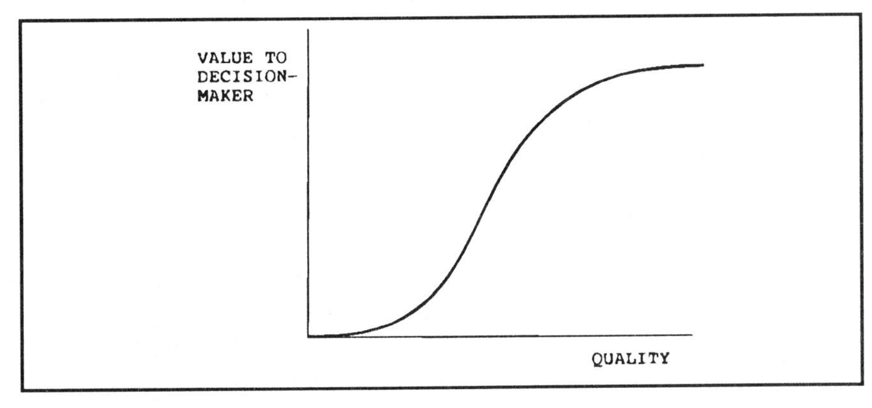

Figure 6: Value and quality of information — patent search

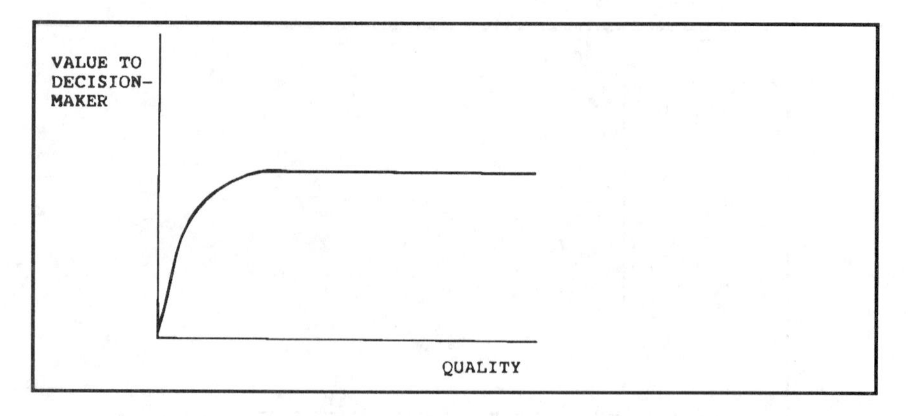

Figure 7: Value and quality of information — non-critical search

low value then, as the quality improves, we would expect a slow increase in value followed by a steep increase. Eventually, a plateau is reached where further increases in quality yield only minimal improvement in value. The value of the information is asymptotic to a maximum level (i.e. minimum risk) which is dependent on the end use of the information. For instance, this maximum value may be higher for an infringement search than for a novelty search.

Figure 7 shows, for comparison, the possible relationship between value and quality for a search whose results are non-critical: typically, the research scientist's request for 'a few references' on a topic. Comprehensiveness is not required and no major decision will be based on the results. In this case, the results of even a fairly low quality search are of high value (with minimal associated risk); beyond a certain point, it is not worth improving the quality of the search (i.e. expending any more resources) as the end use does not justify this.

In Figure 8, the curves for cost/quality and value/quality are superimposed. Point A is where the cost incurred balances the value of the information, that is the benefit to be gained from obtaining it. For a search of lower quality, the value is outweighed by the cost and it is possible to identify a point of maximum lost to the decision-maker. For a search of quality higher than shown at A, the cost is outweighed by the value and it is possible to identify a point of maximum benefit to the decision-maker. From a decision-maker's point of view, therefore, minimum risk is achieved by obtaining a search of good quality in this upper region. Searches of quality lower than A will always lead to some reduction in risk but may prove to be a false economy as the cost of performing them exceeds their value to the company. Sometimes, it would be better to save the resources than to spend them on a search which can only be of poor quality because of the constraints imposed.

Figure 9 shows the curve for cost/quality with non-critical value/quality curve superimposed. Here, the cost/value intersection occurs at such a low quality that any subsequent increase in quality gives a sharp improvement in value. The plateau is reached very quickly, showing that the search need only be of mediocre quality

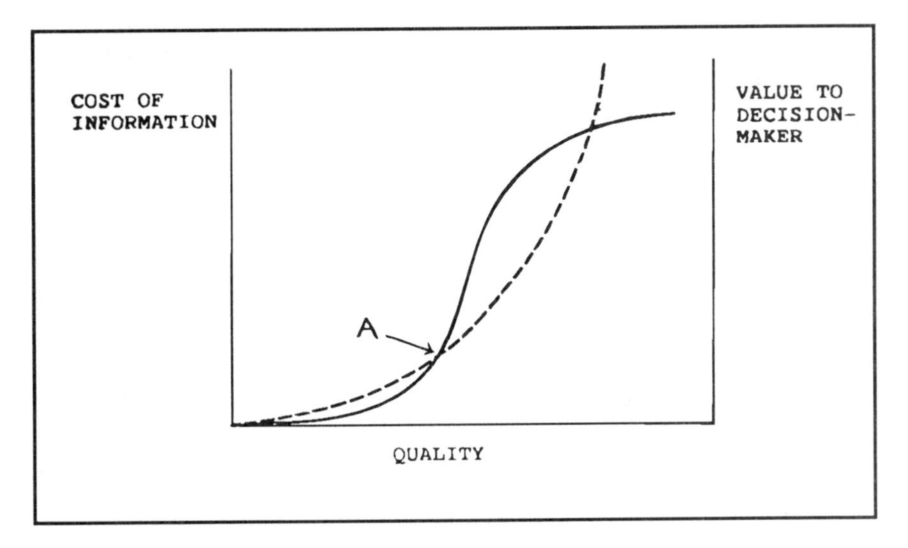

Figure 8: Cost, quality and value of information — patent search

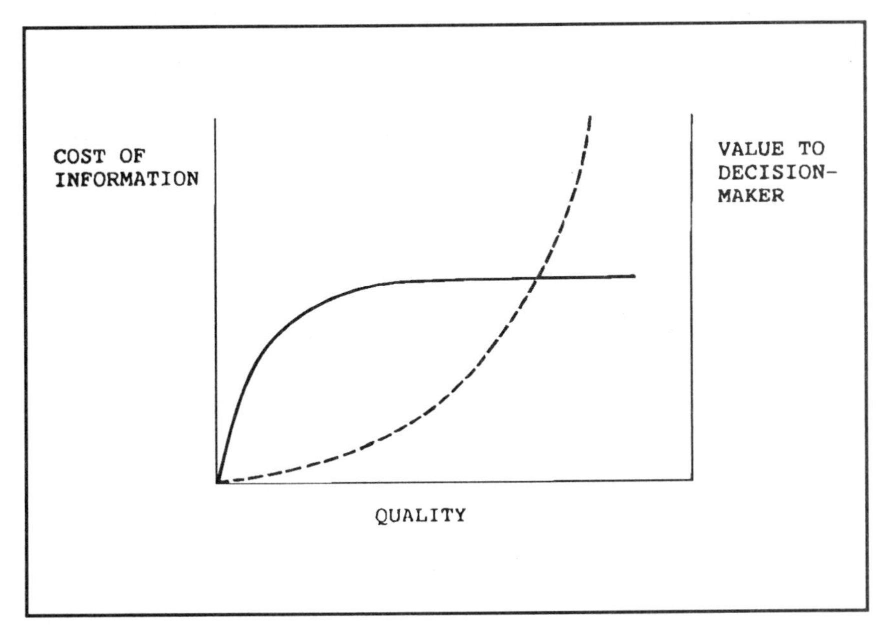

Figure 9: Cost, quality and value of information — non-critical search

to be useful for the requester's purposes. This is not typical of the patent information field, in my experience, but is included for purposes of comparison.

Conclusions

This model shows the influence that good quality patent information can have on decisions that are made in the patents area. It also shows the futility of basing decisions on poor quality information in situations where the penalties are great and the risks high. By considering where the cost/value trade-off occurs, we can make more sensible judgements about the quality of our patent information and thus about the risks of our decisions.

To return to Huxley, it is rare to be out of danger in the patents field, but good quality patent searching and analysis can supply the knowledge required to understand that danger and to minimise it.

Acknowledgements

I should like to thank my fellow professionals for their comments on the draft version of this paper and for discussions of these issues over a number of years of searching for quality.

References

1. T H Huxley '*On Elementary Instruction in Physiology*' (1877).
2. A R Tenner 'Quality Management Beyond Manufacturing', *Research-Technology Management* **34***(5)*, 27–32 (1991).
3. P B Crosby '*Quality Without Tears*'; McGraw-Hill (1986).
4. Based on: T Lucey '*Management Information Systems*'; D P Publications, Eastleigh, Hants. (1987).
5. E S Simmons 'The Paradox of Patentability Searching', *J. Chem. Inf. Compu. Sci.* **25**, 379–386 (1985).

USPTO Automated Patent System: a critical discussion of APS and its use for full-text searching

Lucille J. Brown and Robin M. Jackson

LJB Associates, 464 Inverrary, Deerfield, IL 60015, USA

In 1980 the United States Patent Office (USPTO) was given a mandate by the US Congress to automate the agency's operations. The Automated Patent System — APS — is one aspect of the automation project. APS allows full text searching of US patents issued since 1971 and Japanese abstracts in English since 1981. Using a workstation consisting of dual high resolution monitors and standard person computer architecture, examiners and members of the public can query the system, display documents of resultant sets in ASCII text, or view the image of each document as is currently done in the USPTO search room, as well as augment searches done for certain kinds of art. The focus of this paper will be the mechanics and practicality of using the system as it exists today and a brief discussion of the utility of using full text systems for searching the patent literature.

Background of the automation project

The United States Congress mandated the United States Patent Office (USPTO) in the early 1980s to automate the entire agency operation. The goal of the project was to improve the quality of patent and trademark information dissemination and to improve the integrity and quality of USPTO records. The automation project included the Trademark Examination Operation as well as the patent system. The Automated Patent System — APS — is a result of the congressional directive.

APS is designed to manage patent office operations including office automation, application processing and the activities that support patent examination such as technical searching. Initially, the search portion of APS contained the full text of US patents issued from 1971 and English abstracts of Japanese patents from 1981. The files, USPATS and JPOABS, were full text searchable using Chemical Abstracts' designed and modified Messenger software and also image searchable using Image Search, a protocol created by the Planning Research Corporation and Chemical Abstracts. The full text version of the system was available to all patent examiners. In addition, the image portion was accessible by Art Group 220 (Figure 1) at dual screen workstations (Figure 2). The image search capability was limited to the technology assigned to Group 220, approximately 163,000 documents.

Class	Sub-class	Title
42	all	Firearms
44	varied	Fuel and related compositions
60	varied	Power plants
75	varied	Specialised metallurgical processes, compositions for, therein, consolidated metal powder compositions, etc.
86	varied	Ammunition
89	varied	Ordinance
102	all	Ammunition and explosives
114	varied	Ships
136	varied	Batteries, thermoelectric and photoelectric
148	varied	Metal treatment
149	all	Explosives and thermic compositions or charges
178	varied	Telegraphy
181	varied	Acoustics
204	varied	Chemistry, electrical and wave energy
244	varied	Aeronautics
252	varied	Compositions
264	varied	Plastic and non-metal article shaping or treating; processes
310	varied	Electric generator or motor structure
315	varied	Electric lamp and discharge systems device
330	varied	Amplifiers
340	varied	Communications, electrical
342	varied	Communications, directive radio wave systems and devices
356	varied	Optics, measuring and testing
359	varied	Optics, systems and elements
367	varied	Communications, electrical: acoustic wave systems and devices
375	varied	Pulse of digital communications
376	all	Induced nuclear reaction systems and elements
380	all	Cryptography
419	all	Powder metallurgy processes
420	varied	Alloys or metallic compositions
422	varied	Process disinfecting, deodorising, preserving or sterilising and chemical apparatus
423	varied	Chemistry, inorganic
424	varied	Drug, bio-effecting and body treating compositions
427	varied	Coating processes
428	varied	Stock material or miscellaneous articles
434	varied	Education and demonstration
436	varied	Chemistry: analytical and immunological testing
455	varied	Telecommunications
534	varied	Organic compounds - part of class 532-570 series
558	varied	Organic compounds - part of class 532-570 series
568	varied	Organic compounds - part of class 532-570 series
976	all	European Patent Office Classification of selected US patents in the nuclear engineering field

Figure 1: Group APS coverage

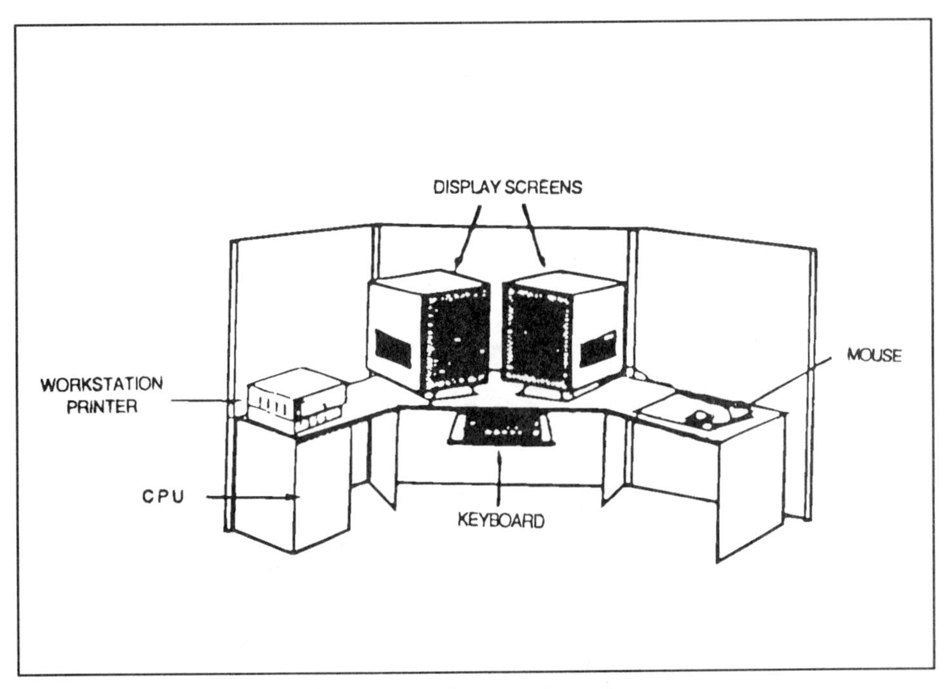

Figure 2: CSIR Workstation

System architecture is shown in Figure 3. The system consists of two NAS/9080 mainframes which provide optical and magnetic storage for text search retrieval and support. The entire system is connected via a local area network based on fibre optic digital switch high density storage devices. Rapid access devices also store optical disk information for enhanced user access. For example, examiners have local storage at each workstation which allows an average size subclass of US patents to be downloaded as images. In addition to interaction with the internal system, various interfaces permit interaction with external systems such as commercial databases.

The documents available for searching were expanded in the late 1980s and segments of the system were opened to the public in the Washington, DC USPTO Public Search Room (PSR). A simplified version of the system architecture open to the public is shown in Figure 4. The PSR is used by inventors, patent attorneys and agents to search for art relating to potential patent applications, patent infringement, litigation, etc. APS text search terminals for full text searching could be used by members of the public trained in basic Messenger software and were also made available to untrained searchers assisted by a USPTO search analyst. Image search capability was made accessible on a limited basis in 1990 using the Classified Search Image Retrieval (CSIR) workstations similar to the workstation in Figure 2. Document coverage for image searching was expanded to the technology assigned to examining groups 210 and classes 364, subclasses 200 and 900, part of Group 230, Electric Computers and Data Processing.

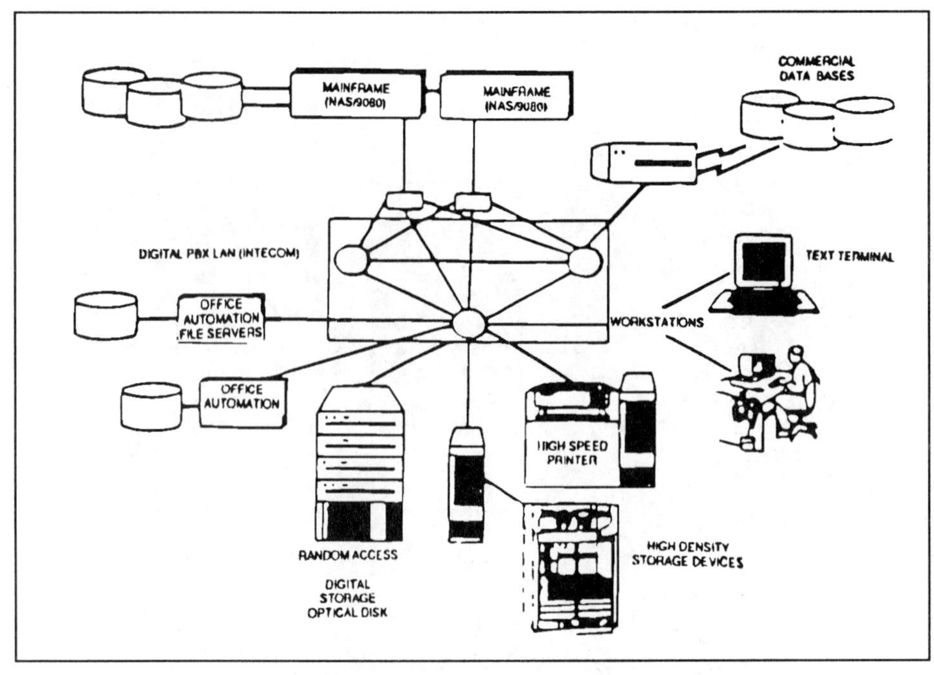

Figure 3: APS system architecture

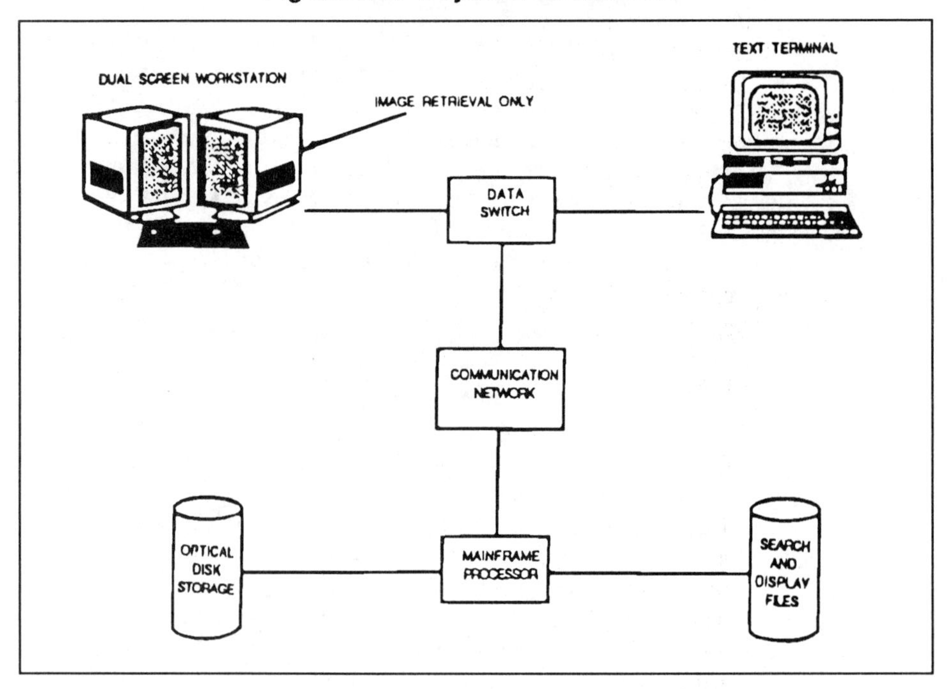

Figure 4

Library	State
Auburn University	Alabama
Boston Public Library	Massachusetts
Cleveland Public Library	Ohio
D.H. Hill Library	North Carolina
Dallas Public Library	Texas
Engineering Library	Nebraska
Los Angeles Public Library	California
Marriott Library	Utah
Miami-Dade Public Library	Florida
Milwaukee Public Library	Wisconsin
Minneapolis Public Library	Minnesota
New York Public Library	New York
Noble Library	Arizona
Oregon State Library	Oregon

Figure 5: Patent Depository Libraries with APS access

Access to APS was extended to the public outside the Washington DC area in 1991 when the system was connected to selected members of the Patent and Trademark Depository Library system (PTDL) (Figure 5). The PTDLs in the United States are designed to facilitate widespread public access to patents and patent information documents. Many PTDLs have complete collections of all US patents issued since 1790 and trademarks published since 1872. All libraries have patent and trademark search systems on CD-ROM to improve use and access of patent material. In addition, PTDLs provide patent and trademark classifications to assist patron searches. The APS systems in the PTDLs are similar to those located in the PSR and are connected to the Washington DC- based system via telephone lines. Unfortunately, PTDL staff were not trained in the use of the system, although its use was free to library patrons.

Searching APS

APS is searchable by Messenger software modified by Chemical Abstracts Service to meet the special needs of the PTO, such as forward and reverse browse capability and unique patent document fields. The file set-up is similar to most commercial full text and abstracted computerised files, although programmed function key use is necessary for complete user system interaction (Figure 6).

Search capability is also similar to that found in most computerised text information sources. The majority of data, patent titles, abstracts, specification and claims, are indexed in the basic index. The system is unique, however, in that much of a patent document is searched using very specific qualified fields. Figures 7 and 8 show which components of the front page of patent documents are available for specific retrieval. All numerical fields are range searchable. The range search capability has been modified for the United States Patent Classification (USPC) to include subclasses in the range which are not in the subclass numerical sequence, but whose subject contents are included in a class range hierarchy. The modification was necessary because of the unusual structure of the USPC. The classes and subclasses

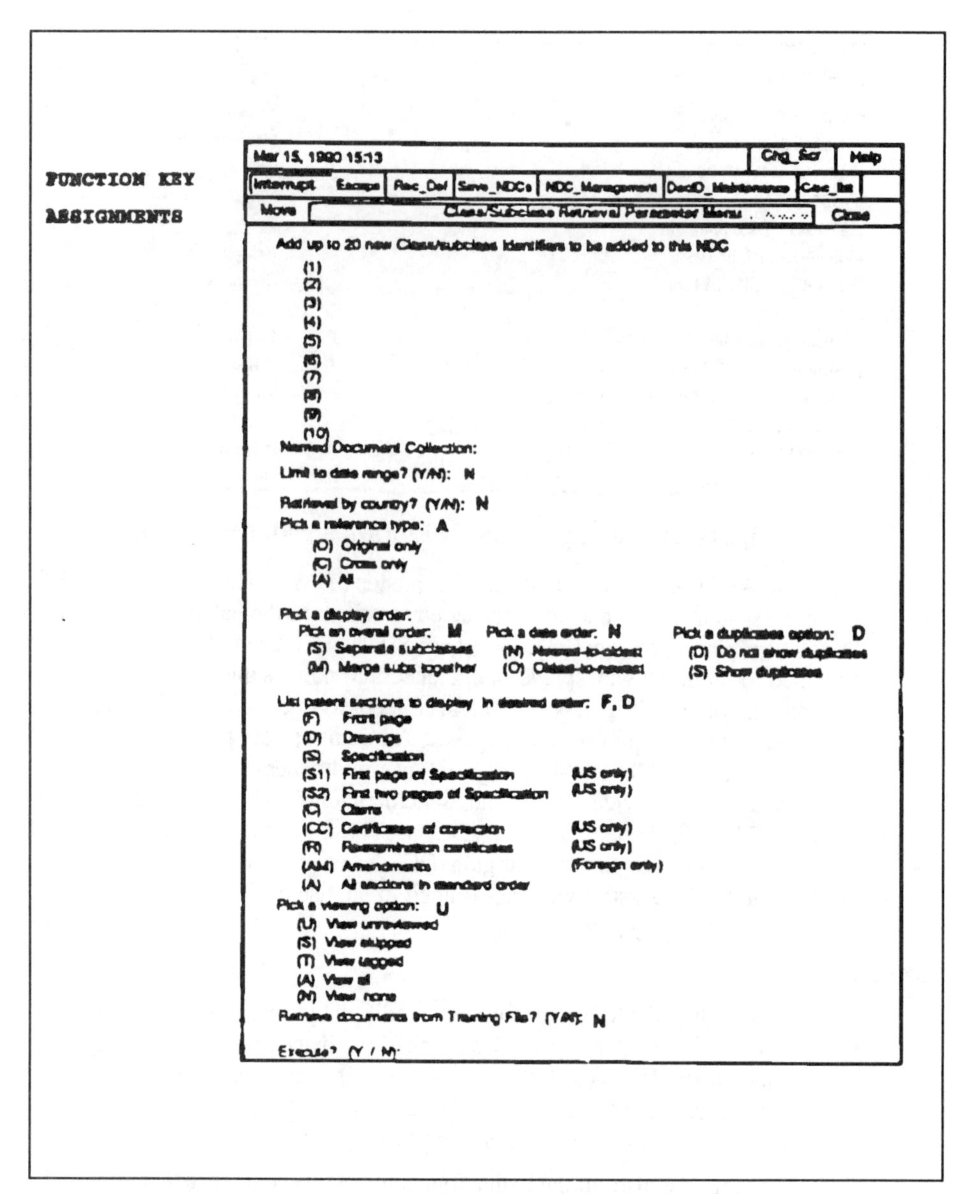

Figure 6: the class / sub-class retrieval parameter menu

which constitute the index are listed in the Manual of Classification (MOC) as groups, but not necessarily in numerical order. Thus searching a strictly numerical range may not retrieve all the subclasses appropriate to certain class–subclass series that are topically related but not in a consecutive numerical range. Class 436 and several of its subclasses are examples of this classification peculiarity (Figure 9).

Proximity searching is not particularly unique, although implied proximity when entering two or more consecutive terms without a proximity operator implies a (w)

United States Patent [19]
Malabarba et al.

[11] **Patent Number:** **5,132,286**
[45] **Date of Patent:** * **Jul. 21, 1992**

[54] **COMPOSITION OF ANTIBIOTIC L 17054 AND A METHOD FOR THE TREATMENT OF BACTERIAL INFECTIONS USING THE ANTIBIOTIC**

/TI

[75] Inventors: Adriano Malabarba; Angelo Borghi, both of Milan; Paolo Strazzolini, Fiume Veneto; Bruno Cavalleri; Carolina Coronelli, both of Milan, all of Italy

[73] Assignee: Gruppo Lepetit S.p.A., Milan, Italy

[*] Notice: The portion of the term of this patent subsequent to Feb. 24, 2004 has been disclaimed.

[21] Appl. No.: 702,797

[22] Filed: May 17, 1991

Related U.S. Application Data

[63] Continuation of Ser. No. 929,040, Nov. 10, 1986, Pat. No. 5,041,534, which is a continuation of Ser. No. 591,096, Mar. 19, 1984, Pat. No. 4,645,827.

[30] Foreign Application Priority Data

Mar. 22, 1985 [GB] United Kingdom 8507347

[51] Int. Cl.⁵ A61K 37/10
[52] U.S. Cl. 514/8; 530/322; 514/25
[58] Field of Search 530/322; 514/8, 25; 536/16.8, 18.1

[56] References Cited

U.S. PATENT DOCUMENTS

4,645,827 2/1987 Malabarba et al. 530/322

Primary Examiner—Johnnie R. Brown
Assistant Examiner—Elli Peselev
Attorney, Agent, or Firm—John J. Kolano

/BI

[57] ABSTRACT

The present invention is directed to the essentially pure preparation of an antibiotic substance arbitrarily designated antibiotic L 17054. This antibiotic substance is obtained from the known antibiotic substance named teicoplanin (formerly teichomycin) by chemical treatment. The new compound and the pharmaceutically acceptable salts possess antimicrobial activity.

2 Claims, 3 Drawing Sheets

/BI

Also contained in the Basic Index

Body of Document, Including:

Parent Case text
Summary
Drawing Description
Government Interest statement
Detailed Description
Claims

Figure 7

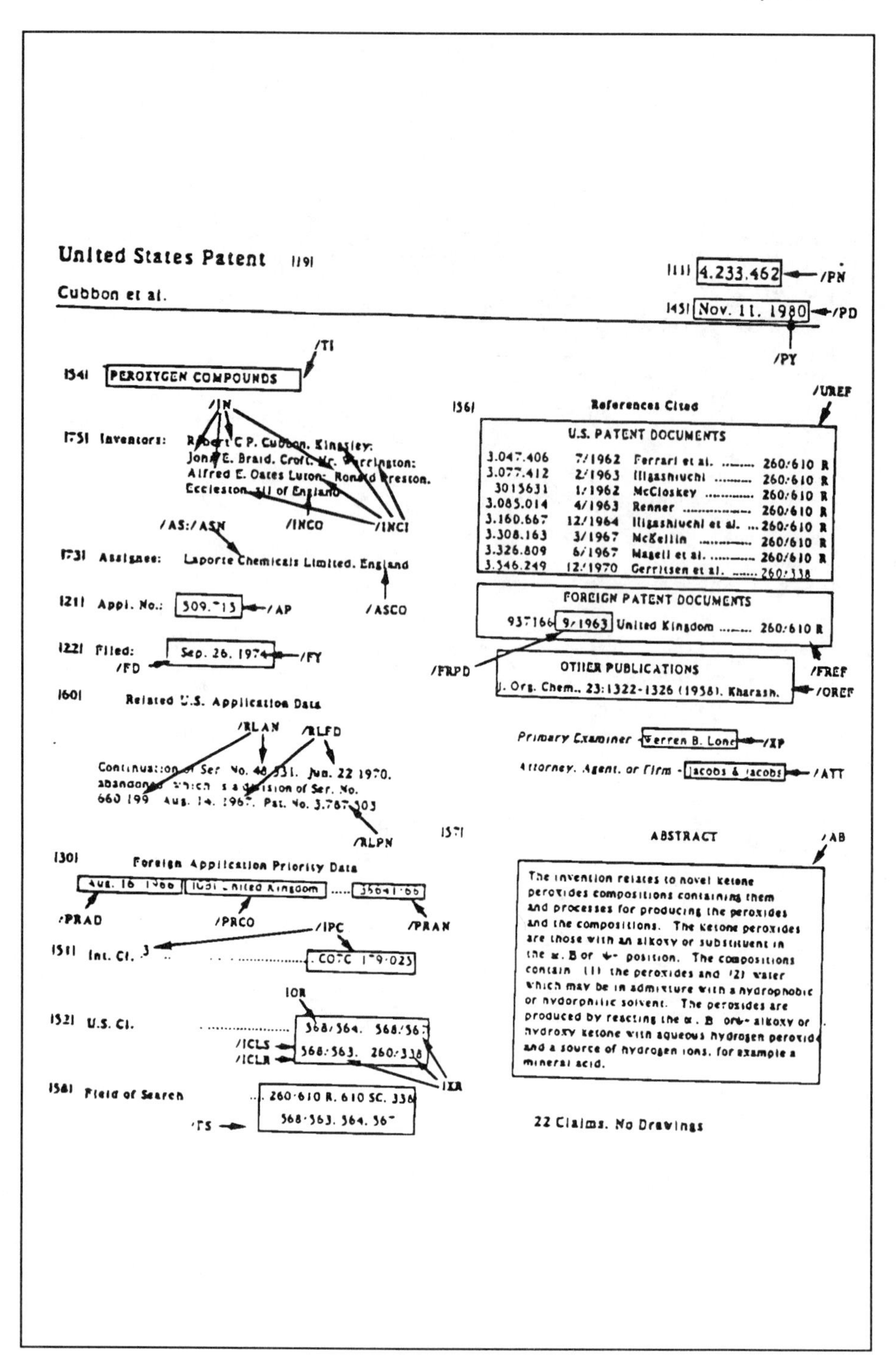
United States Patent [19]
Cubbon et al.
[11] 4,233,462
/PN
[45] Nov. 11, 1980
/PD
/PY
/TI
[54] PEROXYGEN COMPOUNDS
/IN
[75] Inventors: Robert C P. Cubbon, Kingsley; John E. Braid, Croft, Nr. Warrington; Alfred E. Oates Luton; Ronald Preston, Eccleston, all of England
/AS:/ASN
/INCO
/INCI
[73] Assignee: Laporte Chemicals Limited, England
[21] Appl. No.: 509,713
/AP
/ASCO
[22] Filed: Sep. 26, 1974
/FY
/FD
[60] Related U.S. Application Data
/RLAN
/RLFD
Continuation of Ser. No. 48,531, Jun. 22 1970, abandoned, which is a division of Ser. No. 660,199 Aug. 14, 1967, Pat. No. 3,787,503
/RLPN
[30] Foreign Application Priority Data
Aug. 16 1966 [GB] United Kingdom 35641/66
/PRAD
/PRCO
/PRAN
/IPC
[51] Int. Cl.3 C07C 179/025
IOR
[52] U.S. Cl. 568/564; 568/567; 568/563; 260/338
/ICLS
/ICLR
IXR
[58] Field of Search 260/610 R, 610 SC, 338; 568/563, 564, 567
/FS
[56] References Cited
/UREF
U.S. PATENT DOCUMENTS
3,047,406 7/1962 Ferrari et al. 260/610 R
3,077,412 2/1963 Higashiuchi 260/610 R
3015631 1/1962 McCloskey 260/610 R
3,085,014 4/1963 Renner 260/610 R
3,160,667 12/1964 Higashiuchi et al. 260/610 R
3,308,163 3/1967 McKellin 260/610 R
3,326,809 6/1967 Magelli et al. 260/610 R
3,546,249 12/1970 Gerritsen et al. 260/338
FOREIGN PATENT DOCUMENTS
937166 9/1963 United Kingdom 260/610 R
/FRPD
/FREF
OTHER PUBLICATIONS
J. Org. Chem., 23:1322-1326 (1958), Kharash.
/OREF
Primary Examiner - Werren B. Lone
/XP
Attorney, Agent, or Firm - Jacobs & Jacobs
/ATT
[57] ABSTRACT
/AB
The invention relates to novel ketone peroxides compositions containing them and processes for producing the peroxides and the compositions. The ketone peroxides are those with an alkoxy or substituent in the α, β or γ- position. The compositions contain (1) the peroxides and (2) water which may be in admixture with a hydrophobic or hydrophilic solvent. The peroxides are produced by reacting the α, β or γ- alkoxy or hydroxy ketone with aqueous hydrogen peroxide and a source of hydrogen ions, for example a mineral acid.
22 Claims, No Drawings

Figure 8

operator (Figure 10) is available. Internal and external truncations are allowed. Among the more unusual APS search features are special character searching using either the actual character or a word string that represents a character, and the searching of tabular information. The feature that distinguishes APS, however, is the image search facility. Images are searchable by patent number or by class. Thus, using CSIR workstations, the image of a patent document can be displayed or an entire subclass of patent images can be retrieved and browsed online.

Sub-class	Title
260	.preserving or maintain microorganisms
261	.Separation of microorganism from culture media
320.1	VECTOR, PER SE, E.G., PLASMID, HYBRID PLASMID, COSMID VIRAL VECTOR, BACTERIOPHAGE VECTOR ,ETC.
262	PROCESS OF UTILISING AN ENZYME OR MICROORGANISM TO LIBERATE SEPARATE, OR PURIFY A PREEXISTING COMPOUND OR COMPOSITION THEREFOR: CLEANING OBJECTS OR TEXTILES
263	.Textile treating
264	Cleaning using a microorganism or enzyme
265	Depilating hides, bating or hide treating using enzyme or microorganism
266	.Treating gas, emulsion or foam
267	.Treating animal or plant material or microorganism
268	..Treating organ or animal secretion
269	..Treating blood fraction

Figure 9: Excerpts — Class 435 Chemistry: Molecular Biology and Microbiology

Search retrieval is sortable by bibliographic parameters. In addition, records can be sorted by USPC classification as the classes and subclasses appear in the MOC. As previously mentioned, the classification may not be in strict numerical order, but the sorting operation maintains the subject integrity of the hierarchy. Queries may be saved, by an entire session or by specific L numbers. Documents retrieved may also be saved. Document display is similar to standard Messenger displays and also includes unique displays such as the entire front page of the patent or specific patent sections, e.g., patent summary, drawing description and patent claims.

Each workstation, both the text and CSIR, has peripheral printers to print retrieval as requested by the user. The entire print session may also be printed continuously at the text terminals.

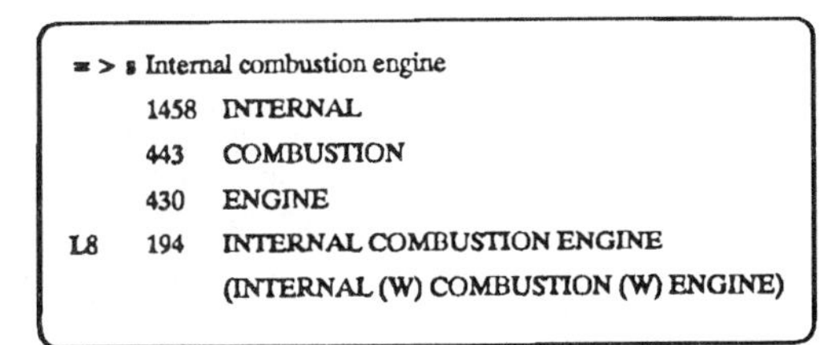

Figure 10: Implied proximity

Practical advantages

APS provides the searching community with significant advantages. Cost is probably the most obvious difference between APS and commercially available resources. The USPTO compared its cost of using fulltext vendors prior to its institution of full text APS. The cost savings were estimated at $1.5 million, as compared to using current full text systems for an analogous time period. Comparable fulltext systems charge by the search. If a query can be represented by a single search statement, the cost is $40.00 for the search. In the case of a complex search where the searcher is highly interactive with the system and numerous search statements are executed, the cost would be $40 per search statement. A session with 10 search statements would be $400 plus displays. As compared to non-full text systems, APS also presents a positive cost differential for special search activities such as citation searching. Commercial systems offer patent citation searching at $20–50 if a document has been cited subsequent to issuance, plus online charges. APS has a flat $40/hour online charge and a hit charge only if a reference is printed, and no per-hit display charge if the citations are not printed. In addition, citations may be limited to patent documents or journal references. Not all systems which allow citation searching conveniently search both journal and patent citations.

Other search capabilities of APS also distinguish it from commercially available systems. Special character, table content searching, sorts by USPC hierarchy are not present in other systems, or are present on a limited basis. For example, other databases allow numerical range searching, but will not allow for subclasses which may fit in the subject hierarchy but are out of numerical order. Special character searching is also limited. If a Greek letter is printed as a symbol and not spelled out, a document using that symbol is not retrieved in most commercial systems.

Image capability is another obvious advantage over other databases and information sources. The entire document is searchable and the output is displayable as an image. The nearest available commercial source of computerised, image-based patent information is the CD-ROM products which are searchable by bibliographic parameters. However, bibliographic searching cannot be compared to searching the entire text of a document. In addition, while manual searching is necessary, it is tedious and hampered by the reality of missing and worn documents. Conducting a manual search using a scanned image assures one of doing a complete search of all the documents that should be in a subclass.

Practical disadvantages

APS is only available in the PSR at the USPTO or in selected depository libraries throughout the United States. The authors know of no plans to make the systems more widely accessible. Professional searchers must travel to Washington DC or local PTDLs to use the system. Arbitrary time limits have been assigned to the APS terminals ranging from 15 min in the PSR to unlimited time in the PTDL if no one else wants to use a terminal. 15 minute searches are probably not worth the time and effort. Unlimited time, of course, may be useful if there is not a patron waiting list. In addition, assessing and maintaining a connection with the system,though a PTDL is difficult between 11:00 AM and 4:00 PM, apparently a peak user period.

The scope of coverage does not compare favourably to some online vendors. Claims go back to 1950 for chemical patents and cover patent citation activity from 1790 to date. While APS offers full text capacity, certain types of searches, e.g., assignee searches, often need to cover greater than a 20 year period.

Current limitations of the image capability are also a disadvantage. The system provides good documentation listing which patent images are displayable, but the technology represented by these displays is so varied that it is difficult for searchers to be sure what has been viewed on a terminal and what must be viewed manually. Other problems include the inability to conveniently download retrieval from specific search statements. Downloading captures an entire session each time a download command is executed. When the output from a search statement is displayed and saved, when a subsequent statement is displayed and downloaded, the previous search statement displayed is downloaded again. 20 average sized full text documents will fill a 1.2 byte disk very easily. The citations from two full text 20-hit search statement would not fit on one disk. In order for the system to reach its significant potential for industry, these concerns must be addressed.

Fulltext searching vs abstracted indexed databases

The majority of industrial patent searchers use abstracted, indexed systems to search the patent literature. Manual searching is a clear advantage when the subject being searched is based on diagrams or engineering drawings. Searching an actual document is also desirable for other types of topics where the technology is described in words and not diagrams. However, workload and the amount of information that must be examined to complete a search make it prohibitive for most searchers to routinely engage in manual searching. Computer access to the full text of patent documents, on the other hand, makes it a more reasonable task and allows searchers to be less dependent on an abstract or an indexer to retrieve relevant citations.

Abstracted, indexed systems such as CLAIMS, Derwent and Chemical Abstracts, of course, are useful in the interpretation and retrieval of the art presented in patents. Unfortunately, many details of an invention must be left out and in certain instances, these details may be required to fulfil a search request. Several examples of search questions where abstracted systems cannot provide the amount of detail required are chemical processes, medical tests on specific biological samples and particular reagent value ranges. A chemical process invention disclosure may claim the use of a special solvent at one step in the reaction. An abstracted system may or may not list solvents if the solvent is cited in a subsidiary claim. In the case of biological samples, some systems will list a group of sample types, e.g., sera or sputum, as 'biological samples', if the sample specimen is cited at all. The searcher must view the actual documents to determine the sample type when sample type is germane to the search. The use of special reagent value ranges, i.e., the amount of the reagent used, will almost never be discussed in an abstracted database. Searching full text systems can retrieve documents which cite special aspects often omitted from abstracts systems or systems which selectively list portions of a document. Full text image systems can also replace traditional manual searching. The problem of lost and damaged documents could be eliminated and also augmented by text search capabilities.

Future of APS

The USPTO automation project continues to evolve. The ultimate search objective is to have the full text of all US and selected foreign patent documents fully searchable and displayable at no cost to the searcher. In addition, the entire USPTO will run a completely computerised operation — electronic processing of patent applications from filing to examination to issuance — to electronic mail and all activities in between.

In order to accomplish such an ambitious goal, The USPTO must address the problems previously discussed as well as consider staffing and training issues incumbent with replacing manual resources with computerised ones. The search system is good, and with the appropriate study and input from information professionals, it can become superlative.

Two into one must go — development of an interface for searching fragmentation code and Markush structure files

Gez Cross and Jenni Mountford

Technical Product Development, Derwent Publications Ltd., Rochdale House, 128 Theobalds Road, London, England, WC1X 8RP.

In this paper, we are going to discuss the challenges encountered in developing a single interface to build queries for two totally different types of chemical indexing system. Each of these systems, for indexing both specific and Markush structures, has a number of structure drawing conventions to minimise the number of structural permutations a user needs to search. Due, at least in part, to the different nature of the two systems, the conventions in each are not identical, although there is a good deal of overlap. Since there was a need to be able to generate strategies for both systems with a single structure drawing, this raised a number of design problems, as did the separate need to allow access to the features unique to each system. These problems will be examined, together with the reasons why it was necessary to undertake the development of the interface. The paper will conclude with a look at the solutions devised to overcome these difficulties and incorporate all the necessary features into a single interface, and an update on the current status. This will include a review of initial testing by users and further developments resulting from these tests.

Background

Since 1963, chemical structures found in patent documents have been indexed in Derwent's 'World Patents Index' database (WPI/L) using a fragmentation code based system. Originally based on 80 column punch cards, this code works by splitting chemical structures into discrete structural fragments, each of which is represented by an alphanumeric 'code'. This code has been through several changes since its introduction, with new codes being added in 1970 and 1972 to improve the relevance of the search results. In 1981, it underwent a complete revision which expanded the number of structural based codes from around 650 to over 1100, and also enabled the non-structural activity and process codes to be expanded greatly. In order to avoid having two separate coding systems, the 'old' codes from records indexed prior to the introduction of the new code were converted into the new format automatically. Whilst this enabled users to search the entire span of data using the same system, it did not remove all the problems as it introduced another time slice into the file, making it necessary to accommodate codes introduced on four separate occasions. As the use of a code valid from, say, 1981, linked with a code valid from 1972 onwards, would only retrieve records from the later period, users had to be very careful to avoid losing valid references from earlier time periods. Various printed user aids were produced to help users produce valid strategies for searching the data online, but with the growing popularity of the PC, there was a call for more computer based help.

Consequently, a PC based offline interface, TOPFRAG1, was created and released in 1987 in order to facilitate online searching and help users to cope with the complex time ranging necessary as a result of the changes described above. Initially, this program could only create strategies based on single compounds but was very popular with users. Despite, or maybe because of, its popularity, a number of improvements were suggested by users to increase the capabilities and user-friendliness of the interface.

In 1989, after gauging user needs via a questionnaire which had been sent out to all users of the original program, a new version (Generic TOPFRAG) was produced. The main features were the ability to use generic terms and variable groups in the query structure as well as increased hardware, especially printer, support. In addition, several other new features were introduced to make it easier for users to edit the generated strategies manually. However, there was still not the ability to perform sub-structure searches other than by modifying a generated strategy manually. Also, due to the ability to input generic structure queries, some of the strategies produced were very complex and exceeded the capacities of the online hosts that mount the WPI/L files. A further improvement was therefore seen to be necessary.

Whilst these software developments were proceeding, Derwent started production of a chemical structure file, World Patents Index — Markush (WPIM), containing both specific and Markush structures, with indexing starting from 1987. Whilst this gives much more precise retrieval than the fragmentation code, it has resulted in the problem of having two different indexing systems which need to be searched by users. This can not be rectified by converting the backfile into the new format since there is insufficient information in the fragmentation code records to create accurate Connection Tables suitable for searching with the new system.

It therefore became apparent that an offline query building interface was needed that would produce strategies for searching both the backlog of fragmentation coding and the Markush structure file. This would also provide the opportunity to improve the generic version of the software to allow substructure searching and include other improvements requested by users. On superficial consideration, this does not sound too problematical, given that a good part of the work had been done already for the fragmentation code generation. Additionally, taken in isolation, generating a strategy for searching the Markush structure file would not cause many problems. The problems start to become apparent when the different nature of the two files is considered, together with the different structure drawing conventions and the features which exist in one file but have no meaning or relevance in the other. Let us look now in more detail at some of the challenges that needed to be faced, to achieve the aim of the project to obtain a single interface for the two systems.

The challenge

1. Free sites

One of the biggest differences between the two files and the systems used to search them concerns the ability to perform sub-structure searches. These are searches in which a core part of the structure is input and labels or 'free-sites' are placed on parts of the structure to indicate that further substitution is possible at these positions. This enables the retrieval of all references which include the input 'core' or sub-structure somewhere within the full structure, as shown in Figure 1.

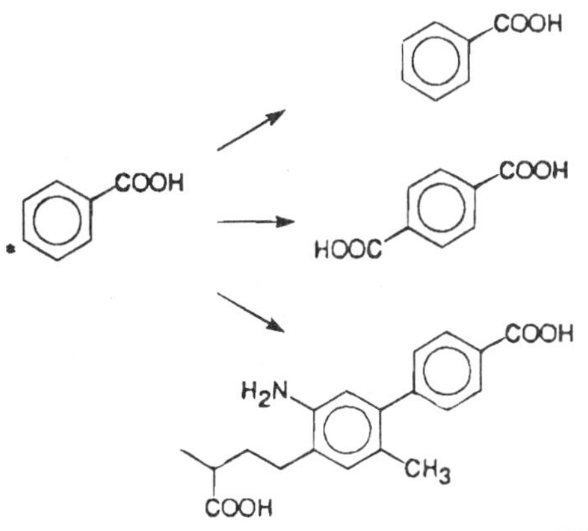

COOH
COOH
COOH
HOOC
COOH
H_2N
CH_3
COOH

Figure 1

For the newer Markush system, this is not a problem. The input structure is just labelled with 'n*' (where n = the maximum number of substitutions permitted at that point) at the position(s) on the structure at which further substitution is permitted. The searchable Connection Table produced includes the number of free-sites, if any, on relevant atoms, and the search system does the rest. This is not difficult for a structural database, even one containing Markush structures.

For structures indexed by fragmentation code, it is not so simple. It is very easy to provide a means within the interface for labelling the positions at which substitution is permitted. However, the nature of the fragmentation code system makes it very difficult to work out the combinations of codes necessary to perform complete sub-structure searches. Let us consider a simple example of this:

To search for benzoic acid would involve using the code for one aromatic acid (J011, J131), the code for benzene (G100), one aromatic ring present (M531) and the code for mono-substitution on a benzene ring (G010) amongst others. For a benzoic acid with one free-site on the ring para to the carboxy group, the codes involved need modifying to incorporate the possibility that the free-site could represent further aromatic acids, further aromatic rings or just H. This is illustrated in Figure 2.

COOH	COOH
G100 *benzene*	(G100 or G111 or G112 or G113) *1 benzene* / *1 benzene (+ other)* / *2 benzenes (+ other)* / *≥ 3 benzenes (+ other)*
J011 *1 COOH (deriv)*	(J011 or J012 or J013 or J014) *1 COOH* / *2 COOH* / *3 COOH* / *≥4 COOH*
J131 *1 aromatic COOH*	(J131 or J132 or J133) *1 arom. COOH* / *2 arom. COOH* / *≥ 3 arom. COOH*
G010 *monosubstd benzene*	(G010 or G013) *monosub.* / *1,4-disub. benzene*
M531 *1 aromatic ring*	(M531 or M532 or M533) *1 arom. ring* / *2 arom. rings* / *≥ 3 arom. rings*

Figure 2

Already the codes involved are becoming more complex. In addition, the program has to take into account the fact that some of the codes involved were introduced at different times. This means that the first code (G100) examined above is used in isolation to search records indexed prior to 1981, as the other codes were not introduced until that date, whereas post-1981 they must be ORed together. Consequently, the strategy which must be produced to retrieve answers from the complete database is further complicated.

If we now go one step further and introduce another free site on the ring, e.g. in the 3 position relative to the carboxy group, we introduce another level of complexity. All of the above changes are relevant, but in addition, there is the possibility of interaction between the two permitted points of substitution. Thus, further rings could be fused onto the benzene ring, making it a fused aromatic or heterocyclic system, as shown in Figure 3. This means we can no longer search using the benzene codes, as there are a large number of alternative ring systems possible and it would be impossible to OR them all together. It is also more difficult to use the ring substitution codes as we must include the possibilities for mono-, meta-, para- and 1,2,4-substituted benzene, beta-substitution on a fused carbocyclic aromatic and beta-substitution on the carbocyclic ring of a fused heterocycle.

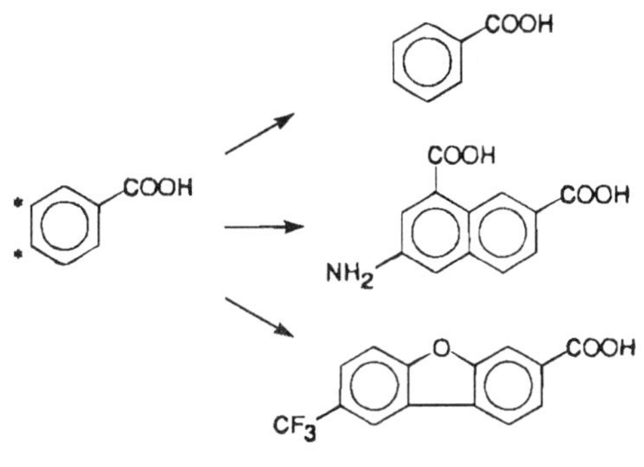

Figure 3

In these cases, we have chosen a simple ring with only one explicit substituent. For other cases we might also have to consider the effect of free-sites on a combination of groups. This is particularly true in cases that have to be considered for potential tautomerism, such as when a free-site is placed on a not necessarily aromatic carbon-carbon double bond. An example of this is a 2-aminocyclohex-4-en-1-one optionally substituted on the 3-position (i.e. represented by a free-site on position 3).

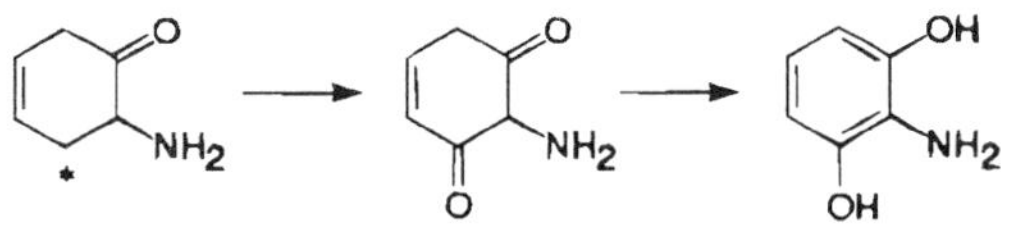

Figure 4

The free-site includes the possibility of a doubly bonded atom e.g. another keto group, at position 3, in which case the structure would be tautomerised according to the database tautomerism rules to a hydroxyaniline derivative, as shown in Figure 4. Therefore, for some possibilities represented by the free-site, the program has to include the codes for 1 or more alicyclic keto groups + 1 or more alicyclic amines + cyclohexene. These have then to be ORed with the relevant codes for 1 or more aromatic hydroxys + 1 or more aromatic amines + benzene, taking into account changes over time. This particular problem also applies when drawing free sites on structures in the Markush file. It is necessary to consider the effect of potential values for the free-site on the drawn structure and allocate bond values accordingly. It is, of course, made worse by the fact that the actual structure drawing conventions are not identical in the two files, and the ideal is to have a single drawing producing two valid strategies, as we shall see in the next section.

Another class of problems to be solved included decisions as to where to draw the line in selecting fragments and their codes that could be included in a particular sub-structure. Should Hal-C2-O* include codes for -C-O-O, COOC(O), C-O-C(O)-N,.. as well as the more straightforward alcohol and ether codes, for example, or should the -O* group only be included as "chain attached to O + Halogen"?

Free-sites are very useful for obtaining a high level of retrieval, but this must be attained without the relevance of the retrieved answers being adversely affected to too great a degree. Ideally, the user should be able to balance retrieval and relevance by appropriate selections. Providing this capability without the user-interface becoming too cumbersome was a not inconsiderable challenge. Let us move on then to look at other aspects involved in constructing an interface for the user which would give the necessary capabilities without being unduly tedious to use.

2. Common drawing for the two systems

As indicated in the discussion on free-sites, one of the biggest problems in the building of a common interface for the two systems is related to the different structural conventions in the respective files. For the most part, similar conventions regarding tautomerism were invoked when the new Markush system was started to those already in place for the fragmentation code. However, the Markush system contains a bond type that does not occur in the fragmentation code system, Normalised bonds. These are used to represent aromatic and/or tautomeric bonds in the Markush structures. Incorporation of these necessitated some differences in the conventions. The primary rule in the Markush file is to normalise a cyclic path to produce a fully normalised (aromatic) ring, if at all possible, for rings containing n atoms and 2n alternating single/double bonds. A second rule relates to keto-enol tautomerism, indicating that keto-enols should be coded in the keto form. Where the two rules are in conflict, as in the case of phenol or the hydroxypyridines, the first rule takes precedence. For the fragmentation code, the keto-enol rule is the same, except that for phenols an exception is made, so that for hydroxy groups directly attached to a carbocyclic aromatic ring the compound is coded as the enol. There is no such exception for heterocyclics however; hence the representations for 3-hydroxy pyridine used for the two files are as in Figure 5.

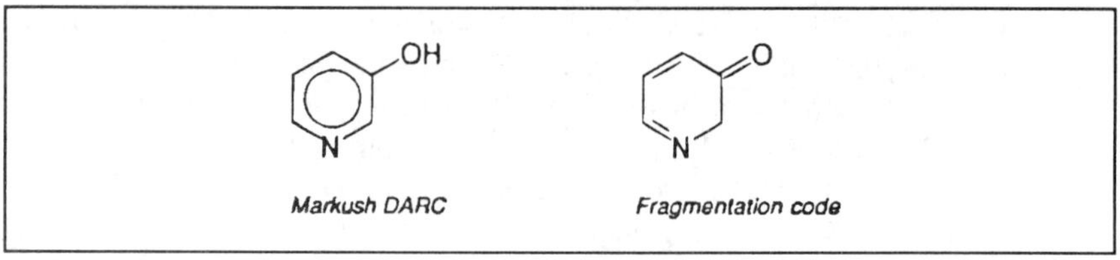

OH N *Markush DARC*

O N *Fragmentation code*

Figure 5

There are a number of other rules used in the fragmentation code to make structures fit comfortably within the codes available and obtain good results on searching. Many of these are irrelevant when searching in a graphical structure file, and as they did not always make chemical sense (as opposed to searching sense) were not used in the Markush file. Whilst this is a reasonable strategy when the two files are being used in isolation, it makes searching the two together, using a single structure drawing, more difficult.

As well as differences in the bond types used, there are also differences in the terms used to represent groups of atoms in the Markush system and in the existing Generic TOPFRAG interface to the fragmentation code. These include the chemical 'shortcuts', where simple alphanumeric text symbols are used to represent clusters of atoms, e.g. Ph for a monovalent phenyl group, tBu for a tertiary butyl group. The shortcuts used in WPIM (and the DARC system overall) are wholly included in the set used in the Generic TOPFRAG interface for searching the fragmentation code. However, the symbols used to represent the same concept were not exactly the same in the two cases. In most cases, the differences are not large, (e.g. TBU vs t-Bu for tertiary butyl, OBE vs o-C6H4 for an ortho-substituted benzene) but one of the aims was to allow users of the previous program to use saved structures in the new program, which prevented use of only the WPIM format. In addition, users would have grown used to a larger set of shortcuts, and would generally be averse to giving up many of them, as it could involve extra effort drawing structures.

A similar situation existed for Generic terms ('Superatoms'). The WPIM database contains a number of Superatoms which are used to represent generic terms found in patents, such as 'alkyl', 'aryl' or 'halogen'. In the fragmentation code interface, it seemed useful to use these symbols in a similar way. Thus HAL was used for halogen and generated the codes for F, Cl, Br, I, which saved inputting the four alternatives as values of a variable group. Because even then we hoped to produce a combined interface, the same symbols were used to represent the same concepts as existed in WPIM. However, the structure of the fragmentation code lent itself to the creation of new Generic terms, so whilst in WPIM there is only one term for transition metal (TRM), in the code interface there are three: TRM1, TRM2, TRM3 to represent 1st, 2nd and 3rd row transition metals. In programming terms, this is a minor problem (not even a real challenge), as it is relatively straightforward to apply TRM in the Markush strategy whenever any of the three coding variations are used. The same applies to the other extra superatoms. The major difficulty lies in the way the superatoms are handled.

In the Generic TOPFRAG interface, and hence the code part of the new program, the superatoms drawn by the user produce either a series of codes which are ORed together, or a truncated code to retrieve all the relevant members of the set. They are thus treated as partially customisable shortcuts, as many of them can be restricted by attributes set by the user. In the Markush strategy, the superatoms are included directly in the strategy and when searched match directly with the corresponding superatoms indexed in the database structures. At present, though, the Markush DARC system which is used to search WPIM, cannot 'translate' between generic terms and specific instances of those terms. So, the term for alkyl (CHK) will only match against a corresponding CHK group and not against a methyl group, for example. In order to retrieve both generic and specific references, it is therefore necessary to construct queries to include both the generic term and any specifics of interest. This is a major difference between the two strategies. Table 1 shows the query input necessary to retrieve all 1-3C alkyl halides.

Fragment code interface	Markush Interface	
CHK1-3c-HAL	G1-G2;	G1=CHK,C,ET NPR, IPR
(produces all relevant codes)		G2=HAL,F,Cl,Br,I

Table 1

This was a significant challenge, even for the superatoms that represent fixed sets of elements. For these, it would be possible to translate the superatom into every member of the set, but this is probably not what every user would want every time. If a user was searching for a specific value, for instance, iron or methyl, they would need to use the generic terms as well in order to retrieve all possible references. In these cases, it would be necessary to input Fe- or TRM-, C- or CHK- respectively, and users would not be happy if the program automatically converted the TRM to every possible transition metal. Similarly, if the program automatically included the generic every time a specific was used, this would solve the above case, but would be inconvenient if ONLY the specific was wanted.

Of course, several solutions are still possible, but it is planned to introduce this translation capability into the Markush DARC system in the near future, which would render some solutions unnecessary. This also raised the question of cost in that it was necessary to consider how much it was worth spending on a capability which could be redundant only a few months after the release of the program.

Finally in this section, it was obviously desirable to have similar limits on the number of variable (Gi) groups, the number of values per group, and the level of nesting permitted. As the Markush DARC system had inbuilt limits, these were the obvious choices. This was not a big problem for the actual drawing interface, but generating fragmentation codes for the possible permutations involved was not simple. Given that the limits are a maximum of 20 Gi groups, with a maximum 20 values per group, the number of permutations that could potentially be involved is huge. Given sufficient memory, a fast processor and time, there is no problem in computer terms in generating codes for any number of permutations, but in a commercial environment, it may not be realistic to expect a user to wait hours for a strategy to be generated. Artificial limits could be imposed, but these would adversely affect users who needed a large number of groups, but with few values in each, or a small number of groups with a large number of values. Questions of this type involving not just technical considerations are difficult to resolve and usually need some input by potential users on the proposed solutions.

3. Drawing for one system

The previous section dealt with the problems associated with generating two strategies from a single structure drawing, which it is believed will be the most common requirement. However, it is likely that, from time to time, users will wish to construct searches just for the new Markush file when performing update searches or just for the backlog information indexed with the fragmentation code when looking for historical information.

There are also other occasions where it might be necessary or advisable to construct the strategies independently. The two search systems are fundamentally different and, for the most part, the new Markush structure file will give good retrieval with much higher precision than is possible by using the fragmentation code. However, each system has certain strengths, and by drawing queries which make use of these strengths, it is possible to obtain better results. Given that to perform sub-structure searches under the fragmentation code it is often necessary to make assumptions about the type of answers required, it may be necessary to use significantly different queries in order to obtain maximum retrieval with high precision in both systems.

To illustrate this, we will consider a simple benzazepine optionally substituted on the N-containing ring. For the Markush system there is no problem: free-sites are placed in the relevant positions on the ring, and all corresponding answers are retrieved, whether they contain simple substituents, further rings fused to the benzazepine, or just H at the relevant positions. However, before a fragmentation code search can be constructed, a decision has to be taken on whether answers containing further fused rings are desired. If the answer is positive, further decisions have then to be taken about the number and nature of the possible heteroatoms and about the number and size of the rings fused onto the benzazepine. All of these will affect the number and actual identity of the codes which need to be ORed together in the final strategy, to maximise precision. For example, in this case, a searcher could decide to look only for ring systems containing a total of two or three fused rings, and containing no more than three heteroatoms including the nitrogen and optionally one sulphur and/or one oxygen atom, as shown in Figure 6.

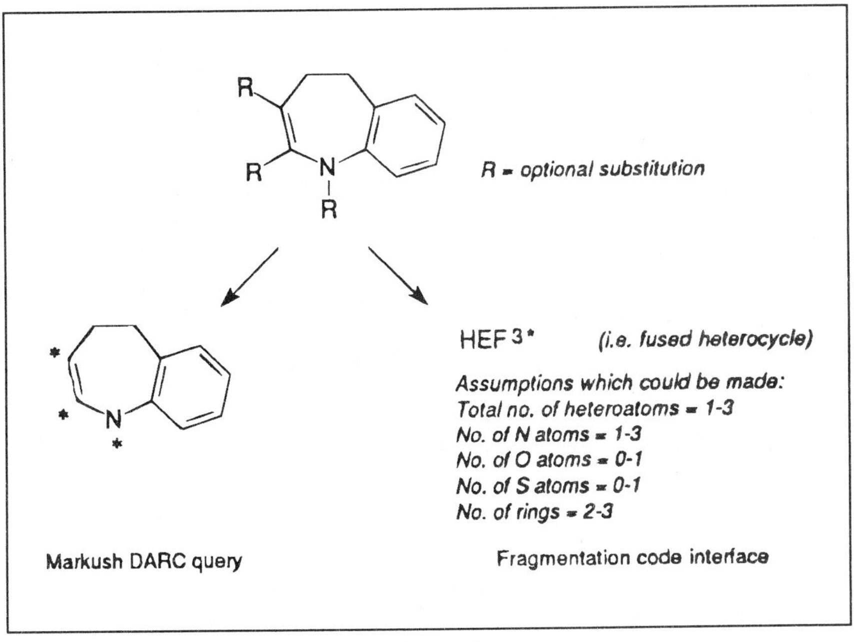

R
R N
R

R = optional substitution

* * N *

HEF 3* *(i.e. fused heterocycle)*

Assumptions which could be made:
Total no. of heteroatoms = 1-3
No. of N atoms = 1-3
No. of O atoms = 0-1
No. of S atoms = 0-1
No. of rings = 2-3

Markush DARC query

Fragmentation code interface

Figure 6

It takes a good deal of intellectual effort whilst preparing the search to think of every possible type of answer that may be required (in terms of ring systems). It is necessary though, since if all possibilities are not included, retrieval will suffer. The situation is simpler if only the code for the base ring is included, but there is the assumption that no further fusion is possible (and indeed the Markush query could be modified to allow for this). The other, simpler, alternative is to leave out the specific ring codes altogether, and use the more general fused heterocyclic ring codes to restrict to relevant compounds. This presupposes that there are other groups present which will serve to define a searchable entity. Although this is simpler, it may well give too many answers of low relevance unless the attached groups are very unusual and limit the search adequately themselves.

To attempt such searches with a common query may still give high retrieval but it may well be at the expense of precision in one or both of the systems. Therefore it should also be possible to construct queries for each system separately, which can take advantage of the inherent capabilities of each, as well as for both systems together.

4. Further considerations

As indicated previously, other improvements needed to be made to the existing capability for generating strategies to search the fragmentation code. Users were able to input very generic queries resulting in some very complex strategies, which did not become apparent until the product was used in the commercial environment. The different online vendors each have various limits on the amount of workspace allocated to users, the number of operations possible in a single search set, the number of answers allowed in each and/or all the search sets and so on. In addition, there appear to be several internal, limits in some of the systems, which are not generally advertised. Due to the problems associated with handling the time-ranging of the various permutations, strategies tended to be complex. The final 'pulling together' of the permutations for each time period was carried out towards the end of the strategy. At this point, many nested Boolean and proximity operations tended to be performed, resulting in various system limits being breached on some occasions. Messages such as "Argument Table full:......", "Too many postings in all search statements" or "Strategy too complex, please simplify" were displayed. The actual reasons for these overflows tended to be rather complex and not always easy to fix. For example, in at least one of the above cases, the problem appeared to be a combination of the complexity of the particular line (number of nested operations) and the number of postings associated with the codes being treated. Other problems only seemed to occur at certain times of day when usage tended to be higher on the host system, indicating that the workspace allocated was less, which is not something that can be programmed easily into an offline interface. However, something clearly had to be done to improve the situation.

There were also problems generating strategies due to insufficient memory available on the PC, as the existing Generic TOPFRAG interface was DOS based and could only use the normal 640K RAM. If this was a problem for the existing interface, there would be increasing problems as more capabilities and greater complexity were added.

Many of the user queries received were related to lack of memory causing one problem or another, and to hardware compatibility. It was also desirable for the program to run on, or with, as great a range of hardware as possible.

Finally, it was necessary to bear in mind that the Markush DARC system for searching the WPIM Markush structure file was still being enhanced, with a number of developments due to be implemented over the next year. Consequently, the interface had to be constructed in such a way that it would be straightforward to upgrade it, when necessary.

Given the problems we've just examined, it might seem more appropriate if the paper was entitled "Two into one WON'T go" — a more common English expression — but as the title of the paper indicates, users had stressed that "Two into one MUST go" so we will now look at how this is being achieved.

The solution

The growing popularity of Microsoft Windows 3.0, coupled with the existing popularity of the Apple Macintosh, provided the solution to several of the problems. These Graphical User Interfaces (GUIs) provided easy access to all the memory that the computer system possessed. In addition, Apple provided support for printers through the system files provided with the computer, or else the manufacturers of the printer supplied the software. A huge variety of hardware was supported by the Windows files, thus resolving one of the most common sources of customer query. Both of these GUIs would facilitate a modular design for the software, which in turn would enable memory to be used to best advantage and make the dual nature of the program more easy to accommodate. Future updating would simply be a matter of updating the relevant module and not necessarily the whole program, and totally new modules could be bolted on if desired as a result of customer feedback.

Therefore, it was decided to produce the new interface ('Markush TOPFRAG') on these platforms. In order to minimise the amount of work, it made sense to create the interface on one GUI first, thus working out most of the problems, and avoiding unnecessary duplication of effort. The Microsoft Windows platform was chosen as the first to be implemented, given its worldwide popularity, with an Apple Macintosh version to follow on as soon as possible.

1. Common Drawing

The ability of the GUIs to call up different modules and pass information between them greatly facilitated the solution to the 'common drawing' problem. It was already necessary to include checks within the program to ensure that users had input structures according to the conventions of the relevant files, and, where necessary, correct them. This was already built into the old TOPFRAG programs, and could easily be converted to the new platform. By having separate modules for the drawing interface, the Markush strategy generation and the fragmentation code generation, the biggest problems can be solved relatively easily.

In principle, the user would draw in one format and the drawing interface would construct the relevant connection table. This information would be passed to a fragmentation code module, which would perform the relevant checks on the structure, converting it where necessary to the correct form, and generating the strategy. The Markush module could then be loaded, the connection table information passed to it and the relevant checks and conversions made for that file, before the strategy was generated.

The format chosen for the common drawing interface was the simpler fragmentation code form without normalised bonds. This was necessary, since whilst it is possible to convert single/double bonds to normalised without problems, it is not always possible to do the conversion unambiguously in the other direction, especially for tautomers. It was also decided that when drawing for both systems, it would only be possible to select those atoms, superatoms and shortcuts and bonds common to both systems. This would avoid the need for an excessive number of conversions at the strategy generation stage, which would otherwise slow down that step and increase the memory requirements. In fact, it would have been difficult to allow the use of shortcuts that were not valid in one system, as the only way to deal with them would have been to expand them out during the strategy generation procedure. As this would have involved altering the connection table substantially while the structure drawing module was not active, this would not have been trivial, and the generated strategy would have been difficult to relate back to the original query for checking.

On starting a structure query, the user is presented with the option to draw for one or both of the indexing systems. If the choice is made for a particular system rather than 'Both', all the atoms, shortcuts, superatoms and bonds relevant for that system will be allowed. Those not permitted in the chosen system will not be selectable from the menus and will be precluded from manual input. Frequently, the reason for selecting a single drawing would be to take advantage of all the strengths of that system, and then to do the same for the other system. In these cases, it would be desirable to have some way of basing a structure drawn for the Markush file on a structure previously input (and saved) for the fragmentation code. We will now look at how this was achieved.

2. Co-ordinating single system drawing

There are three main differences that need to be taken into account when converting a structure drawn for one to the other. These are bond types (e.g. normalised or cyclic bonds), shortcuts and superatoms.

When converting a fragmentation code structure to a Markush structure, the bonds are automatically converted as appropriate on generation of the strategy, so no further routines were necessary here. When converting in the opposite direction, however, the program can not unambiguously convert normalised to single and double bonds. Similarly, if a Markush structure has been drawn with cyclic bonds, the likelihood is that it will be necessary to alter the drawing significantly for the fragmentation code. Consequently, it was felt that it would be best not to attempt an automatic conversion of such bonds, but to leave it to the user. Obviously, checks had to be built in to trap any of these bonds that the user had not modified manually and warn the user about them.

The second conversion needed was for shortcuts, which was a two part problem. The first part was a simple conversion of shortcuts having the same meaning but different symbols to the appropriate shortcut for the other system. The second involved expanding out shortcuts that did not exist in the other system into the corresponding multi-atom fragment. This was achieved without too many problems, although it was less straightforward when the offending shortcuts were included as values of variable (Gi) groups.

The conversion of TOPFRAG type superatoms into the correct Markush superatoms, with appropriate specific alternatives, was more of a challenge. However, a relatively simple and elegant way was devised, which also gave the user a good deal of control over the conversion. The answer was a simple conversion table, which the program would read and use to convert the superatoms into Gi groups consisting of the corresponding superatom and specific alternatives. Defaults could be incorporated, which could be amended by users, as the table would be a simple ASCII text file. Hence, HAL could translate to G1, where G1 = HAL, F, Cl, Br, I. This also gave the opportunity for users to have their favourite metals as alternatives to the corresponding superatoms, e.g. TRM1 converting to TRM, Fe, Co, Ni. Where no sensible conversion could be included by the program developers, a warning was incorporated, to alert users that they needed to consider adding specific alternatives to the generic term, due to the lack of such translation in the Markush system at present.

A sample of how the table could be used is given below:

Code	**Markush**	
CHK=	CHK+C+Et+n-Pr+i-Pr	
ARY=	ARY+o-C6H4\+m-C6H4\+p-C6H4\	{"\" *indicates mono/fused options*}
HEF=	HEF+WARNING	
TRM=	TRM+Fe+Pd+Pt	
LAN=	LAN+La+Ce+Pr+Nd	
HAL=	HAL+F+Cl+Br+I	
TM1=	TRM+Fe+Co+Ni	

In the initial version, the conversion was limited to terms which could be entered in the table in a text form, e.g. atoms, superatoms, shortcuts. It is hoped to be able to incorporate multiatom fragments as well, possibly by drawing out the fragments, saving them as a type of template and incorporating the saved name(s) in this table.

Figure 7 illustrates the latter two types of conversions.

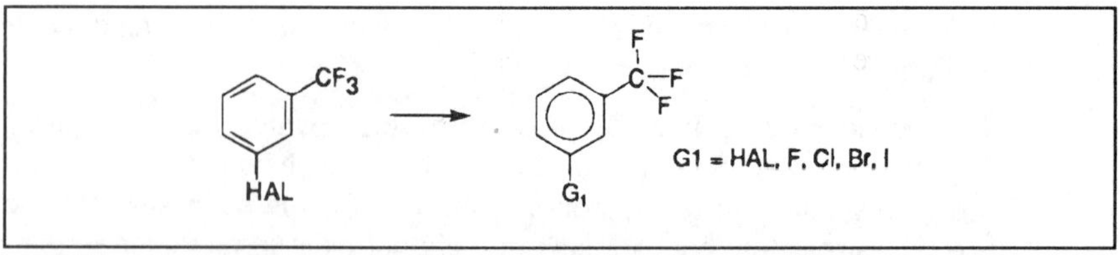

Figure 7

3. Optimised strategies

One of the biggest problems in developing the new interface was to 'improve' the actual strategies produced for searching the fragmentation code from generic structures. By 'improve', it was generally meant that the strategies should run online without exceeding any of the limits set by the various hosts. Ideally, they should be formatted in such a way as to facilitate editing of the strategies by knowledgeable users.

One of the biggest sources of problem was the limit on the overall number of (intermediate) answers stored in each search set or statement. In order to overcome this problem, a reformatting of the strategy was necessary. The original method had involved generating the lines for the essential part of the structure, then the lines for the permutations resulting from the various alternatives, generic terms etc., and then pulling them altogether at the end, at which point the time ranging took place. The new method was first to construct a strategy for the essential part of the structure and then to time range the strategy so far. This gave a number of references, all of which contained the basic, unchanging part of the structure. The variable parts of the structure could then be included, but limited to those references retrieved for the essential part.

On one host this consisted of using an existing LIMIT command to limit all subsequent lines to only those documents retrieved by the essential part. For the other two hosts, it involved linking subsequent lines with the restricting line, which was a little more clumsy but adequate. This would serve, it was believed, to reduce the number of references stored overall as well as potentially improving the overall performance of the search.

Other changes were made, especially in the way codes with a large number of references were handled, further improving the performance.

Also, some effort was made to avoid using redundant codes — codes that whilst accurate, were irrelevant for restricting the search, due to the presence of other codes. So, if an alkyl chain could be present 0, 1, 2, or 3 times, there was no point including any of the codes for these concepts, since all possibilities from zero to maximum were covered. As all these codes have a very large number of references (or 'postings'), it could make a significant difference to the search performance if they were left out, without any effect at all on the retrieval or relevance of the final answers. Finally, in order to overcome some of the complexity problems occurring on some of the hosts, an artificial limit was imposed on the level of nesting of proximity and Boolean operations. This matched a real limit on the other host which had not had the same problem. The result of this was that some lines of the strategy which were very long and complex were split up onto two lines of lesser complexity. The hope is that this will ameliorate the problem, if not solve it altogether.

4. Free-Sites

The biggest problem by far, and the greatest challenge to the ingenuity of both designers and programmers, was the incorporation of a free-site capability for the fragmentation code module. Users had previously indicated in the TOPFRAG questionnaire that they would be happy if the program popped up context-sensitive questions which would enable the free-site capability to generate appropriate codes. Therefore, this approach was taken, programming in situations which required user input and leaving it to the user to define their requirements.

To achieve this, upon generation of the strategy, the program would highlight each atom containing a free-site in turn, and ask the user to make appropriate selections for what was allowed at each position.

The selections available were principally H, AA (any atom), OH, SH, N, attached via single, double or triple bonds. The exact selections available would depend on the context, so that if there was no possibility of tautomerism rules being called into operation, the choices for OH, SH, N would not be presented. If the free-site was applied to a carbon-carbon double bond, however, it would be necessary to find out whether the user envisaged the possibility of a hydroxy group being there, which would bring in the keto-enol rules. Having determined what was allowed at each position, the atom is converted to a variable group having the relevant groups attached, as shown in Figure 8.

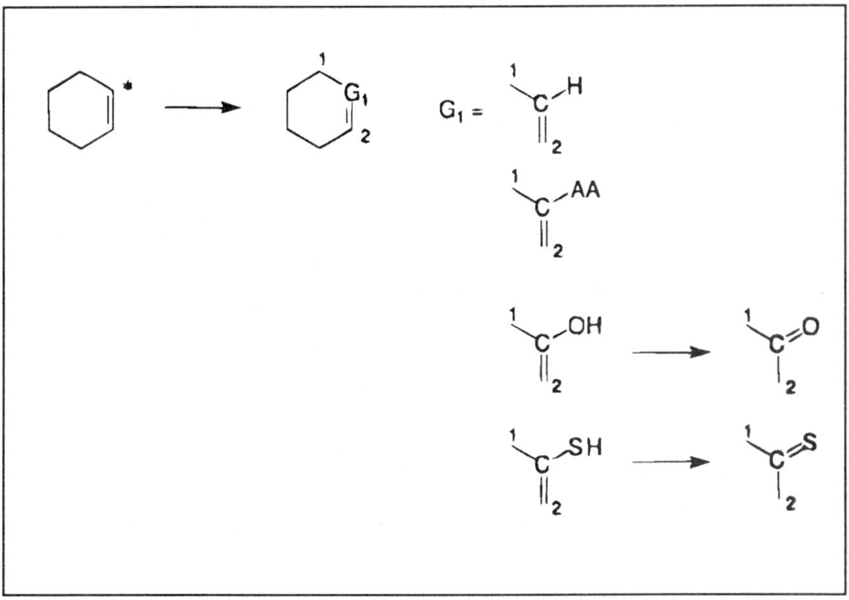

Figure 8

This allows all the relevant permutations to be taken into account for groups which could be attached to the position represented by the free-site.

If there are 2 or more free-sites attached to a ring, it is important to find out whether the user wishes to include the possibility that the ring might be fused to additional rings. Therefore, in this situation, an appropriate question, with a Y/N answer is asked. If the answer is 'Yes', the whole ring is replaced by a ring superatom (or several superatoms), chosen by the user, to indicate the type(s) of rings permitted. These superatoms can be qualified according to size, number of rings, heteroatoms etc. to achieve the desired results.

After this, the user would be allowed to make certain specifications about the overall structure, before the strategy was generated. In order to set (or ignore) various other codes, it was then felt necessary to allow the user to specify whether further ring systems could be present in potential answers, other than those drawn out in the base structure.

Finally, the Fragmentation Code includes a number of codes which can be used to prohibit the presence of certain groups. These are very useful for eliminating 'noise' from answers sets, but cannot be set automatically if free-sites are used. Free-sites, by definition, imply ANY substituent at the marked position, which means that these codes cannot be used automatically. However, users need to be able to set these manually if required, so the most common ones, which would make the biggest difference are presented for selection.

All that has been described so far, is a means of specifying what the individual user means by the free-site(s) placed on the structure. The program then has to provide a sensible strategy from this information.

Free-sites have certain implications for Fragmentation Code strategies. If any of the codes which would have been set for the specifically drawn out structure relate to the number of times a substituent can be present, these have all to be ORed with their incremented values, as illustrated in Figure 2. If a hierarchically-based code is included in the strategy, the codes from higher in the hierarchy should also be applied as alternatives, unless prohibited by the choices specified above. Apart from this, a decision was made to treat the variable groups artificially created by the program just as other groups would be and work out the various permutations involved. It was felt that this was an area where users involved in testing the program would have a large number of comments, both on the application and restriction of the free-sites and the resulting strategies. This feeling was accurate, as we shall see in the next part of this paper.

Success?

It is general policy at Derwent to allow testing to take place by knowledgeable users after a new product has been tested internally. Given the complexity of the development, it was especially necessary in this case. The testing took place mainly during April and May 1992, with testers in Europe, Japan and the USA.

The overall reaction was favourable, but there were a number of suggestions for improving the program as well as a number of bugs found. These included suggestions for making the drawing interface more friendly, features which needed improving and even some ideas not previously suggested. It was not possible to agree to all of the suggestions, but most of them have been incorporated. Some of them were very small improvements, but the small improvements combine to make a big difference to the overall feel of the interface.

As expected, there was a good deal of comment on the free-site capability and it became obvious that improvements needed to be made in the strategies produced when free sites were used. In general, the generated strategies were much more complex than similar manually created ones, and it was felt that the program was actually doing too much. One user commented that rather than taking a strategy and adding codes to cover all possibilities, we should be taking them away, to make the strategy more generic. Consequently, a way had to be found of using the information given by the user, to allow for potential tautomerism changing the codes used for the basic system, but not generating codes for the 'potential' group itself. The main problem with this procedure, was that at the point in the strategy generation where this was taken into account, there was no way of identifying whether the information was from a definite group or from a free-site specification. It has been suggested that a way round this could be to use a 'Pseudoatom' in the connection table which would be used for tautomerism purposes, for example, but would not generate the actual codes for the attached group that the real atom would have. This approach looks promising and is currently being followed up.

The other main comment concerned the inclusion of a terminal emulator for sending the text strategies and viewing the results on the structure file. In the past we have always taken the view that most of our users already had their own favourite communications and/or terminal emulation software packages and there was no point in trying to force another package onto them. However, it seems that a number of users have problems with existing emulator packages and the setting up of them, which admittedly can be complicated. We are currently investigating the possibility of including this as an optional extra, in response to these requests.

So, have we been successful in achieving our aim of creating a single interface for the two different systems? The initial response, from the testing would have to be 'partially'. However, as a result of the testing we believe that we can improve on the test version and achieve our aim.

Conclusion/ summary

We have looked at the reasons for creating a single interface for searching the Derwent Fragmentation Code on WPI/L and the Markush Structure file, WPIM. The main obstacles and challenges have been examined and it has been found that these could be overcome. Testing of the potential interface has taken place and the comments of the testers are being implemented. Whether we have achieved our aim to the satisfaction of the users who requested the interface at the beginning, only time will tell. Reactions so far seem promising, so we are confident that our aim will be achieved. Further updates to the program are already planned, as the capabilities of the systems for which it is designed to provide search strategies are improved. We look forward to many more such challenges in the future.

Acknowledgements

This software has been developed for Derwent by Hampden Data Services Ltd., and under the aegis of the CEC DG XIII B2 IMPACT programme, which has provided support and partial funding.

References

(1) Meyer, D.E. 'Special Application software for Chemical Structures'. In *Chemical Structure Software for Personal Computers*; Meyer, D.E., Warr, W.A., Love, R.A., Eds.; American Chemical Society, Washington D.C., 1988, pp 73,81

The use of computer-assisted molecular modelling in pharmaceutical research

K. Gubernator, H.J. Ammann, C. Broger, D. Bur, D.M. Doran, P.R. Gerber, K. Müller, T. Schaumann

Hoffmann - La Roche AG, Pharma Research, Basel, Switzerland

Studies of drug-receptor interactions at a molecular level have become possible in recent years because experimental structures of biopolymers are now available at atomic resolution. These structures include proteins, protein-inhibitor complexes, DNA fragments, DNA-drug complexes, proteins bound to DNA, carbohydrates alone and in complexes with proteins. Major targets of today's pharmaceutical research are inhibitors of enzymes which belong to classes for which experimental data are now available. Detailed investigations of enzyme inhibition using these data require powerful computer-methods due to the complexity of the systems considered: Databases of X-ray structures of proteins and small molecules together with efficient retrieval software, real-time three-dimensional display systems, efficient force-field methods and fast computers, tools for visual analysis of interactions, manipulation and modification tools for molecules of any kind [1,2].

Serine Hydrolases

Serine proteases and serine esterases are enzymes that play a key role in diverse physiological systems: Trypsin digests peptides and proteins, thrombin initiates blood coagulation, betalactamases mediate bacterial resistance to antibiotics, lipases digest nutrition fat and acetylcholinesterase degrades the synaptical neuro-transmitter acetylcholine [3-7]. They all use a serine sidechain hydroxy group as a nucleophile in their enzymatic reaction. These serine enzymes belong to three families that share no sequence homology or structural similarity. A detailed study of the catalytic mechanisms of these enzymes based on their X-ray structures and on the structures of several complexes with substrates, substrate analogues and inhibitors revealed a common stereo-electronic mode of action.

Thrombin

The natural substrate of thrombin is fibrinogen.This protein is cleaved C-terminal to a specific arginine residue to fibrin which then crosslinks the blood clot during coagulation. The mechanism of this cleavage reaction is understood in detail due to crystallographic investigations of thrombin complexed with fibrinopeptides [8] and by analogy with trypsin [3]. The substrate binding mode is shown schematically in Fig. 1. Key components are a nucleophilic serine-OH (S), activated by an adjacent histidine (H), a recognition pocket for the basic arginine side chain, and two neighboring hydrophobic pockets recognising side chains N-terminal to the cleavage site (P- and D-pocket).

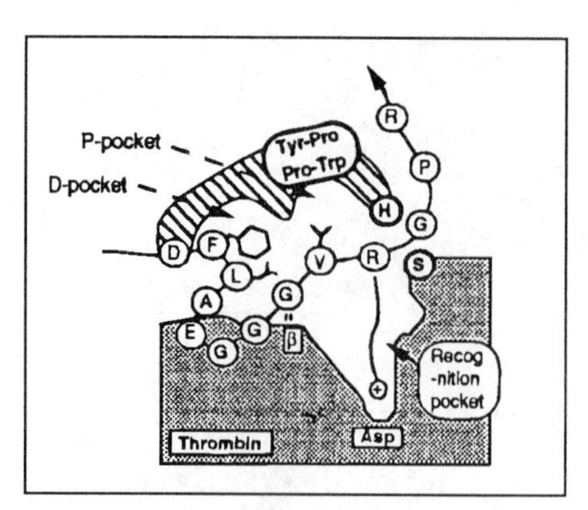

Figure 1

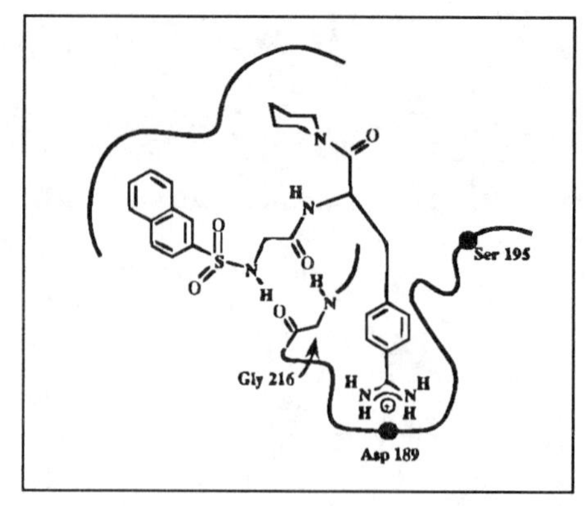

Figure 2

X-ray structures of several complexes between human thrombin and thrombin inhibitors revealed a binding mode in which the recognition pocket and the two hydrophobic pockets are occupied by the inhibitor, but there is no interaction with the catalytic serine [4,9]. This binding mode is exemplified in Fig. 2 by the complex of thrombin with NAPAP : The amidine group forms a salt bridge with Asp189; the piperidine and the naphthyl units occupy both hydrophobic pockets and the glycine subunit of NAPAP forms two hydrogen bonds to Gly216 of thrombin similar to an antiparallel beta-sheet.

These observations have lead to a detailed understanding of the mechanism of inhibition of thrombin and, based on that, contributed to the discovery of novel classes of inhibitors [10,11].

Betalactamases

Bacteria can acquire resistance to antibiotics by producing large amounts of betalactamases. These enzymes cleave penicillin and cephalosporin type antibiotics before these reach their targets, the cell wall synthesizing enzymes. Interference with these enzyme systems is lethal to the bacteria. Antibiotics that are not susceptible to betalactamase cleavage or compounds that inhibit betalactamases are possible means to circumvent this type of resistance. The target enzymes of betalactam antibiotics all belong to the same superfamily and most likely share the same catalytic mechanism. Therefore, classes of inhibitors might be discovered that inhibit several or all members of that family of enzymes. This could open new opportunities for advances in antibacterial therapy.

As a first step in this direction, the structure of the Class C betalactamase from C. Freundii and its complex with the monobactam antibiotic aztreonam has been studied [5]. In the complex structure the reaction between enzyme and inhibitor has been trapped at the stage of the acyl enzyme, after the ring-opening acylation. The arrangement in space of the functional groups around the serine is similar to that in trypsin or thrombin. However, the serine activation is performed by a tyrosine flanked by two lysine $NH3^+$ which as a group are deficient by one proton (Fig. 3 and 5). Based on this experimental structure, a hypothetical model of the acyl enzyme stage of Penicillin G was built (Fig. 4 and 6). The orientation of the acylamino side chain and of the negatively charged group ($-SO3^-$ in aztreonam, $-COO^-$ in penicillin) are similar, but the substituents that could be in the way of the incoming water are different. This structural detail could help to rationalise the observed differences in deacylation rate which is slow for aztreonam and fast for Penicillin G.

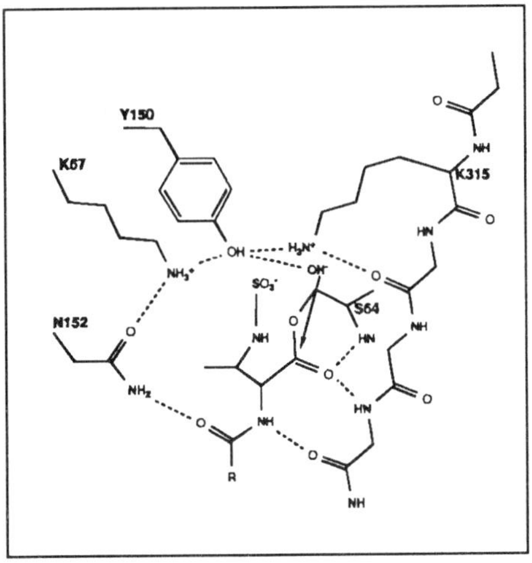

Figure 3

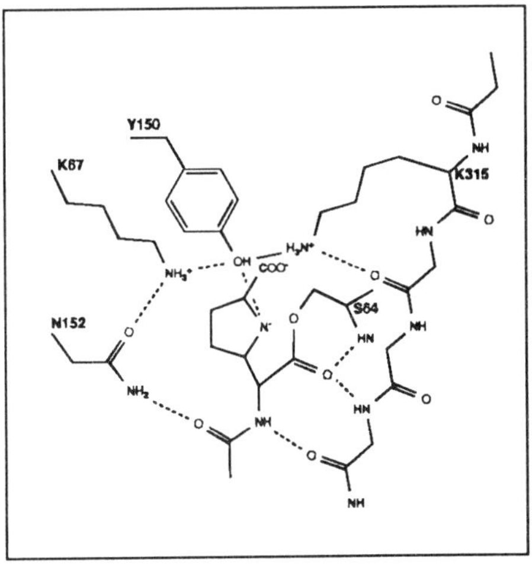

Figure 4

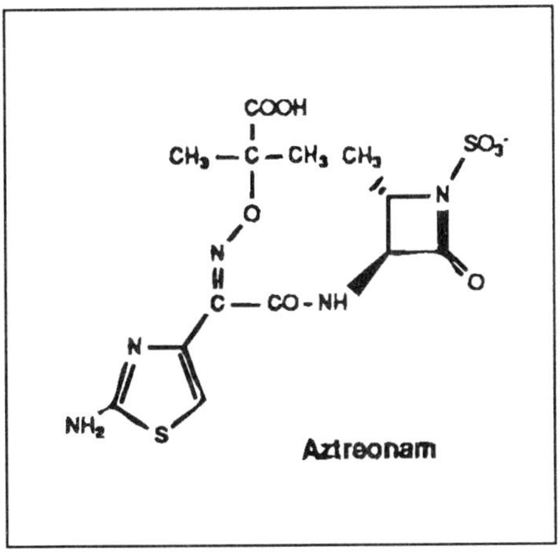

Aztreonam

Figure 5

Penicillin G

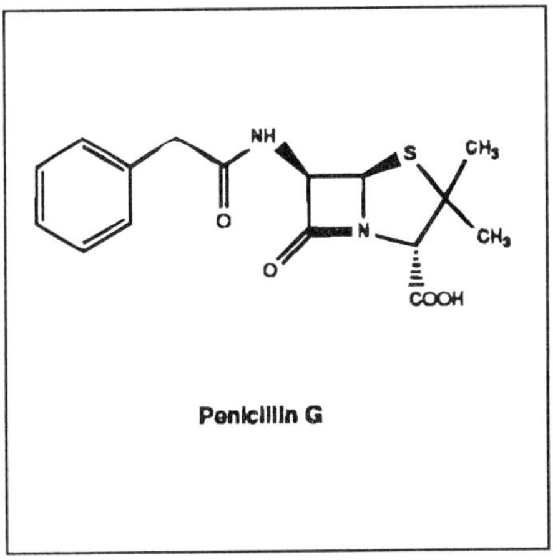

Figure 6

Interestingly, in the case of the class A betalactamases, exemplified by the known structure of the B. licheniformis enzyme [12], there exists an alternative deacylation pathway. A water molecule can approach the serine ester from the opposite side; it is activated by a glutamic acid side chain which is unique to the class A enzymes. These mechanistic details will have an impact on ongoing research directed towards the discovery of betalactamase inhibitors to be combined with potent third generation cephalosporins or, as a long term goal, towards betalactamase-insensitive antibiotics.

Lipase

The structure of the human pancreatic lipase as observed in the crystal [6] represents a catalytically inactive form of the enzyme. In order to become catalytically active, it has to undergo interfacial activation by interaction with its substrate in a fat droplet [13,14].

A model of the active form of the enzyme as well as its complexes with a substrate triglyceride and the inhibitor tetrahydrolipstatin (THL) were built. Sketches of the transition state models are shown in Fig. 7 and 8, respectively. These models allow the formulation of a consistent pathway from the Michaelis complex to the deacylation by water and also provide an explanation for the inhibitory properties of THL: the alcohol group formed after lactone ring cleavage remains in place and prevents deacylating water from attacking the acyl enzyme ester group. When comparing this hypothetical model of the catalytic pathway with trypsin-like proteases, which share the Asp-His-Ser catalytic triad, two marked differences emerge: First, the handedness of the tetrahedral carbon in the transition state is reversed (considering the four substituents of the carbonyl carbon: sidechain

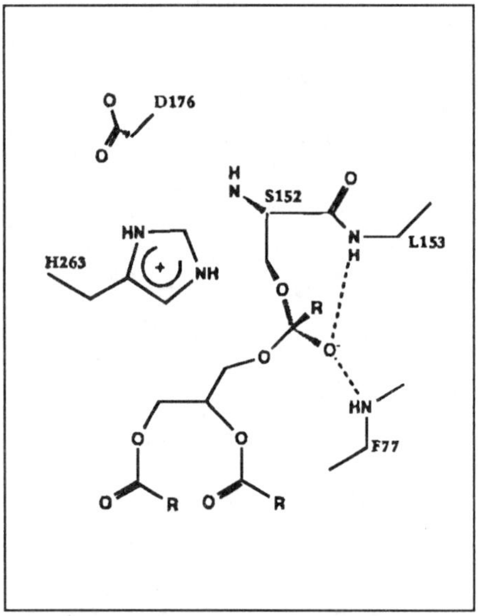

Figure 7

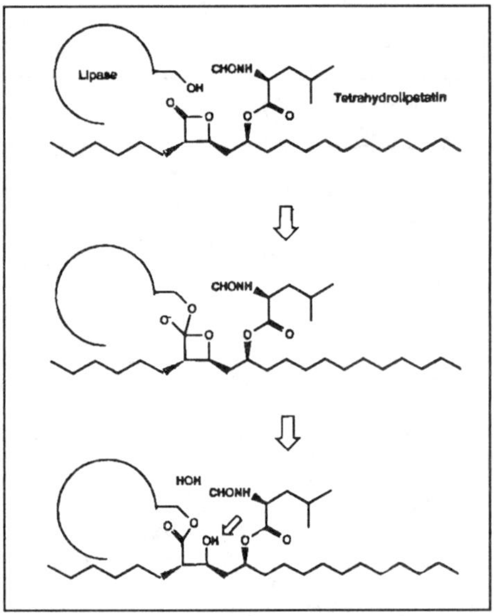

Figure 8

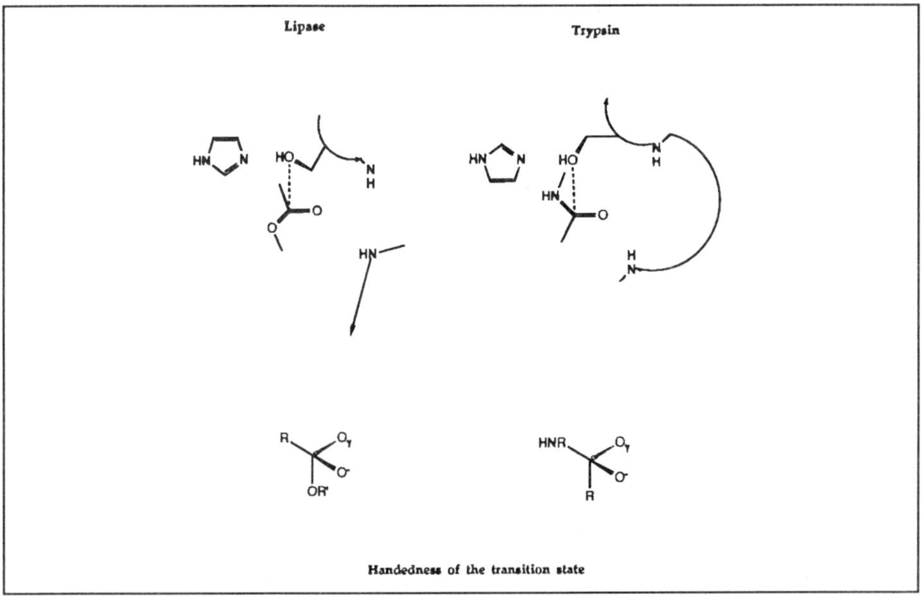

Lipase

Trypsin

Handedness of the transition state

Figure 9

carbon, serine Oγ, oxy anion and amine or alcohol leaving group, Fig. 9); and secondly, the peptide chain direction of the serine containing strand is reversed. As a result, the 3-dimensional arrangement of the functional groups involved in the mechanism of lipase is similar to the mirror image of the corresponding groups in trypsin.

Acetylcholinesterase

The structure of acetylcholinesterase of the electric fish Torpedo californica has been determined recently [7]. The active site of this 537 amino acid protein is situated at the bottom of a deep gorge. The overall fold and the active site residues resemble those of the human pancreatic lipase structure, although no sequence homology can be detected. In contrast to lipase, the third residue of the catalytic triad is changed to glutamate and is part of a different secondary structure element. An oxy anion hole formed by a similar arrangement of backbone NH groups is present in both enzymes; a third NH at appropriate position contributes to the anion stabilization.

Based on these similarities, it is tempting to translate the hypothetical models of the lipase mechanism to acetylcholinesterase (Figure 10). The orientation of the substrate in the active site, the handedness of the transition state and the direction of the incoming water can be used to construct consistent models of the acetylcholine cleavage. Interestingly, the positively charged quaternary ammonium group of acetylcholine in the extended conformation is in contact with the indol ring of a tryptophan at the bottom of the active site gorge.

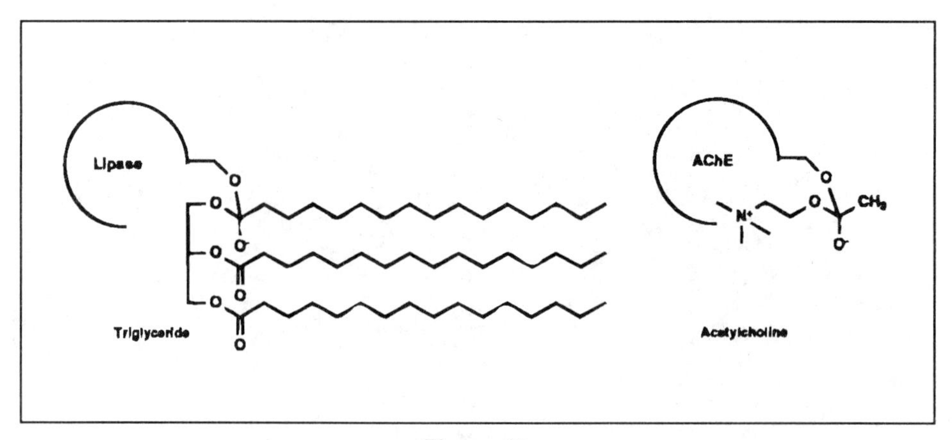

Figure 10

A common stereo-electronic mode of action

The hydrolytic reaction of all of these enzymes is a two-step reaction; an activated catalytic serine is acylated and a leaving group is expelled; the negatively charged oxygen in the transition state is stabilised by appropriately positioned hydrogen bond donor groups. The second deacylation step is a reversal of the acylation step with a water molecule approaching the serine ester. When the pathway of the incoming water is blocked, the reaction stops at the acylenzyme stage and the hydrolase is inhibited.

The detailed knowledge of the mechanism of action and the mechanism of inhibition is the basis for novel approaches in medicinal chemistry in several pharmaceutical research areas and might contribute to improved therapeutic methods in the future.

References:

[1] Müller, K., Ammann, H.J., Doran, D.M., Gerber, P.R., Gubernator, K., Schrepfer, G., *Bull.Soc.Chim.Belg.* **97**, 655 (1988).

[2] Gerber, P.R., Gubernator, K., Müller, K., *Helv.Chim.Acta* **71**, 1429-1441 (1988).

[3] Huber, R., Kukla, D., Bode, W., Schwager, P., Bartels, K., Deisenhofer, J., Steigemann, W., *J.Mol.Biol.*, **89**, 73-101 (1974)

[4] Banner, D.W., Hadvary, P., *J. Biol. Chem.* **266**, 20085- 20093 (1991).

[5] Oefner, C., D'Arcy, A., Daly, J.J., Gubernator, K., Charnas, R.L., Heinze, I., Hubschwerlen, C., Winkler, F.K., *Nature* **343**, 284- 288, (1990).

[6] Winkler, F.K., D'Arcy, A., Hunziker, W., *Nature* **343**, 771-774 (1990).

[7] Sussman, J.L., Harel, M., Frolow, F., Oefner, Ch., Goldman, A., Toker, L., Silman, I., *Science* **253**, 872-879 (1991).

[8] Stubbs, M.T., Oschkinat, H., Mayr, I., Huber, R., Angliker, H., Stone, S.R., Bode, W., *Eur.J.Biochem.* **206**, 187-195 (1992).

[9] Martin, P.D., Robertson, W., Turk, D., Huber, R., Bode, W., Edwards, B.F.P., *J. Biol. Chem.* **267**, 7911-7920 (1992).

[10] Banner, D.W., Ackermann, J., Gast, A., Gubernator, K., Hadvary, P., Hilpert, K., Labler, L., Müller, K., Schmid, G., Tschopp, T., van de Waterbeemd, H., *Proceedings of the 12th Int. Symp. Med. Chem.* 1992 Basel.

[11] European Patent Application 0468231.

[12] Moews, P.C. Knox, J.R., Dideberg, O., Charlier, P., Frere, J.M., *Proteins* **7**, 156-171 (1990).

[13] Gubernator, K., Winkler, F.K., Müller, K., *Proceedings of the CEC-GBF Workshop 1990 in Braunschweig:* 'Lipases: Structure, Mechanism and Genetic Engineering'; Alberghina, L., Schmid, R.D., Verger, R., (Ed.), GBF Monographs, VCH Publishers, Weinheim, pp. 9-16.

[14] Winkler, F., Gubernator, K.,: 'Structure and Mechanism of Human Pancreatic Lipase.' in P.Woolley (Ed.): 'Lipases — their biochemistry, structure and application.', Cambridge University Press, in preparation.

The development of meaningful 3-D search queries for drug discovery

Steven L. Teig

BioCAD Corporation, 1390 Shorebird Way, Mountain View, CA 94043, USA

1. Introduction

Over the last twenty years, computational searching techniques have been developed to permit the rapid searching of molecular databases for compounds containing a given substructure. [1,2] Determining the maximal common substructure of a set of molecules is typically straightforward, [1] so it is easy to construct a superficially reasonable substructure query (also called a '2-D query') from a congeneric series of molecules. Nevertheless, substructure search is of only limited value in general, for exploring only congeneric molecules prevents the chemist from even considering structures that might be more active than those with the specified substructure, have fewer or less severe side effects, or simply broaden a patent position. Further, natural product screens and peptide assays can turn up molecules with highly dissimilar structures but similar IC_{50}'s, making substructure search inappropriate for unearthing new leads in these cases.

Since molecular behavior such as ligand-receptor binding is strongly influenced by the three-dimensional locations of functional groups in space, it is reasonable to want to search databases for molecules possessing a particular set of three-dimensional characteristics. Often, many of the fragments in a small molecule function mostly as structural scaffolding and can be replaced by isosteric fragments without reducing activity. Three-D database searching holds the promise of unearthing such structurally diverse active molecules, and computational techniques for finding compounds that satisfy a set of geometric constraints have been researched actively in recent years. [1,3]

Although the problem of finding molecules that satisfy a geometric query is non-trivial, there is a far more difficult problem that must be solved if 3-D database searching is to be a useful part of molecule discovery — finding a reasonable 3-D database query! This problem has been largely unaddressed in the literature, [4] yet its solution is crucial to making 3-D structure databases useful in small molecule discovery. Readily extracting information from 3-D structural databases necessitates a system that integrates 3-D database searching with 3-D query generation under a simple user interface.

Catalyst is an easy-to-use, integrated system for drug discovery aimed specifically at medicinal chemists as end-users. It contains rapid 2-D and 3-D database searching capabilities and tools for the manual construction of queries. Catalyst also includes a unique facility called 'hypothesis generation' that automatically derives a three-dimensional 'hypothesis' from a set of 2-D drawings of molecules and their experimentally measured activities. A Catalyst hypothesis can be used for several purposes including 3-D database searching using the hypothesis as a query.

The remainder of this paper focuses on the definition and purpose of hypotheses and the issues surrounding their automatic generation.

2. What is a hypothesis?

The concept of a Catalyst hypothesis is predicated on the more general notion of hypothesis prescribed by the scientific method. A scientific hypothesis is a model designed to describe or to 'explain' a set of data. The scientific method consists of constructing a hypothesis to describe a set of phenomena and then attempting to falsify the hypothesis by trying to predict new phenomena. If the predictions are repeatedly validated by experiment, then the hypothesis comes to be considered 'true.' An observation that contradicts the hypothesis forces the abandonment of the hypothesis and motivates the construction of a revised hypothesis that reflects the new counter-example as well as the data 'explained' by the previous hypothesis. This property of falsifiability is the essence of a meaningful notion of 'hypothesis' and that the scientific method is thus a fundamentally iterative process. [5]

A good, scientific hypothesis has the characteristic that it is representative: it fits the data that it purports to explain. In order to be useful, the hypothesis must also be falsifiable by some deterministic procedure. There might be many hypotheses that meet these two characteristics; in an effort to maximize predictiveness, Catalyst is biased towards simpler and more concise hypotheses over more complex explanations of the same data. Finally, among those equally concise, representative and falsifiable hypotheses, hypotheses that are intuitive and visualizable are considered preferable to those that are just computational aids.

A Catalyst hypothesis is a particular sort of scientific hypothesis: the distinguishing characteristic of a collection of molecules. To formalize this notion a little, searching a database of all possible molecules with a hypothesis as a query should cause only those molecules with the specified characteristics to be returned as hits. A hypothesis should be able to contain 1-D (i.e., alphanumeric, such as molecular weight), 2-D (topological or substructure), and 3-D (geometric) criteria. This definition of 'hypothesis' unifies quantitative structure-activity models, pharmacophores, and substructure and other database queries under one general notion. A Catalyst hypothesis takes the intuitive concept of pharmacophore and makes it quantitative, so that it can be used to evaluate new molecules. Within Catalyst, a hypothesis is considered to be as concrete as a molecule.

Hypotheses can be constructed manually or generated automatically. In either case, hypotheses can be used as database search queries. Manually generated hypotheses may contain an essentially arbitrary mixture of substructures, chemical functions, distance constraints, and location constraints: Examples are shown in Figure 1.

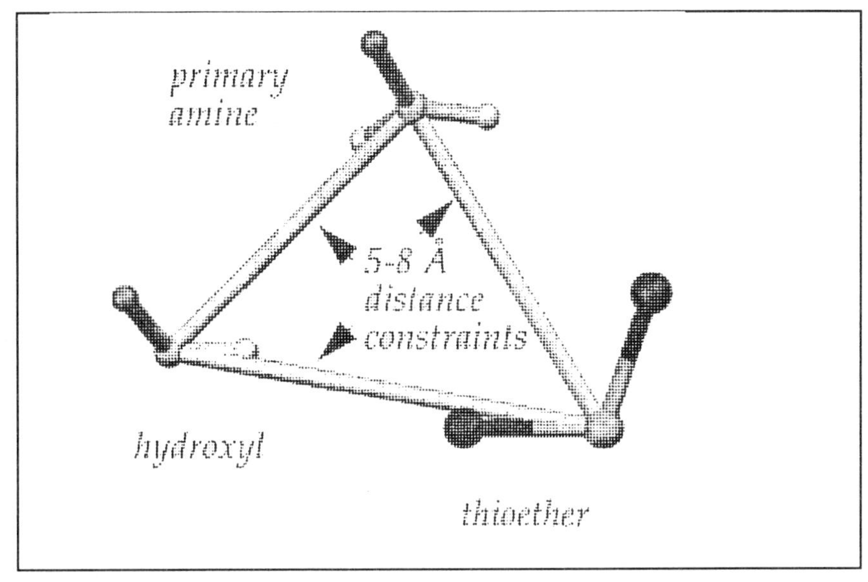

Figure 1a: Hypothesis consisting of three substructures (or fragments) connected by distance constraints. The three substructures are a primary amine, a thioether, and a hydroxyl, all separated by 5-8 Å

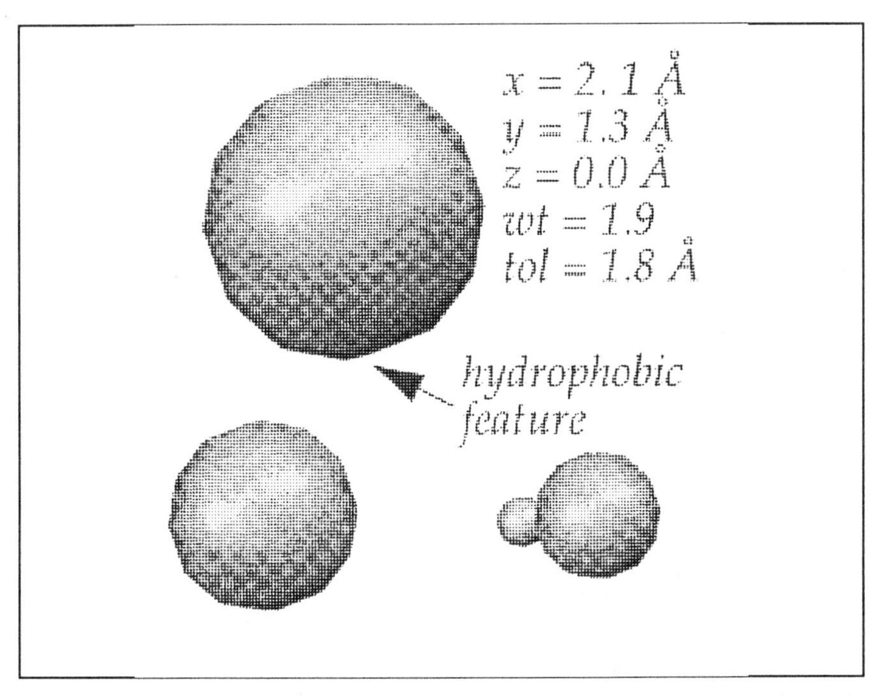

Figure 1b: Hypothesis consisting of four features, each with a chemical function and a location constraint including location, weight, tolerance. The large sphere is a hydrophobe with a particular location in 3-D space. The radius of the sphere is the tolerance

Substructures (or fragments) contain atom and bond specifications, specifying topological connectivity of molecular groups.

Chemical functions are descriptions of chemical properties and, as such, are more general than substructures: for example, the function 'negative charge' would include all negatively charged substructures.

Distance constraints represent distance requirements between two substructures or two chemical functions.

Location constraints are spheres with absolute locations in space in which particular chemical functions must appear in order to satisfy the hypothesis. A location constraint includes the location of the sphere's center, the radius of the sphere (representing the tolerance for geometric location of the desired chemical function), and an associated weight representing the contribution that the presence of the function makes to estimated activity; it is essentially the number of orders of magnitude this feature contributes to estimated activity. The relevance of weights and tolerances will be made clearer by the discussion of flexible fitting description later. The combination of a chemical function and a location constraint is termed a *feature*.

The use of the general word 'hypothesis' in place of the more traditional 'pharmacophore' or 'query' reflects the use of hypotheses for purposes other than database searching, and the ability of hypotheses to represent more general chemical properties than just *in vitro* binding of drug molecules. Currently, though, automatically generated hypotheses are typically used to describe features believed to be relevant to ligand-receptor binding. They are generated to model a set of structure-activity data, such as IC_{50}'s. Such hypotheses consist of four or fewer features (functions and location constraints). Currently, the functions are restricted to come from the following list:

- hydrophobic group
- hydrogen bond acceptor
- hydrogen bond donor
- positive ionizable
- negative ionizable
- positive charge
- negative charge.

In addition to its location constraints, an automatically generated hypothesis carries with it a regression line that maps a geometric fit to the hypothesis into an activity estimate (Figure 2).

3. Why is hypothesis generation difficult?

The automatic generation of three-dimensional hypotheses that reasonably relate molecular structures to activities has historically proved to be very challenging. Just determining which features of an active molecule are relevant to the activity can be subtle, as is mapping corresponding features in dissimilar active molecules. The previous approaches to this sub-problem have made the assumption that there is always a 'pharmacophore': that is, every active molecule possess every feature that

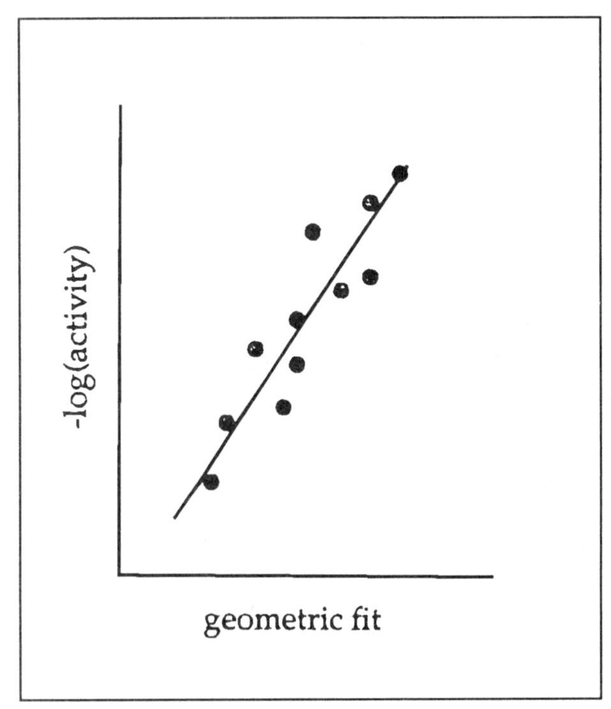

Figure 2: A regression equation that expresses a relationship between geometric fit and -log(activity) is contained within each Catalyst-generated hypothesis

is crucial for activity [6] (ensemble distance geometry). [7] Yet, captopril and enalapril are a demonstration that this assumption is sometimes false. The phenyl ring in enalapril is responsible for over an order of magnitude in activity over enalapril minus the phenyl ring, yet there is no corresponding feature in captopril, although captopril is undeniably a potent ACE inhibitor, too. [8]

In addition to the difficult problem of identifying the key features, there is the problem of aligning molecules so that the corresponding features superimpose geometrically. The molecules might be structurally dissimilar from each other, and they might be highly flexible, too (e.g., peptides). Alignment presents the chemist with a very large combinatorial problem to solve, yet contemporary 3-D pharmacophore generators based on field- [9] or fragment-based [10] methods require the alignment as input!

Finally, it has proved challenging to generate hypotheses that are intuitive to chemists. The chemists' intuition and knowledge are key assets in generating chemically meaningful hypotheses, yet they have frequently been ignored in historical structure-activity methods. Classical QSAR equations [11] and neural networks [12] sometimes produce predictive models, but it is difficult to see what the models 'mean', what assumptions the models are making, and when new candidate molecules are too different from the training molecules for the model's estimated activities to be considered meaningful.

4. Hypothesis generation in Catalyst

Catalyst generates hypotheses from a set of 2-D drawings of molecules (i.e., topologies or connection tables), their experimentally measured activities along with the experimental uncertainties, and a chemical function dictionary, which describes the types of molecular features (e.g., hydrophobic group or hydrogen bond acceptor) that can matter in principle for relating structure to activity. Catalyst attempts to find the simplest hypothesis that adequately 'explains' the data. It tries to minimize the mean squared error of activities estimated with the hypothesis against the experimentally measured activities, while trying to honor a set of criteria for chemical reasonableness. A chemist is likely to be surprised by a feature that is responsible for more than two orders of magnitude in activity, for example. Also, it is expected that most of the features are shared by the most active molecules, even if not all of them are. A chemist can control the extent to which hypothesis generation pays attention to these 'reasonableness' considerations in trying to find a hypothesis that fits the data.

To address the alignment problem, Catalyst takes a novel approach to conformational analysis. Rather than attempting to enumerate all of the low-energy local minima that a molecule can assume [13,14], Catalyst instead tries to maximize the coverage of the bio-accessible (i.e., low-energy) conformational space with a small number of carefully chosen conformers. Since a receptor can readily donate five or ten kilocalories of energy to the ligand during binding, all conformations within that range of the globally minimal energy are possible bio-active conformations. Local energy minima are completely irrelevant; the goal is to produce a compact description of the infinite number of candidate bio-active conformations. Catalyst constructs a collection of conformations that are collectively 'close enough' to all of the low-energy conformations that no interesting conformation will be missed given this collection as a hint. (Figure 3) Hypothesis generation requires only that the conformational model be good enough to allow it to extract the structure-activity signal from the noise.

To try to construct a predictive hypothesis, Catalyst employs an information-theoretic formulation of Occam's Razor as follows. Hypothesis generation seeks the simplest description of the activities within their experimental errors. A formal 'description' consists of a structural hypothesis plus the deviation in each activity from the hypothesis' estimate.

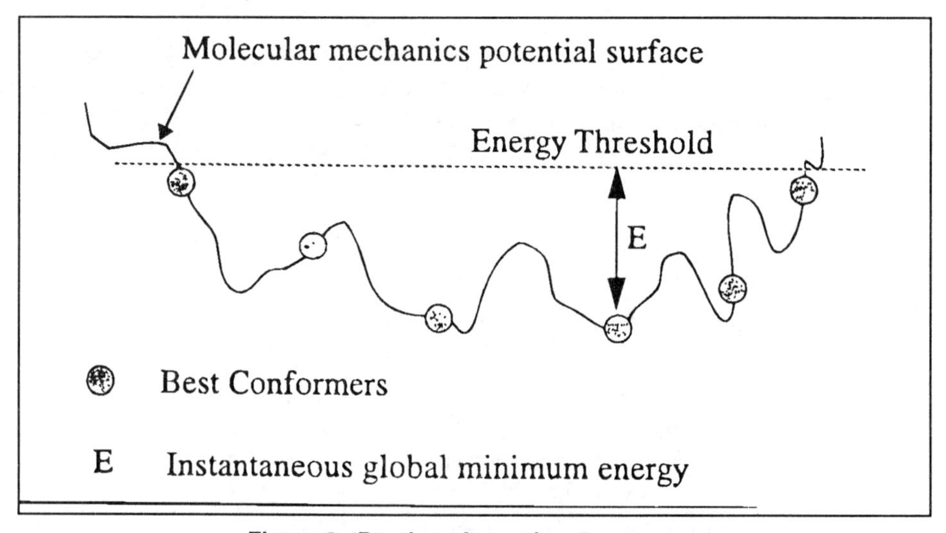

Figure 3: 'Best' conformational coverage

Consider the null hypothesis, which has no structural component at all and estimates that each molecule has an activity that is the average of the activities of the training set molecules. The information-theoretic complexity of the structural component of the hypothesis is zero, but the complexity of the deviations of the training set molecules' activities from their mean is likely to be quite large. Catalyst is willing to increase the complexity of the structural portion of the hypothesis only if doing so provides a more than corresponding decrease in the complexity of the deviations portion of the hypothesis. It is worth enriching the structural model of activity only if the overall information-theoretic description of the activities goes down. This mathematical formulation of the 13th century principle that the simplest hypothesis is best is how Catalyst tries to produce a predictive model and tries to avoid overfitting the data.

An example will clarify the process of hypothesis generation in Catalyst.

5. An automatically generated hypothesis for ACE inhibition

In the 1970s, inhibition of ACE offered a new approach toward controlling high blood pressure. [9] Information about potential inhibitors came from investigation of peptides found in snake venom and from related studies at companies such as Squibb and Merck. Finding non-peptides that could be developed into orally active ACE inhibitors remained the major challenge.

The training set consists of 20 di- and tri-peptides known to Ondetti and Cushman at Squibb in 1972. [15] The peptides, their activities, and the uncertainties in their activities are shown in Table 1. The resulting Catalyst hypothesis is represented by the 'estimated activities' in Table 1 and is graphically displayed in Figures 4-7, superimposed on several leads. The Error column in Table 1 is the ratio of the estimated activity to the measured activity, or its inverse if this ratio is less than one.

Explaining Experimental Results

The ACE hypothesis is shown in Figure 4 and Figure 5 superimposed against two peptides from the training set: Ala-Pro and Phe-Ala-Pro. Important functions suggested by the hypothesis for interaction with ACE include the proline carboxyl group (negative charge), the Phe-Ala amide carbonyl (hydrogen bond acceptor), and a hydrophobic interaction at the back of the proline ring. An additional hydrophobic interaction is suggested for the phenyl ring of Phe.

Suggesting Synthetic Direction

Succinyl proline is shown in Figure 6 to be a suitable surrogate for Ala-Pro. The fit of succinyl proline is similar to the fit of Ala-Pro, but the side chain carboxyl group acts as the hydrogen bond acceptor. By extending the molecule toward the second hydrophobic group, additional binding should be available. One such molecule, enalapril, maps to all four functions and is predicted to be very active (Figure 7). The 2-D structures of these molecules are found in Figure 8.

Beginning with information from small peptides, Catalyst generated a hypothesis that points to non-peptidic compounds that are potent ACE inhibitors.

Name	Measured activity (IC_{50}) nm	Uncertainty	Estimated activity (IC_{50}) nm	Error
nLeu-Ala-Pro	700	3	2,300	3.2
Val-Trp	1,700	3	4,900	2.9
Leu-Ala-Pro	2,300	3	2,300	1.0
Ile-Tyr	3,700	3	3,700	1.0
Phe-Ala-Pro	4,200	3	5,700	1.4
Arg-Ala-Pro	16,000	3	85,000	5.2
Phe-Pro-Pro	78,000	3	18,000	4.3
Ile-Pro	150,000	3	79,000	1.9
Ala-Pro	270,000	3	1,800,000	6.7
Ala-Val	300,000	3	1,800,000	6.0
Glu-Ala-Pro	360,000	3	84,000	4.2
Val-Pro	420,000	3	260,000	1.6
Gly-Phe	450,000	3	120,000	3.6
Ala-Leu	1,600,000	4	290,000	5.5
Ala-Gly	2,500,000	4	1,800,000	1.4
Gly-Glu	5,400,000	4	8,800,000	1.6
Gly-Lys	5,400,000	4	1,900,000	2.8
Pro-Pro	7,500,000	4	1,800,000	4.1
Ala-His	9,000,000	4	8,800,000	1.0
Gly-Asp	9,200,000	4	8,800,000	1.0

Table 1. ACE Inhibitor Activity Table. These are the di- and tri-peptides used to construct an ACE inhibition hypothesis. The *Uncertainty* is the multiplicative uncertainty in the measured activity. The *Estimated Activity* is the activity that Catalyst estimates for the compound by comparing it to the hypothesis. The values in the *Error* column represent the ratio of the estimated activity to the measured activity, or its inverse if this ratio is less than one

6. Searching a database with a hypothesis

The ACE hypothesis above (and any other hypothesis generated with Catalyst) can be used to search a database. Three-D database searching with hypotheses makes possible the rapid 'electronic screening' of hundreds of thousands of molecules in the search for patentable leads or novel, structural motifs that point the way to new leads. To try this with the ACE hypothesis, ten diverse, low-energy conformers for each of the molecules in the Pomona Database [16] were generated with Catalyst. The use of multiple conformers functions here as a first approximation to a flexible

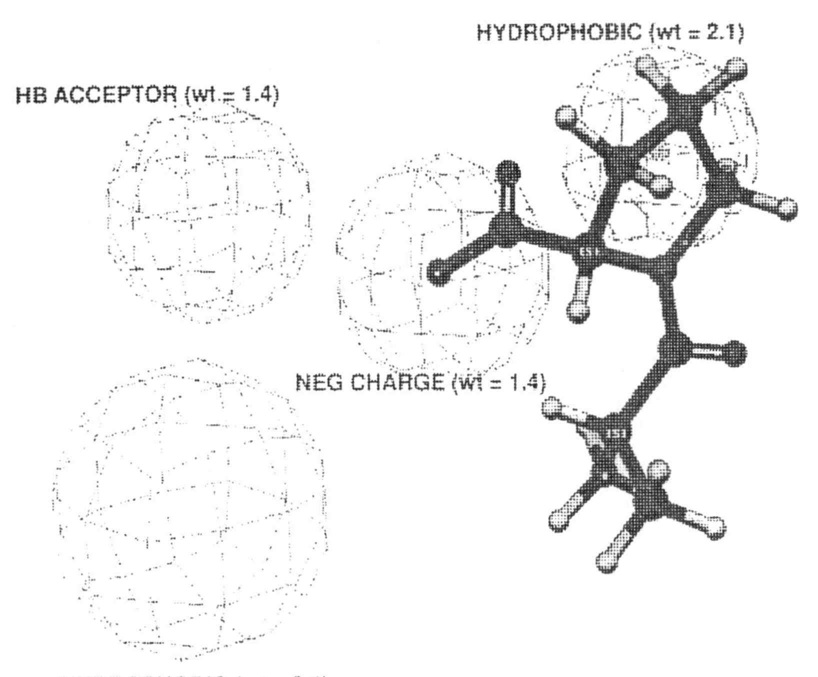

Figure 4: Ala-Pro superimposed on the ACE hypothesis

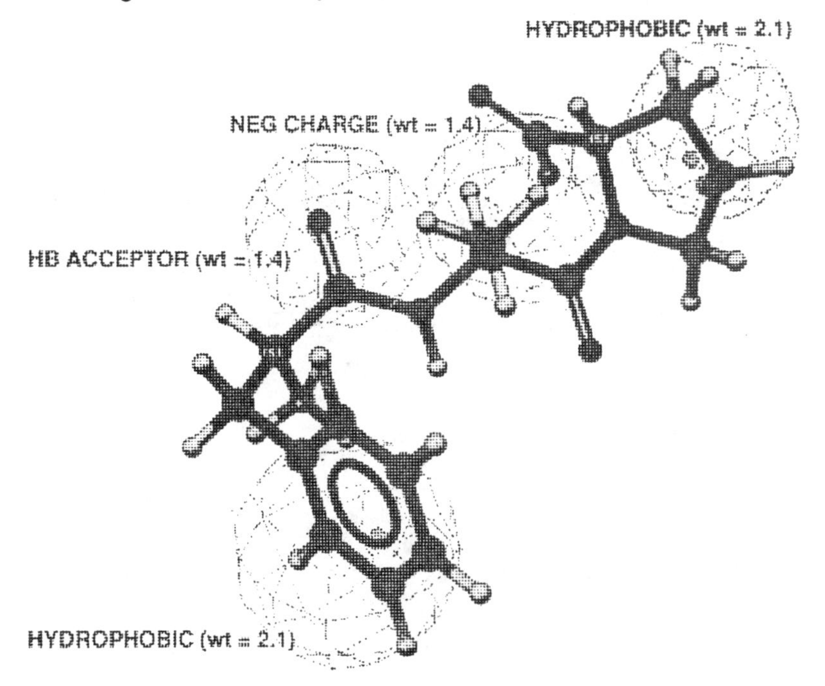

Figure 5: Phe-Ala-Pro superimposed on the ACE hypothesis

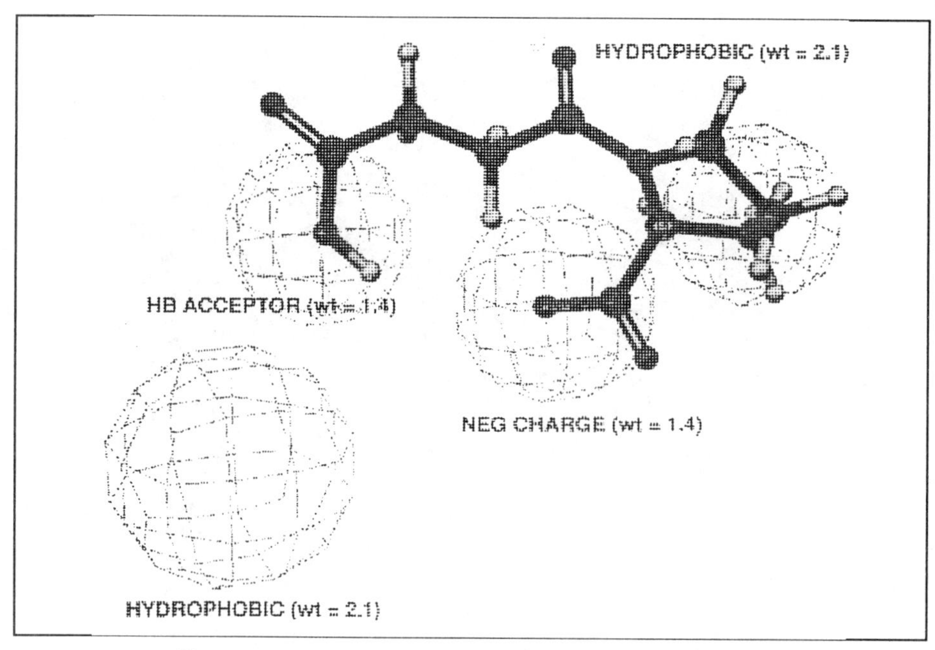

Figure 6: Succinyl proline, an early peptide surrogate

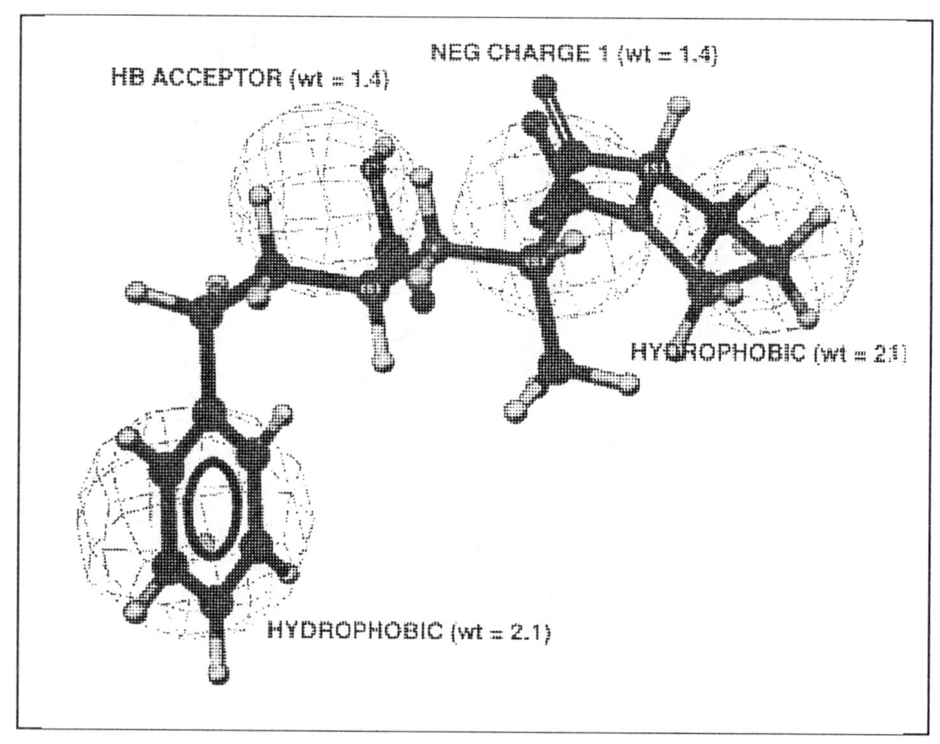

Figure 7: Enalapril mapped to the ACE hypothesis

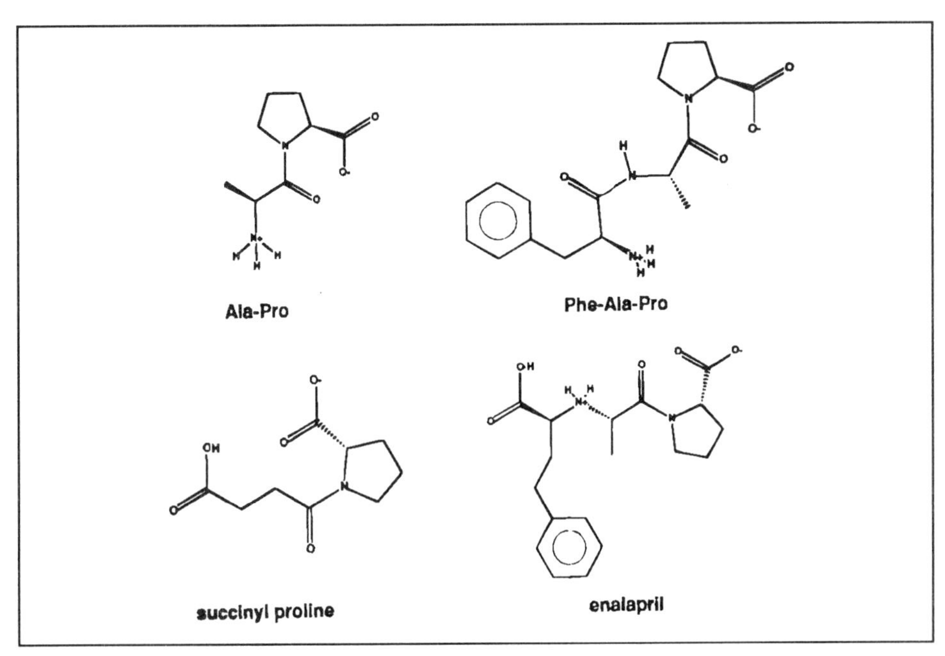

Figure 8: 2-D structures of ACE inhibitors

database search, since the conformers were generated with conformational coverage in mind, as described above in section 4. There are about approaching 200,000 conformers in the multi-conformer Pomona Database being searched here.

Searching on an R3000 Indigo from Silicon Graphics, Catalyst returns 45 hits in less than 30 seconds. Many of the hits are pharmaceutically inappropriate, but a few of them are somewhat interesting. Table 2 is a list of the hits from this search. Certainly, the search has returned structurally diverse hits all of which match the hypothesis-as-query. The use of chemical functions (e.g., hydrogen bond acceptor) instead of purely atomic fragments (e.g., carboxylate) in automatically generated hypotheses greatly increases the likelihood of finding hits that are structurally very different from the training set molecules, yet which have a reasonable chance of binding to the receptor of interest. The hits of the ACE search include sulfones used as hydrogen bond acceptors and isopropyl groups used as hydrophobes even though none of the di- or tripeptides have either substituent. These highly non-peptidic molecules could lead to ACE inhibitors that are quite different from the peptides and enalapril structurally, potentially broadening a patent position or possessing different side-effect profiles.

11-DEHYDROCORTICOSTERONE,GLUC.AC.CJ.(BU)4N+
11-DEHYDROCORTICOSTERONE,GLUC.AC.CJ.(PEN)4N+
11-DEOXYCORTISOL,GLUC.AC.CONJ.(BU)4N+
11-DEOXYCORTISOL,GLUC.AC.CONJ.(PENT)4N+
ACETYLCHOLINESALT-BROMTHYMOLBLUE
AGENTITSALT-BROMTHYMOLBLUE
BENHEPAZONE
BENZILPENICILLIN,TETRABUTYLAMMONIUM SALT
BESULPAMIDE
BETHANECHOLSALT-BROMTHYMOLBLUE
BU (4) AMMONIUMSALT-BROMTHYMOLBLUE
CARBACHOLSALT-BROMTHYMOLBLUE
CEFEMPIDONE
CEFPIROME
CEFTAZIDIME
CEPHALORIDINE
CETIPRINSALT-BROMTHYMOLBLUE
CHOLINESALT-BROMTHYMOLBLUE
CICLOSIDOMINE
CILADOPA
CORTISONE,GLUCURONICAC.CONJ.(PENT)4N+
DECAMETHONIUMSALT-BROMTHYMOLBLUE
ET(4)-AMMONIUMSALT-BROMTHYMOLBLUE
ET-18-OCH3
INDOMETHACINANION
ME(4)-AMMONIUMSALT-BROMTHYMOLBLUE
METHANTHELINESALT-BROMTHYMOLBLUE
METHYLATROPINE-BROMTHYMOLBLUE
METHYLSCOPOLAMINE-BROMTHYMOLBLUE
NADIDE
OXACILLIN,TETRABUTYLAMMONIUM SALT
PENTOLINIUMSALT-BROMTHYMOLBLUE
PHENETHICILLIN,TETRABUTYLAMMONIUM SALT
PHENOXYMETHYLPENICILLIN, TETRABUTYLAMMONIUM SALT
PHENYLALANYLPHENYL ALANINE,TETRAPENTYLAM.SALT
PIPERIDINIUM,N,N-DIMETHYL-2-HYDROXYMETHYL
PIRIBENZIL--BROMTHYMOLBLUE
PR(4)-AMMONIUMSALT-BROMTHYMOLBLUE
PROPICILLIN,TETRABUTYLAMMONIUM SALT
TETRABUTYLAMMONIUMSALT-FASTRED-A
TETRAETHYLAMMONIUMSALT-FASTRED-A
TETRAPENTYLAMMONIUM,BROMTHYMOLBLUE SALT
TETRAPENTYLAMMONIUMSALT-FASTRED-A
TETRAPROPYLAMMONIUMSALT-FASTRED-A
THIAMINESALT-BROMTHYMOLBLUE

Table 2: Hits returned using ACE hypothesis on multiconformer Pomona database *

**The multiconformer database was comprised of 20,303 compounds and a total of 163,046 conformers*

7. Summary

Three-D database searching is proving to be a valuable adjunct to drug discovery efforts. Database searching technology is improving rapidly, but effective techniques to generate meaningful 3-D database search queries have been essentially non-existent so far. A new system for medicinal chemists, Catalyst, integrates rapid 2-D and 3-D database searching with a facility called 'hypothesis generation' that automatically extracts the important 3-D characteristics of a set of molecules from 2-D structures and activity data.

Catalyst's hypothesis generation automatically addresses the problems of determining the important molecular features and aligning flexible molecules. It tries to produce a predictive hypothesis by using an information-theoretic mathematization of Occam's Razor to find a simple hypothesis.

An example of hypothesis generation applied to ACE inhibition has been presented along with the results of a database search using the automatically generated ACE hypothesis as a search query. The search was conducted on a representative sample of conformers generated for each molecule in the Pamona database and found novel structures which are structurally very different from the training set used to generate the hypothesis, yet have a reasonable chance of showing activity.

8. References

[1] S. E. Jakes and P. Willett, *J. Mol. Graph.* **4**, 1, 12-20 (1986).

[2] M. Hicks, In *Software-Entwicklung in der Chemie 3*, Berlin: Springer-Verlag 9-16 (1989).

[3] R. P. Sheridan, R. Nilakantan, A. Rusinko III, No. Bauman, K. S. Haraki, R. Venkataraghavan, *J. Chem. Inf. Comput. Sci.* **29**, 4, 255-260 (1989).

[4] O. Güner, D. Henry, R. S. Pearlman, *J. Chem. Inf. Comput. Sci.* **32**, 101-109 (1992).

[5] K. Popper, 'Conjectures and Refutations: The Growth of Scientific Knowledge'. New York, Harper & Row (1965).

[6] D. Mayer *et al., J. Comp. Aided Mol. Des.* **1**, 3-16 (1987).

[7] M. Bures, Y. Martin, Abstract in *ACS Proceedings*, Washington, D. C. (1992).

[8] J. H. Laragh, *Prog. Cardiovasc Dis.* **21**, 159 (1978).

[9] CoMFA, R. Cramer *et al., J. Amer. Chem. Soc.* **110**, 5959-5967 (1988).

[10] V. Golender, A. B. Rozenblit, 'Logical and Combinatorial Algorithms for Drug Design', Wiley & Sons (1983).

[11] C. Hansch (ed.), 'Comprehensive Medicinal Chemistry, 4', Oxford, Pergamon Press (1990).

[12] M. Lipton, W. S. Still, *J. Comp. Chem.* **9**, 4, 343 (1989).

[13] G. M. Crippen, T. F. Havel, 'Distance Geometry and Molecular Conformation', Chemometrics Research Studies Series 15, New York, Wiley Pub. (1988).

[14] M. Saunders, *J. Am. Chem. Soc.* **109**, 10, 3150-3152 (1987).

[15] D. W. Cushman, et al., *Prog. Cardiovasc Dis.* **21**, 176 (1978).

[16] A. Leo, C. Hansch, Pomona Medicinal Chemistry Project.

Large chemical structure and reaction databases: challenges and solutions

William G. Town

William Town Associates Ltd, Abingdon Road, Clifton Hampden, Abingdon, Oxon, England OX14 3EG

Introduction

We are entering an exciting new era in chemical information. Over twenty-five years of computerised database building have resulted in the creation of large chemical structure and chemical reaction database systems (some now consisting of several millions of machine-readable records) the design of which includes the challenging task of avoiding information overload for the user. Fortunately, novel approaches to information handling, new software tools, and more powerful computing environments interconnected through high bandwidth networks are leading to improved information retrieval and data exploration methods which use increasingly sophisticated analytical tools to process the stored data and deliver information to the end user. To understand these developments, it is necessary to review the currently available databases and their delivery systems, explore how algorithmic selection of data can enhance the usefulness of the existing delivery systems, and finally examine some of the new data exploration systems and their potential for revolutionising the way we think about chemical information retrieval.

Large chemical reaction databases

In reviewing the available chemical structure and chemical reaction databases, various approaches are possible. One classification of chemical reaction databases (Judson, 1990) divided these databases into three categories according to their intended purpose:

1. *comprehensive databases* which contain almost all the reactions included in the journals covered by the database (e.g. CASREACT, InfoChem/ZIC)
2. *thematic databases* which are smaller subset reaction databases with a common theme (e.g. reactions of epoxides, steroid reactions)
3. *selective databases* which are subset databases whose purpose is to provide comprehensive coverage of all known reactions by the inclusion of typical reactions chosen from the examples present in the literature. Until recently, selective databases were produced by costly and subjective manual selection methods (e.g. Comprehensive Heterocyclic Chemistry, Theilheimer's Synthetic

Methods). Algorithmic selection methods have been introduced recently to create new selective databases (e.g. *ChemReact, ChemSynth, ChemSelect*).

An alternative classification (Zass, 1991) differentiates chemical reaction databases by size and by their annual rate of growth. Figure 1 shows an updated version of Zass' reaction database diagram showing database size versus database annual growth plotted on a log/log scale. If the three algorithmically selected databases, *ChemReact*, *ChemSynth* and *ChemSelect*, are omitted from the diagram (Figure 2), the reaction databases are seen to fall into three main groups:

1. *high volume/high growth databases* represented by the Beilstein, CASREACT and InfoChem/ZIC databases
2. *low volume/high growth databases* such as ISI's Current Chemical Reactions (CCR) and FIZ Chemie's ChemInform RX (REACCS version)
3. *low volume/low growth databases* including Theilheimer, Pergamon's Current Heterocyclic Chemistry, Derwent's Journal of Synthetic Methods (REACCS version), MDL's Current Literature File (CLF), CHIRAS and Metalysis (all now discontinued) and Derwent's Chemical Reaction Documentation Service (CRDS) on ORBIT.

Current Synthetic Methodology, which is a new current awareness database derived from ChemInform RX, falls into a class of its own. The role and purpose of the algorithmically selected databases will be discussed further in a later section.

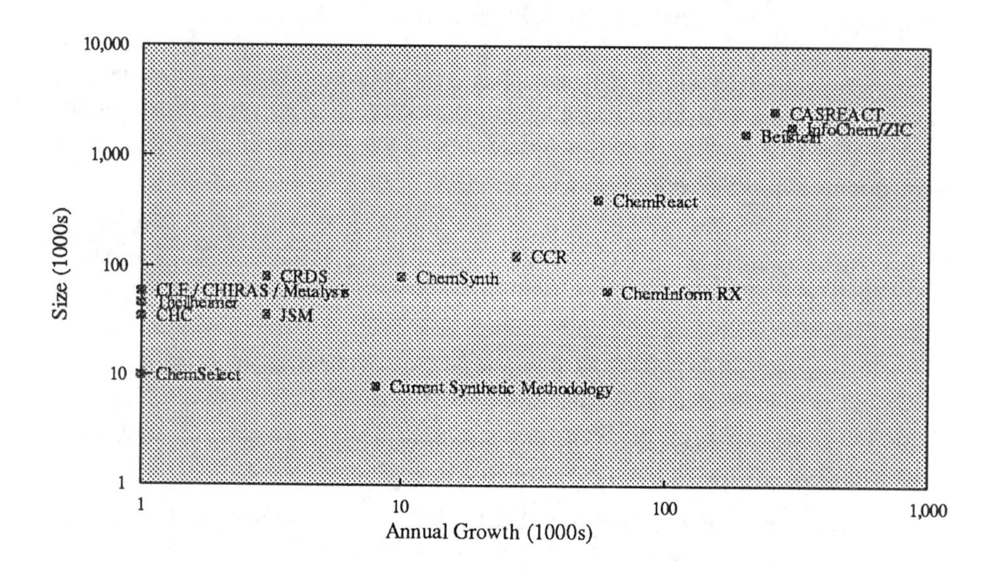

Figure 1: Reaction databases 1992: an overview

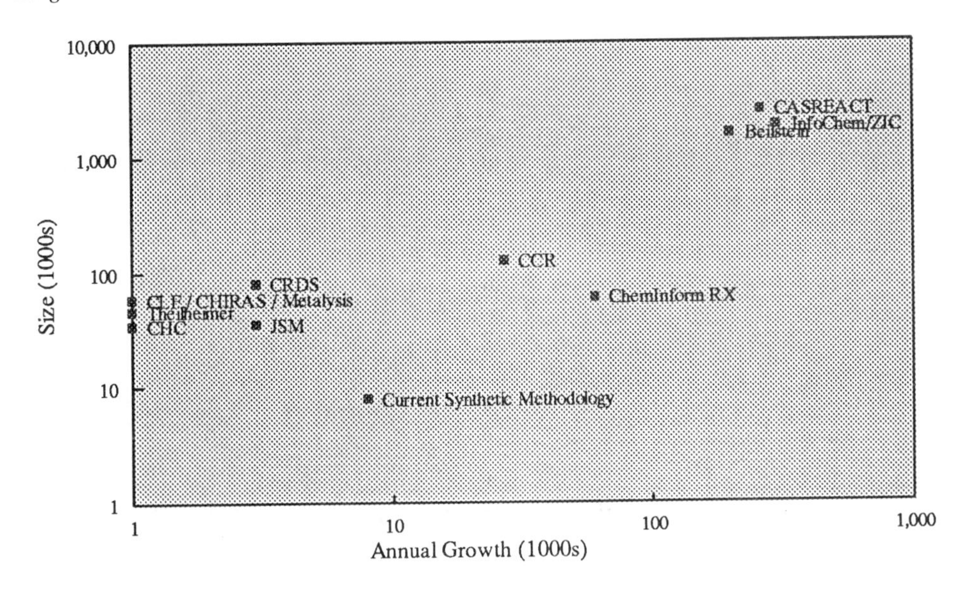

Figure 2: Manually created reaction databases

Large chemical structure databases

The classifications applied to chemical reaction databases by Judson and Zass can be applied to chemical structure databases in an analogous manner. Classification of chemical structure databases by intended purpose leads to the following results:

1. *comprehensive structure databases* which include all chemical compounds from the selected journals, are represented again by Beilstein, the Chemical Abstracts Registry file and the VINITI/ZIC databases

2. *thematic structure databases* include drug databases (Derwent's Standard Drug File, PJB's PharmaStructures, Prous' Drug Data Report), fine chemicals databases (MDL's ChemQuest), crystallographic databases (CCDC's Cambridge Structure Database), spectroscopic databases (Sadtler's spectral libraries, Chemical Concepts' Specinfor), and environmental or toxicological databases (National Chemical Emergency Centre's CHEMDATA, NIOSH Registry of Toxic Effects of Chemical Substances (RTECS), and the CEC Environmental Chemicals Data and Information Network (ECDIN)).

3. *selective structure databases* are less common but could be argued to include chemical dictionaries (Chapman & Hall Chemical Database, Merck Index).

Figure 3 shows the same chemical structure databases plotted by size and rate of growth. Two clusters of databases are apparent:

1. *high volume/high growth databases* represented by Beilstein, the Chemical Abstracts Registry File, and the VINITI/ZIC databases

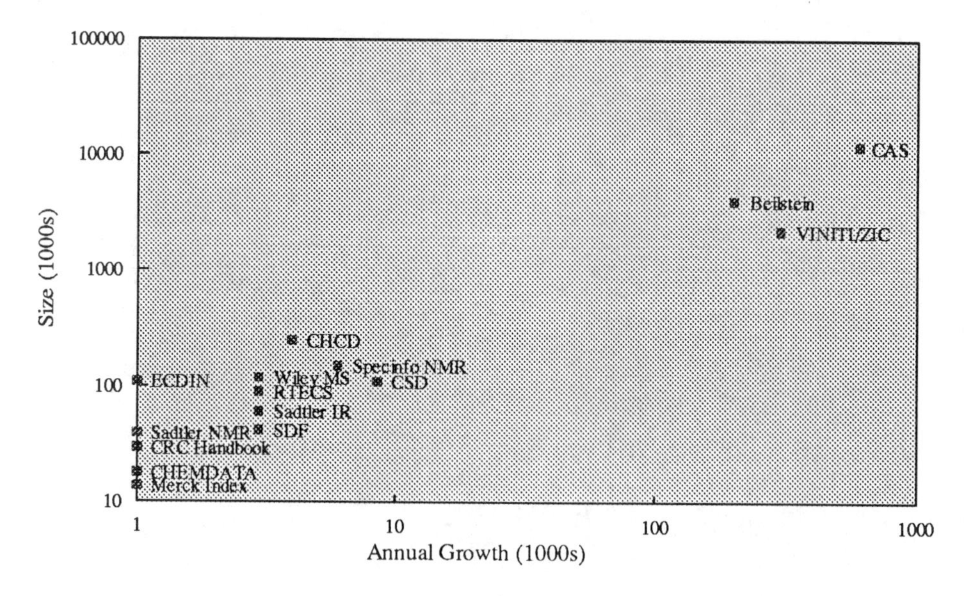

Figure 3: Chemical structure databases (1992)

2. *low volume/low growth databases* which include virtually all other databases.

All selective chemical structure databases are produced by manual methods at present.

Delivery systems for chemical structure and chemical reaction databases

Until recently, delivery systems for chemical structure and reaction information could be classified into three main groups:

1. *personal database systems* which operate on personal computers (mostly IBM PC compatible) and permit (sub)structure and/or reaction searching of databases up to say 100,000 records in size with the presently available hardware technology. Software products which fall into this category include *ChemBase* (MDL), *PsiBase* (HDS) and *S4* (Softron). Recently, structure-searchable CD-ROM products for use on personal computers have appeared. These have effectively raised the upper size limit of databases which can be handled on personal computers by facilitating the delivery of the database to the user. Products of this type include *Current Facts in Chemistry* (Beilstein/Springer Verlag) which uses *S4*, and the *CHCD Dictionary of Natural Products* on CD-ROM (Chapman & Hall) which incorporates *PsiBase for Windows*.

2. *in-house or corporate database systems* which operate on mini- or mainframe computers (mostly Digital VAXs) and which provide online access to users with dumb terminals or PCs emulating dumb terminals. This sector of the market is

dominated by MDL's *MACCS* and *REACCS* products but other systems in use include *DARC-SMS* (Questel SA), *ORAC* and *OSAC* (ORAC Ltd). The apparent effective maximum database size for these systems is conditioned by the database building limitations of *MACCS* and *REACCS*. *MACCS* appears to search 500,000 record databases effectively, but the time required to load a record seems to grow as the size of the database grows. Similarly, the effective maximum size of individual *REACCS* databases seems to be of the order of 100,000 records.

3. online database systems which operate on mainframe computers or specially designed networks of computers and which provide online access to remote users with dumb terminals or PCs emulating dumb terminals. The most familiar system of this type is the *Messenger* system developed by Chemical Abstracts Service for the CAS ONLINE service and now used at all three STN International nodes. The *Messenger* system is being continually improved and upgraded and, in addition to structure searching, now handles reactions for CASREACT and generic structures for MARPAT. CAS databases are also structure searchable on the Questel host using *Generic DARC* software. The Beilstein database is searchable using Messenger on STN but is also structure searchable on DIALOG using elements of *S4* and on Orbit using *HTSS*. Online database searching is improved by the use of a personal computer 'front end' software package such as *STN Express* (CAS) or *Molkick* (Beilstein/Springer) for the offline preparation of structure queries and post-processing of results. Online database systems effectively permit searching of the largest multi-million record chemical structure and reaction databases. However, the presentation of the retrieved information to the user has hardly progressed since the first systems were introduced in the 1970s. Results are presented on a virtual, 80 character-wide, 'wallpaper roll' reminiscent of the TTY (Teletype) terminals of the 1970s using fixed-width characters and a single typeface and point size. Use of a package like *STN Express*, permitting reverse scrolling and the capturing and post-processing of the 'transcript' of the search session, improves the usability of the online service but it cannot overcome the basic limitations of the system which are in part due to the transmission speed limitations of the public data networks.

In 1991, a new generation of delivery systems appeared when MDL introduced the first release of the components of its new ISIS distributed information system based on a client/server architecture. The PC or workstation components, *ISIS/draw* and *ISIS/base*, provide the query formulation, local database and results processing elements of the system. *ISIS/base* interacts with *ISIS/host* (running on a mini-computer or UNIX workstation) which mediates the search request and interacts with the chemical and relational databases to obtain the search results. Eventually, it is planned that *ISIS/host* will also provide a gateway to CAS databases on STN International. At present, the structure and reaction search engines in *ISIS* are identical to *MACCS* and *REACCS* respectively and therefore suffer all the drawbacks of those two systems with respect to large databases mentioned above.

Algorithmic selection of databases

The development of multi-million record databases has presented a challenge to the designer of search and retrieval systems. In-house reaction database systems have had to become more sophisticated in their approach as databases increased in size, and they now offer features such as reaction site searching, atom/atom mapping and stereospecific searching, as a matter of course. As Zass (1990) has indicated, it is even more important that online systems, such as CASREACT, adopt these features to avoid expensive and annoying false hits. Fortunately, Chemical Abstracts Service has recognised this need and the latest releases of *Messenger* and *STN Express* now incorporate many of these features (STN Technical Note 91/04). Buntrock (1992) has reviewed chemical reaction searching systems and has found that many of his earlier criticisms have been dealt with.

However, with the increasing size of databases, it is inevitable that, for any given search, many examples of a given reaction type will be retrieved even from a perfect retrieval system. This increases the cost of the search and makes the review of the search results more tedious. Unless a comprehensive search was the original intention, the user would be better served by good representative examples of each reaction type rather than being overwhelmed by every recorded example. This was the intention of the thematic and selective databases such as Theilheimer's Synthetic Methods mentioned earlier. The large chemical reaction databases now available to us present new opportunities for the creation of selective databases by algorithmic or automatic methods whilst retaining links to all the examples found in the comprehensive database.

As a result of a series of fortuitous circumstances, the large chemical reaction database created by VINITI (the All-Union Institute of Scientific and Technical Information of the Academy of Sciences of the USSR) in collaboration with ZIC (VEB Zentrale Informations-Verarbeitung Chemie in East Berlin, now ZIC-GmbH, Berlin) has been made available to InfoChem GmbH in Munich for marketing and further development. The 1.8 million reactions in this database, covering the period 1975–88, were derived from preparation information contained in the 2.2 million chemical structure database resulting from the VINITI/ZIC collaboration. This chemical structure database has now also been made available to InfoChem. A series of selective chemical databases have been developed by InfoChem from the large InfoChem/ZIC database and the relationship of these selective database to each other and the original comprehensive file is shown in Figure 4.

ChemReact contains one example reaction for each reaction type found in the comprehensive database and the 1.8 million chemical reactions in the InfoChem/ZIC database are represented by 414,142 typical reactions in *ChemReact*. The *ChemReact* selection process consists of a sequence of five steps:

1. *atom-atom mapping for each reaction.* This step determines which atom in the reactant molecule corresponds to each atom in the product molecule. Corresponding atoms are assigned equal numbers.
2. *identification of reaction centres for each reaction.* Using the atom–atom mapping, the reacting atoms and bonds are identified in this step. An atom is

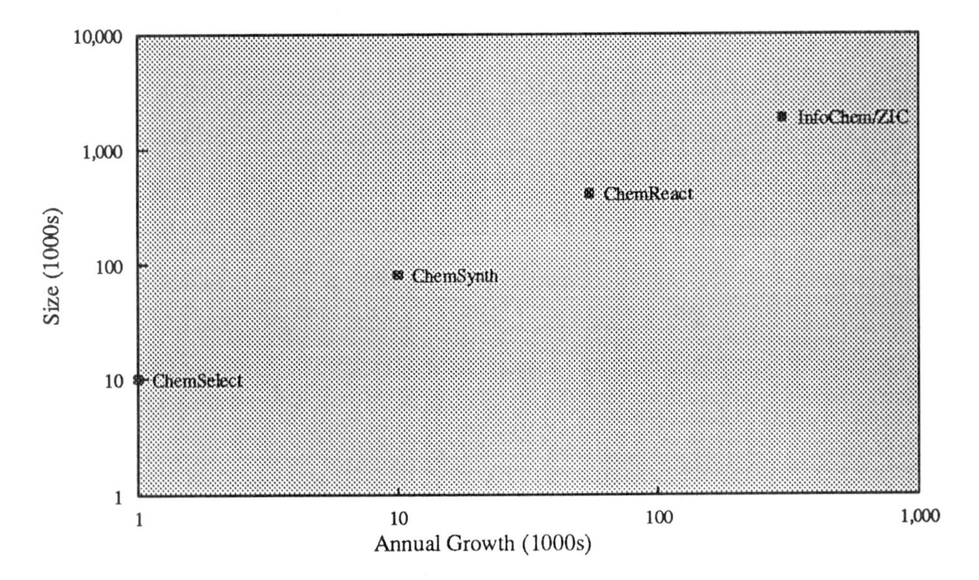

Figure 4: Automatically selected reaction databases

defined to be a reacting centre if its degree of unsaturation, its charge, unpaired electrons, or the number of attached hydrogens change during the reaction. A bond is considered a reacting centre if it is either completely formed or cleaved in the reaction or if its bond order changes together with the degree of unsaturation of the participating atoms.

3. *identification of reaction type codes.* The information on all reacting centres is combined into one numerical reaction type code which also includes information on the chemically relevant environment. Reactions with the same code are considered to be identical.

4. *selection of the best example in each document.* For each reaction type in a document with more than one example, the 'best' reaction is selected using yield, known spectra, etc. as selection criteria.

5. *selection of the best example in the database.* For those reaction types represented by more than one document, a further selection is made using again yield, known spectra and publication year (most recent) as selection criteria.

A further selection was made for *ChemSynth* by excluding any typical reaction in *ChemReact* which did not appear in one of a set of 'leading journals', any reaction which was found only once in the literature, and any reaction with a yield of less than 50%. The resulting 80,000 chemical reaction database is a comprehensive methodology database for organic reactions encompassing a wide range of different

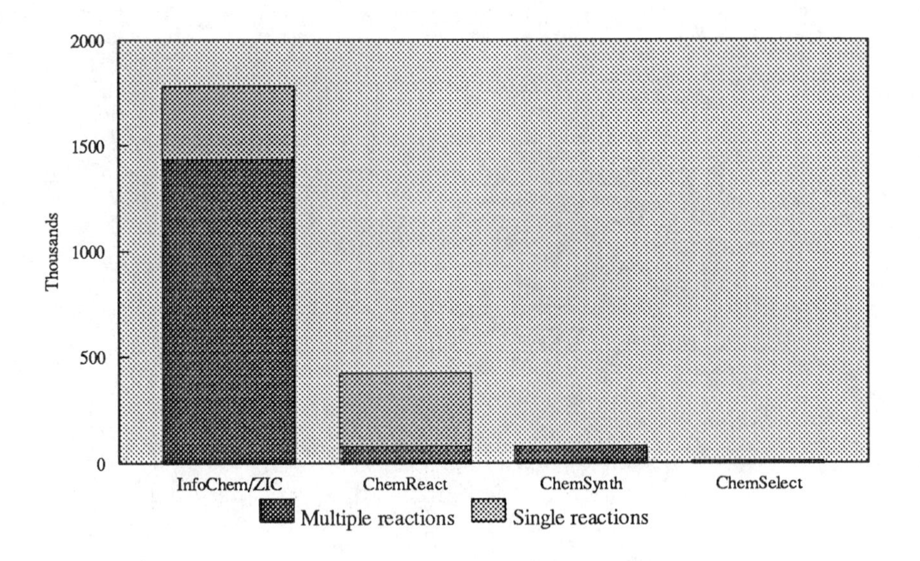

Figure 5: Relationship between databases

and applied synthesis methods for the period 1975–88. Both *ChemReact* and *ChemSynth* are available as loaded databases in *REACCS* and *ORAC* formats. Finally, InfoChem has produced a smaller subset, *ChemSelect,* by excluding from *ChemSynth* any reaction which has not been cited at least five times in five different journals, resulting in a database of 9,922 high quality reactions which is available in *ChemBase* format.

Figure 5 shows the relationship between these products schematically. The 334,142 singly-occurring reactions in InfoChem/ZIC are also included in *ChemReact.* The remaining almost 1.5 million reactions in the comprehensive database are represented by the 80,000 reactions in *ChemSynth.* Hence, as long as the relationship between these two files is preserved, a search of an 80,000 record database in *REACCS* or *ORAC* is equivalent to a search of the larger 1.5 million record file.

Using a compact form of storage, InfoChem has been able to reduce the 1.8 million record InfoChem/ZIC chemical reaction database to a small display file. This file and the corresponding software form part of OSIRIS (Organic Synthesis Investigation and Reaction Information System). The links between the selective database in *ORAC* or *REACCS* are the internal registration numbers displayed in the *ORAC* or *REACCS* record: ZIC reaction number, ZIC document number, and reaction type number. Using these internal registration numbers as search parameters in OSIRIS, it is possible to browse the database by retrieving all examples of a given reaction type in the literature or all reactions found in a particular document. However,

OSIRIS is more than a simple display system. It also contains a novel type of synthesis planning system consisting of a synthesis tree search of known chemical reactions reported in the literature. From a requested target compound, all reactions which produce the target compound directly are first identified and then the starting materials or precursors of these reactions are searched for, in turn, as the products of other reactions. Under control of the user, the shape of the synthesis tree can be controlled (eg maximum number of branches per step) and further levels in the synthesis tree can be explored. This application of the 1.8 million record file is independent of the use of *ORAC* or *REACCS*.

It should be clear that as the comprehensive InfoChem/ZIC database grows, the selective databases *ChemReact*, *ChemSynth* and *ChemSelect* will change, but not necessarily grow in proportion to the growth of the parent database. For any given reaction type, the new information added to the parent database may result in the selection of a newer and better example of the reaction type. New reaction types will be found in the latest literature but, as the database grows the ratio of the total number of reaction types to the total number of reactions and the proportion of new reaction types will both fall. Figures 6 and 7 (Loew, 1992) demonstrate how this ratio and the percentage of new reaction types has fallen during the process of

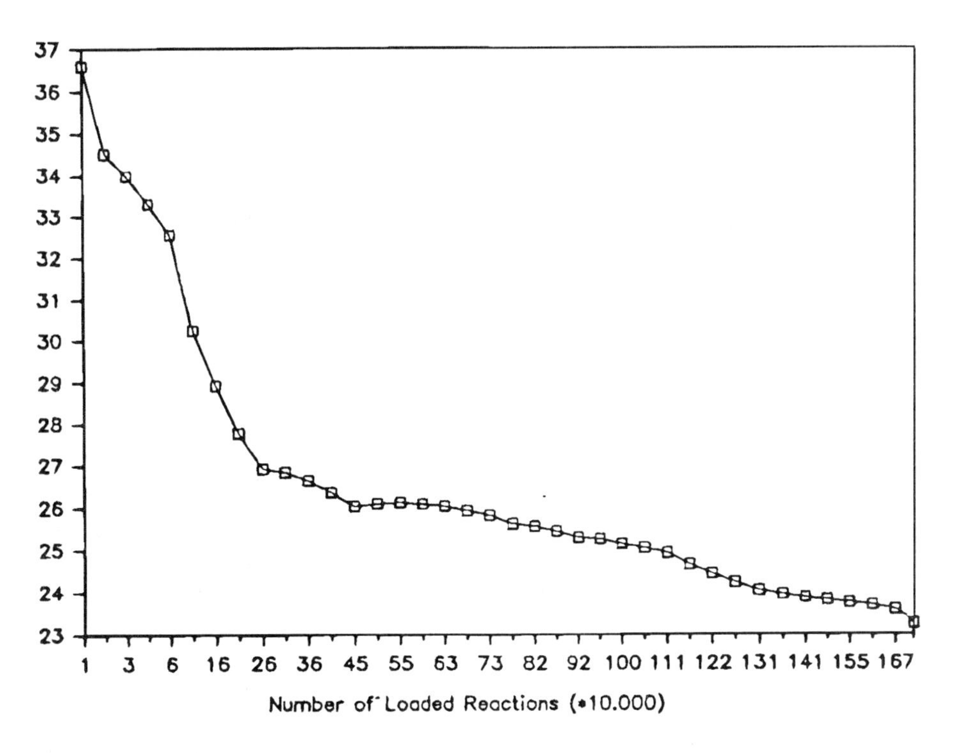

Figure 6: Ratio of the number of reaction types to the number of reactions

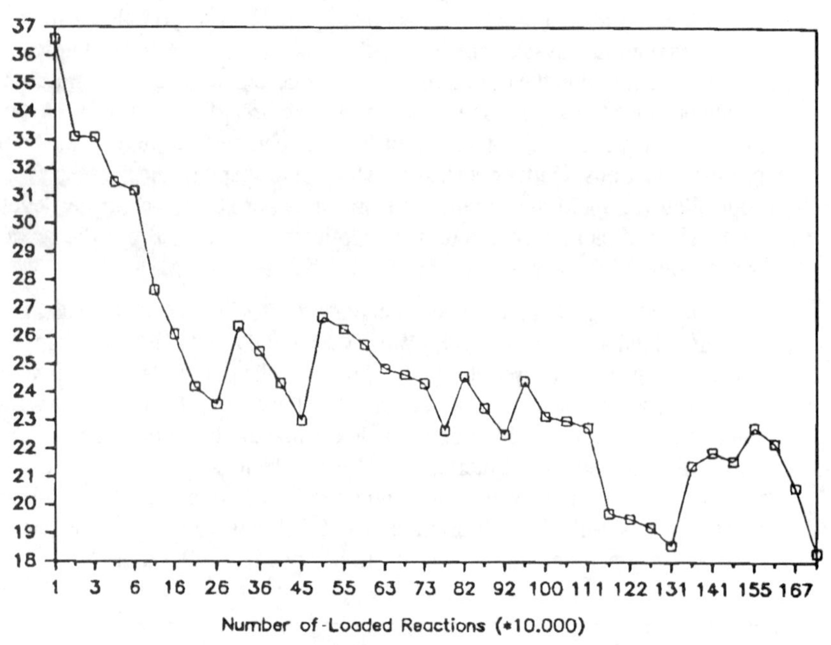

Figure 7: Percentage of new reaction types

building *ChemReact*. At the end of this process, the percentage of new reactions added had fallen to 18.3% (although this percentage does seem to oscillate around 20%). Using this figure and the estimated annual growth rate of 300,000 reactions/ year for InfoChem/ZIC, produces a growth rate of 54,900 reaction types/year for *ChemReact*. These growth rate figures and proportionate growth rates for *Chem-Synth* and *ChemSelect* were used in Figures 1 and 3. Until the family of databases has been brought up-to-date and arrangements have been made for regular updating, these growth rates are somewhat academic. Nevertheless, they demonstrate the principle of algorithmic selection and the possibility of applying similar techniques to other databases.

Novel data retrieval and exploration systems

While the focus of algorithmic data selection has been on improving the usefulness of existing information delivery systems, continuing improvements in the speed and power of computing hardware is making possible the development of new types of information system. Figures 8 and 9 show the trend in hardware over the last five years for central processor units and typical PC or workstation configurations.

Weininger (1992) has been arguing for a number of years that as the cost / performance graphs of storage technologies intersect, new types of information retrieval systems become cost effective. When the cost per byte of disk storage became cheap enough, we saw the transition from batch processing magnetic tape-based systems to online disk-based systems. Although the cost per byte of disk storage has continued to fall, the cost per byte of Random Access Memory (RAM) has fallen even faster. We have now reached a point where RAM-based information retrieval systems are cost effective. Because the access time for disks is of the order of 10 to 20 milliseconds, while for RAM access times are measured in nanoseconds, carefully designed RAM-based information systems can achieve retrieval speeds more than 100,000 times faster than disk-based systems.

Central processor units

	1987	1992
CISC chips	Intel 80386	Intel 80486
	Motorola MC68020	Motorola MC68040
RISC chips		IBM SGR 2564
		MIPS R4000
		Motorola M88000
		Sun SuperSPARC
Clock speeds	16 MHz	25 to 66 MHz
Processor speeds	1 to 4 MIPS	20 to 100 MIPS
		10 to 40 MFlops

Figure 8: Hardware developments

Typical PC or workstation configurations

	1987	1992
CPU	286	486SX, 486DX or RISC
RAM	1 to 4 Mbytes	4 to 96 Mbytes
Disk memory	20 to 100 Mbytes	80 to 2000 Mbytes
Display	EGA (640x350 pixels)	VGA (640x480 pixels) to
		XGA (1280x960 pixels)
		or 1280x1024 pixels

Figure 9: Hardware developments

In 1991, Daylight Chemical Information Systems (Daylight) introduced the first release of its *THOR* and *Merlin* products for UNIX workstations. The initial 4.1 release supported Sun workstations only, but, in subsequent releases, support for Silicon Graphics (SGI) IRIS workstations (v.4.2) and Digital VMS servers (v.4.3) has been added. The application programs have been developed using Daylight's own object-oriented toolkits and are based on client/server principles using TCP/IP to provide transparent interconnection of clients and servers. Both *THOR* and *Merlin* are based on the use of the compact SMILES notation (Weininger, 1988) to represent the chemical structure. Numerical and textual data are held in hierarchical databases in which the relationship between the identifier associated with the data and the data itself is preserved. *THOR* and *Merlin* provide alternate views of a *THOR* database: *THOR* is concerned with the microscopic view (display of all data for a single compound — single compound look-up) and *Merlin* provides a macroscopic view (data exploration of the whole database including substructure, superstructure and similarity searching). Both *THOR* and *Merlin* use RAM-based storage to achieve their remarkable response times.

THOR provides access to all data for a compound through any identifier stored in the *THOR* database. The principal method of access is through a hash table in which the unique SMILES for the compound is the primary key. Other identifiers are also hash coded but are cross-referred to the SMILES notation hash table. Using these techniques a corporate-sized database of, say, 250,000 compounds can be searched in RAM on a workstation with about 32 MBytes of RAM (today, most UNIX workstations have 16 MBytes RAM as standard). As there is normally only one disk access per search to retrieve the data, retrieval times of less than a second are common. In fact, provided there is sufficient RAM, the retrieval time is independent of database size — even the 12 million compound CAS Registry file would be searchable in less than a second given sufficient RAM. Already there are UNIX servers on the market with 640 MBytes of RAM — sufficient for multi-million record structure databases and, with the cost of RAM falling and the amount of RAM per chip increasing, we may expect to see this figure double or quadruple within a few years.

A typical *THOR* screen is shown in Figure 10. The *THOR* control window appears in the top left of the screen. A *THOR* data tree hierarchical display is shown in the bottom left corner. In the top right there is a structure list display (this could be a set of SMILES records saved from a previous search) which enables fast input of query structures. Bottom right contains the Grins structure editor with its tool menu overlaid. In the bottom centre is shown the trackball display which enables viewing of three-dimensional structures (eg CONCORD-generated coordinates or X-ray data). Cross database searching can be achieved by opening several databases and using a database selection window (not shown) to search the same compound across several files which may reside anywhere on the network — even in a different continent. In a later release, Daylight plans to add a virtual database facility to *THOR* which will allow the user to simultaneously search several physically distinct databases as if they were one database.

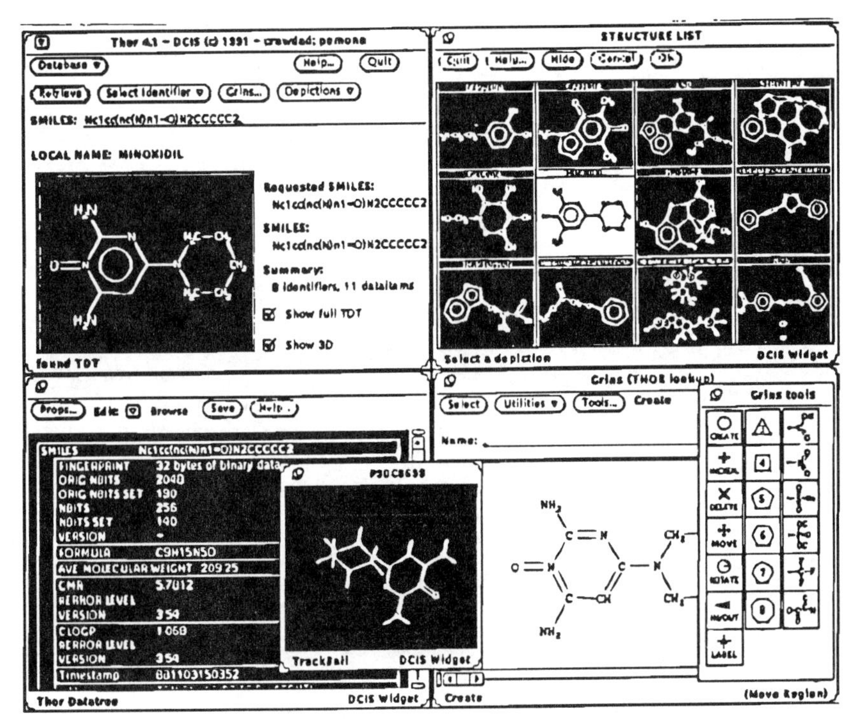

Figure 10: THOR screen

Merlin provides a spreadsheet-like view of the *THOR* database. Usually, not all data fields in the *THOR* database are selected for the *Merlin* pool which is held in memory. A typical *Merlin* screen is shown in Figure 11. The main *Merlin* window is shown in the top left corner of the screen. Each row in the table contains the selected data fields for a given compound and the current hit list (initially the whole database) is shown in the vertical direction. The resizable window is scrollable and, using the numeric keypad (hardware or software version), the user can page up or down through the list or jump to the beginning or end of it. By using the 'adjust' (middle) button on the mouse, data fields can be added or deleted from the display and column widths may be adjusted. An existing column may be redefined to display a different data type by selecting the desired data field from a drop-down menu at the head of the column. *Merlin* uses structure fingerprints to achieve search speeds of tens of thousands of compounds per second while searching by sub-structure or similarity (typical response times are five seconds for similarity search of a 300,000 compound database on a 32Mbyte SGI IRIS Indigo workstation). When combined with *Merlin*'s powerful string searching, sorting and hit list pruning techniques, data exploration of large files becomes easier than ever before. At any time the current view of the hitlist in the main window can be displayed in a hitlist snapshot (right hand side of the screen) and individual structures in the 'paned' snapshot window can be enlarged for easier viewing by clicking with the

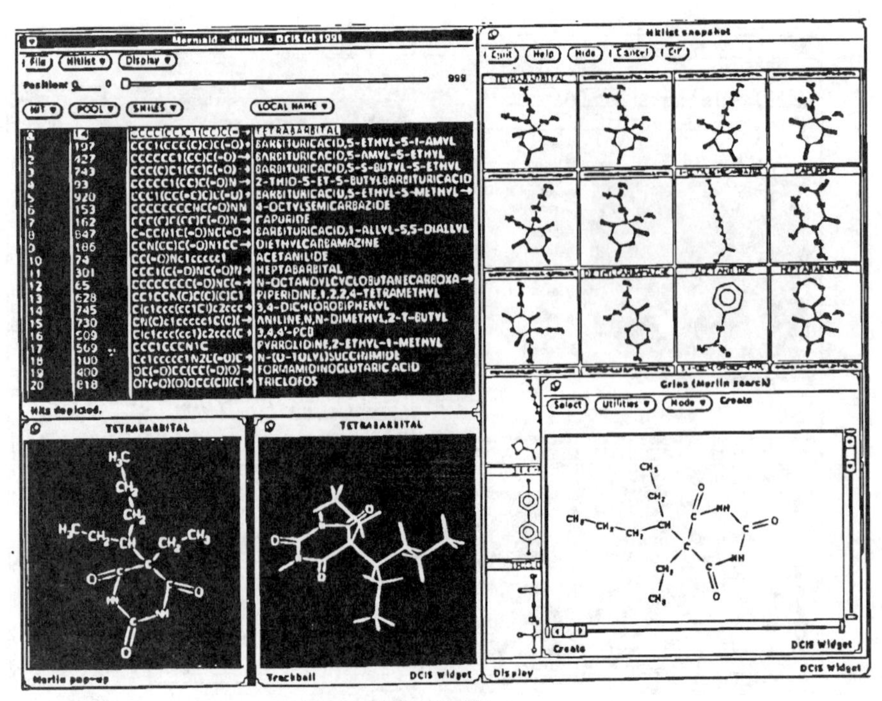

Figure 11: Merlin screen

adjust button. Finally, two- and three-dimensional views of any compound in the hitlist can be displayed (bottom left of the screen).

THOR and *Merlin* are redefining the way in which we will think about chemical information systems in the future. With the current technology it is possible to have the whole Beilstein database mounted on a *THOR* or *Merlin* server on a corporate network or even on an international network such as Internet to provide a new type of online host. The high bandwidths available on this type of network make it possible to search and explore remote databases as if they were resident on one's own workstation. Issues of pricing and security of such services have still to be faced but, providing these commercial questions can be resolved, I am confident that we will see a new generation of online system within the next few years. At present, UNIX workstations tend to be concentrated in the molecular modelling, drug design and spectroscopic groups within chemical and pharmaceutical companies. As the price of these workstations continues to drop, their use will become more widespread. In addition, the future porting of the client portions of *THOR* and *Merlin* to PCs and Macintoshes will make these chemical information systems available to scientists at their workbench with their current hardware.

Conclusions

Large chemical substance and chemical reaction databases will soon become more accessible to scientists through recent developments in hardware and software. Algorithmic data selection opens the possibility of multi-level search systems based on existing information delivery systems such as MACCS and *REACCS*. New generation chemical information systems such as *THOR* and *Merlin* offer more powerful methods for exploring large databases and open up the potential for new types of online hosts.

References

[1] Judson, P.N. 'Recent developments in reaction indexing', *Proceedings of the Montreux 1990 International Chemical Information Conference*, Springer Verlag, Heidelberg 1990.

[2] Zass, E. 'Large reaction databases', *Proceedings of the 15th International Online Information Meeting* , Learned Information, Oxford 1991.

[3] Zass, E. 'A user's view of chemical reaction information sources', *J. Chem. Inf. Comput. Sci.*, **1990**, *30*, 360-372.

[4] Anon. 'New structure search enhancements to CASREACT' *STN Technical Note 91/04* , STN International, Columbus 1991.

[5] Buntrock, R.E. 'Chemical reaction searching — revisited' *Database*, June **1992**, 106-108.

[6] Loew, P., Personal communication, 1992.

[7] Weininger, D. 'Software toolkits for chemical information systems' *Symposium on Software and Data Integration in Chemical Information: Standards, Software Toolkits and Distributed Systems, 203rd ACS National Meeting*, San Francisco, April 1992.

[8] Weininger, D. 'SMILES: a chemical language and information system 1. Introduction to methodology and encoding rules', *J. Chem. Inf. Comput. Sci.*, **1988**, *28*, 31-36. Figure 6: Ratio of the number of reaction types to the number of reactions

A computer system for inputting generic reactions and sequences to produce databases and printed publications

A. Parlow and Ch. Weiske [1] and N. Heß-Pohl, G. Roden, J. Roemelt and C. Zirz [2]

[1] FIZ CHEMIE, Berlin, Germany, [2] BAYER AG, Leverkusen, Germany

Introduction

With the development of reaction databases and retrieval systems, chemists in research and production have become more and more accustomed to such systems. The demand for high-quality reaction databases is growing worldwide. Database users normally need answers to their synthetical problems that satisfy their questions quickly and precisely, preferably by an impressive example; they do not want to be flooded with hundreds of variations of the same reaction.

For producers of chemical information systems, in printed or electronic forms, the major task — and the main cost factor, too — is the intellectual selection and evaluation of the primary literature. From some hundreds of thousand of reactions reported annually in the scientific literature, for instance, only about 60,000 are selected for the ChemInformRX, and from those 8,000 key reactions are selected for the CSM (Current Synthetic Methodology) database. This selection is based on rigid criteria like novelty and applicability, and this is a process that can only be done by experts. On the other hand, it is also not affordable to input each published reaction and to diminish the number of reactions by computer afterwards!

Having the labour cost as the main cost burden, it is essential to get the production process of information systems as economic as possible, otherwise the more or less limited market could not cover the costs. This means looking for solutions which promise a reasonable ratio between expenses and revenue.

The ChemInform traces its roots from the 'Chemisches Zentralblatt' and the 'BAYER-Fortschrittsbericht'. Its predecessors and ChemInform itself have been printed publications for many years. Due to the lack of search capabilities, its carefully produced, high-quality information has been a current awareness service only like a newspaper. To make long-term use of the most valuable information, a research project had been established in 1988 with the generous financial support of the German Ministry of Research and Technology to build up a production system which permits to produce simultaneously the printed information service and a computer-readable reaction database [1]. This project has been accomplished successfully recently. One of the main components of the new production system has been the development of a new module for handling the reaction diagrams appearing in the printed issues. This module was developed by a group of chemists and software specialists at BAYER AG in Leverkusen.

According to the aforesaid, the production procedure should be as economic as possible. This means that it is an indispensable condition to input all needed information only *once*, and to derive from one source the different products — the printed publication including computer-generated indexes and other printed information, *and* the reaction database.

The requirements for a printed publication like ChemInform and a reaction database are in fact quite different: A database that shall be applied with the existing reaction retrieval systems must contain individual reactions with fully defined reactants and products. The easy-to-grasp graphical schemes appearing in ChemInform diagrams receive their value from the Markush-like combination of reaction variations and from the impressive description of multi-step reaction sequences. A system to be used for both product lines must have enough flexibility and functionality to satisfy all those requirements. In the market, systems are available for storing reaction information and for producing printouts [2], but no system existed up to now to allow generic sequences and the necessary flexibility in data handling.

Description of the system

The input module developed for the production of ChemInform has been named CAESAR, which is an acronym for 'Computer-Assistierte Eingabe und Speicherung Allgemeiner Reaktionen' (Computer assisted input and storage of general reactions) [3]. In contrast to other known systems, the geometrical position of a data entry is not pre-determined by means of fixed screens. The role of a data item is instead defined by an additional interactive marking. This allows a totally free and flexible graphical orientation of the entered items, and also a quick and easy change of the data structure.

The main architectural component of the system is the datafield description. It is submitted to the system as a readable text file allowing easy changes of the processed data types. The datafield description consists of the datafield names, their characterisations, and their hierarchical dependencies and relations. A graphical representation of the datafield description used for ChemInform is given in Fig. 1. It will be discussed in detail later. Data types that can be handled are character strings including formulae and numbers, or structures represented as connection tables.

The link between the datafield description and the keyboarder is the main menu (Fig. 2). To the right of the drawing area, the names of the datafields and their hierarchy are depicted. By activating a datafield, either by cursor movement or sequentially, the correspondence between a datafield or record and the subsequent input is fixed. The data item itself can then be placed anywhere on the drawing area according to the requirements of the ChemInform scheme. Corrections and modific-

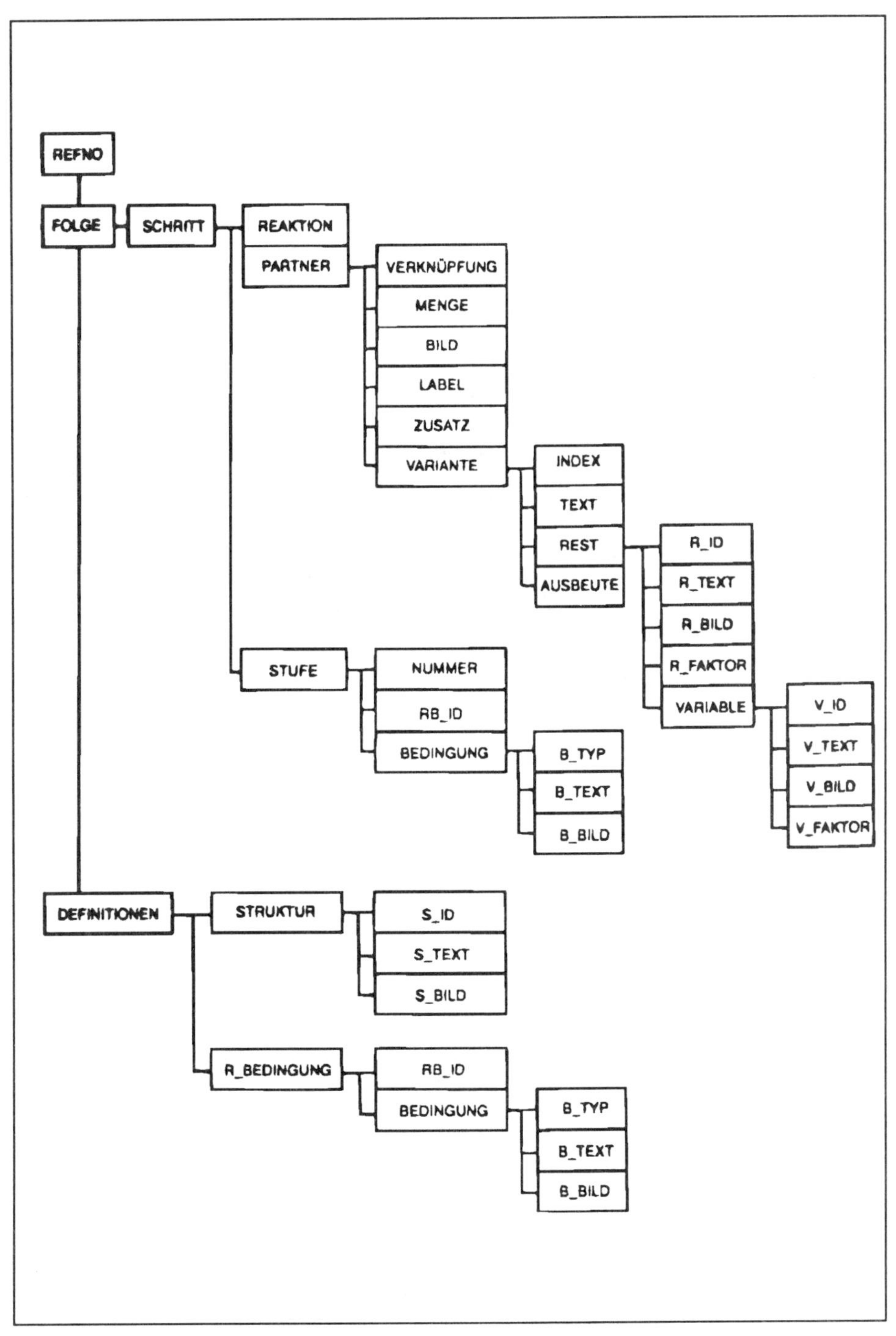

Figure 1: ChemInform datafield description

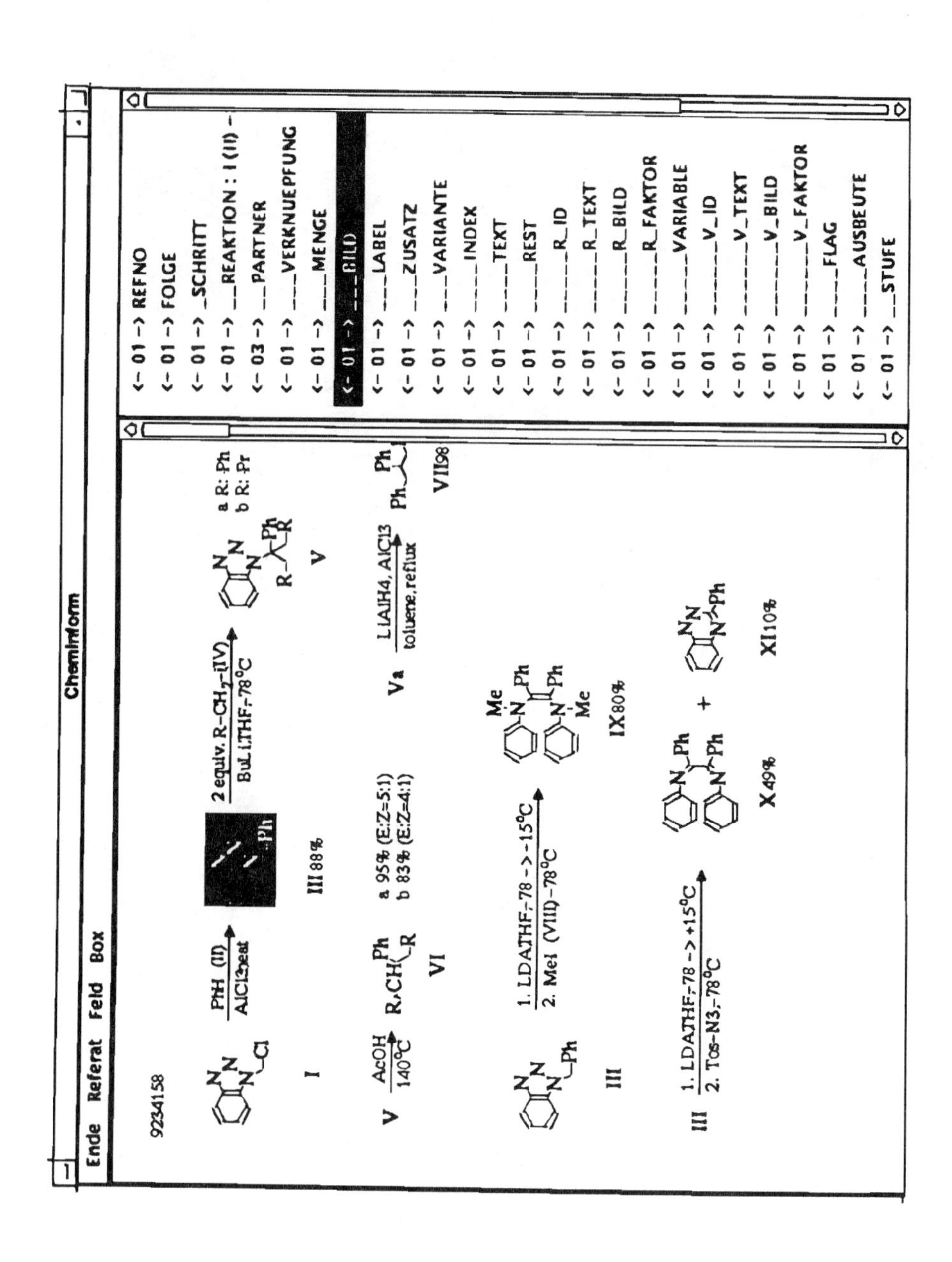

Figure 2: CAESAR main menu

ations can be made in the same way; the relation between field and content is always retained.

If a datafield of type 'structure' is chosen, the structure editor is invoked. This structure editor is an improvement of the editor used in the BAYER system RESY [4] for storage and retrieval of chemical information. It normally uses drawing menus and can additionally be controlled by special shortcut keystrokes to improve input speed. It is capable of handling generic substituents (in several nesting levels) and link groups like -$(CH_2)_7$-. It understands frequently-used abbreviations like 'tBu' and also molecule fragments entered in line formula form like '-CH_2COOH' and resolves them into their connection table. A valence check is included, too. It can handle coordinative bonds or hydrogen bonds (Fig. 3). Stereochemistry is implemented also for centres with coordination numbers greater than 4. In summary, this structure editor is, in its capabilities and functionalities, without parallel worldwide.

After entering the complete information for a ChemInform scheme it is output in the widely used SMD format [5] retaining the hierarchical data structure that is derived from the datafield description. The DDS block is used for this purpose.

Subsequent programs, which have been developed mainly by CHEMODATA in Munich, are used to interpret the SMD files, especially to split up the multi-step sequences into individual steps and to resolve generic variations into the fully defined reactions. Other programs calculate additional data and perform the nec-

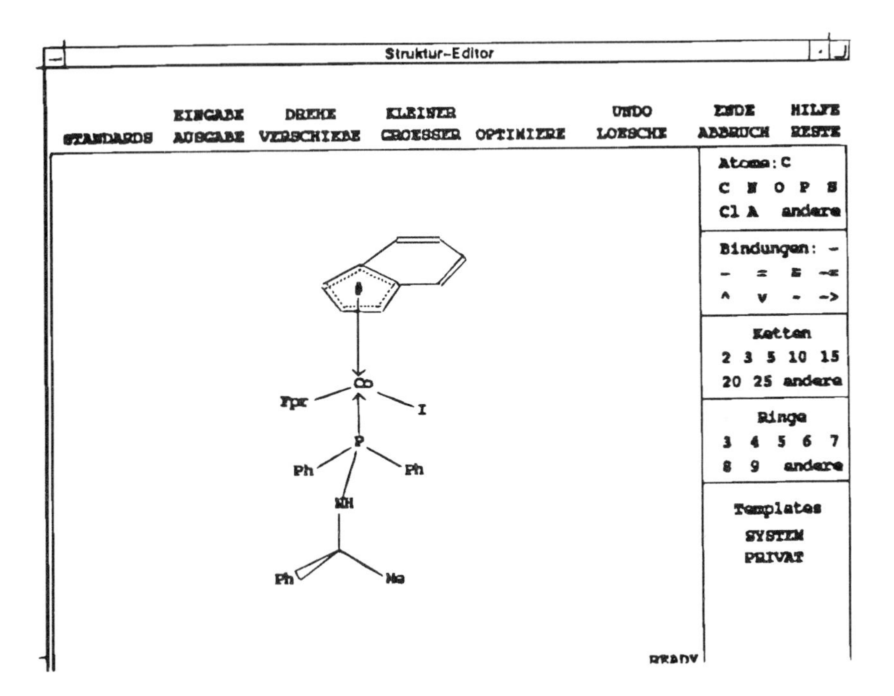

Figure 3: CAESAR structure editor

essary tasks to convert the reaction information into the formats used by the commercial retrieval systems.

The refinement of the graphical scheme and the preparation for the typesetting process for the printout is also accomplished by a special program developed by the company GTS-GRAL in Darmstadt. This program uses only the information that is contained in the SMD file. But it allows for interactive manipulations of the graphical scheme which are sometimes necessary if one wants to include more graphical items in the scheme than are used for the database.

Thus, the SMD file contains in principle all the information that is necessary for both the printed publication and the database.

The ChemInform datafield description

As aforesaid, it is an essential feature of the ChemInform to contain comprehensive and also easy-to-grasp schemes of the synthesis tree of a publication. The reader gains a quick overview of the chemistry dealt with in the publication, a synopsis on the chemical environment. This compactness has been a great challenge for the flexibility and power of the input program. The datafield description has been developed in close cooperation with the editorial offices in Leverkusen and Berlin and with the programming team at CHEMODATA in Munich. It has been improved and adopted to the production process in several cycles. Its hierarchical and manifold subdivided structure will be described in more detail in the following.

The basic information unit on the uppermost level is the abstract. Its identification number REFNO is used as the main access key and also for combining the textual information (bibliography, abstract, system numbers), which is input by text processing systems separately, with the graphic part.

An individual abstract contains sequences (FOLGE) and possibly definitions (DEFINITIONEN), which refer to the whole abstract.

A sequence (FOLGE) is the reaction pathway from the starting materials to the end products of a synthesis procedure. Several sequences may be contained in an abstract (Fig. 4).

A step (SCHRITT) contains all the information that corresponds to a single reaction arrow. A symbolic description of the reacting partners via their labels is keyboarded in the subfield REAKTION as an auxiliary field for the interpreting programs. The reacting molecules are denoted as PARTNERs. Additional information for the reaction, e.g. conditions, are contained in a record for the reaction stage (STUFE).

The PARTNER record is filled with the data for a reacting or produced structure. These are: its label (LABEL), its image or structure (BILD), its connection (VERKNÜPFUNG), and, if necessary, its amount (MENGE), a sub-record VARIANTE and additional data (ZUSATZ). A connection may be a plus sign for the partners on one side of the arrow, or one of the different kinds of arrows that are allowed in the graphical scheme. A LABEL is obligatory for a PARTNER and contains the Roman numeral of the structure, possibly extended by stereochemical information or indices for different variations of the generic structure.

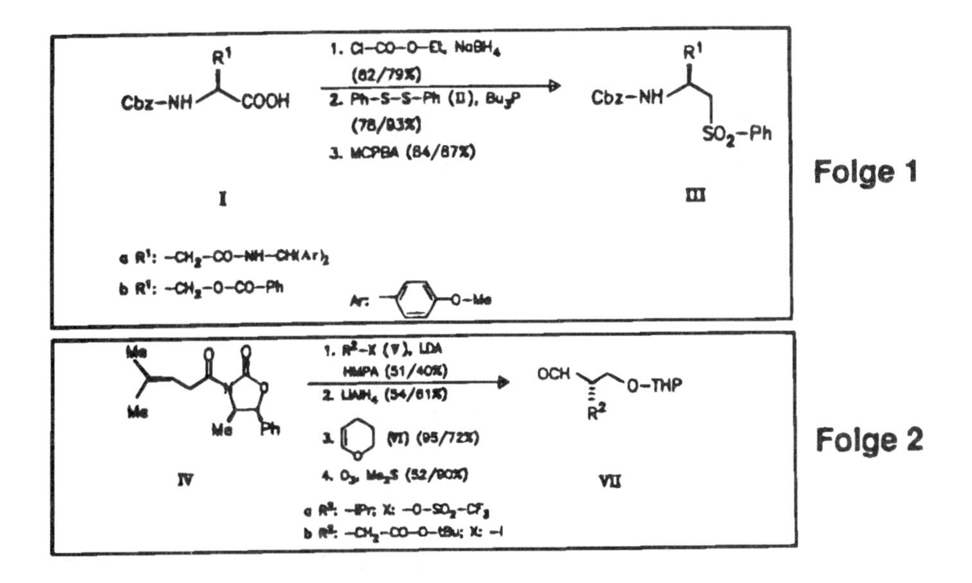

Figure 4: Sequences in a ChemInform abstract

The variation record (VARIANTE) contains the different variations of a generic partner. It is identified by an INDEX and defined by the structural residue(s) (REST) and may additionally contain textual information (TEXT) and information on the yield (AUSBEUTE) (Fig. 5).

The REST record describes the individual residue(s). It contains an identifier (R_ID) and the structural definition (R_BILD) and, if necessary, textual additions (R_TEXT) and multiplication factors (R_FAKTOR) to allow items like -$(CH_2)_n$-. If the residue itself contains variable elements, an additional record VARIABLE is used, which contains the analogous datafields in a lower level.

The stage record (STUFE) consists of the information on reaction conditions for a single reactions step (Fig. 6). It has a number (NUMMER) and several condition

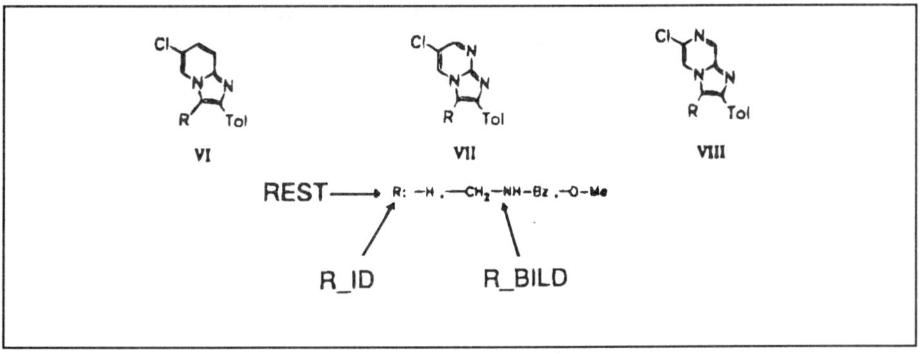

VI VII VIII

REST → R: –H, –CH2–NH–Bz, –O–Me

R_ID R_BILD

Figure 5: REST datafields

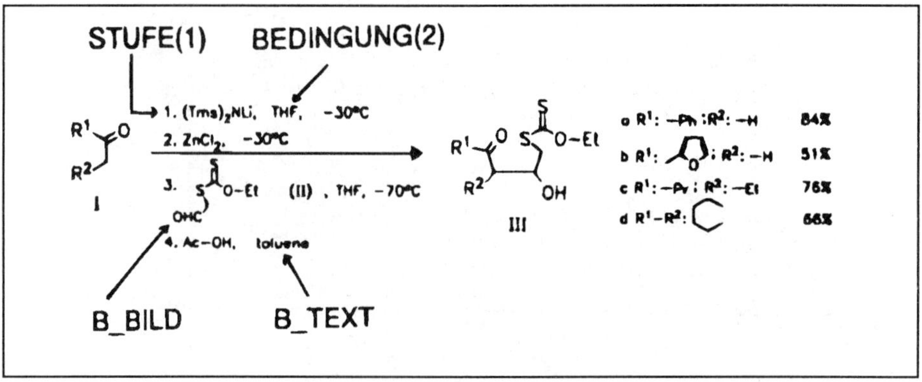

Figure 6: STUFF datafields

records (BEDINGUNG). They in turn contain structures (B_BILD) or textual information (B_TEXT) and a type-descriptor (B_TYP), by which the role of the individual condition, e.g. catalyst, solvent, temperature or others, is defined.

Finally, a definition record (DEFINITIONEN) allows for declarations which are valid for the whole abstract. It may contain structures or structure fragments and reaction conditions; its subfields are analogous to the corresponding main fields.

Conclusion

The input system CAESAR has been used for the production of the ChemInform database and the printed issue routinely and successfully since 1991. All reaction types which can be handled at present by the existing systems can be processed quickly and efficiently. Structures which cannot be described by the common connection tables (e.g. fractional or undefined indices) cannot be handled. On the other hand, for special bond or structure types such as in pi complexes, more information can be input and stored than can be made use of by the existing retrieval systems. But the most innovative development of this system is the ability to input and store generic reaction sequences.

The production of both, a database and a printed issue of high quality, together in one single, highly efficient input process is accomplished by this program, which can easily be adopted to any other application due to the changeable datafield description. Last but not least, it should be mentioned that CAESAR is offered by FIZ CHEMIE.

References

[1] ChemInform — An Integrated Information System on Chemical Reactions; J. Gasteiger, C. Weiske, A. Parlow, *J. Chem. Inf. Comput. Sci.* **30** (1990) 400-402

[2] e.g. CPSS by Molecular Design Ltd., CENTRUM by Polygen

[3] CAESAR — Ein neuartiges System zur Computer Assistierten Eingabe und Speicherung Allgemeiner Reaktionen am Beispiel des ChemInform; W. T. Donner, N. Heß-Pohl, G. Roden, J. Römelt, A. Wagner, C. Zirz, *Mitteilungsbl. GDCh-Fachgr. CIC* **18** (1991) 31-50

[4] CAS Online End-User Searches via Bayer Inhouse System; W. T. Donner, C. Zirz, *Proceedings of 13th International Online Information Meeting* (1989), Learned Information, Oxford, 71-75

[5] The Standard Molecular Data (SMD) Format; H. Bebak *et al., J. Chem. Inf. Comput. Sci.* **29** (1989) 1-5

Draft Specification for Revised Version of the Standard Molecular Data (SMD) Format; J. Barnard, *J. Chem. Inf. Comput. Sci.* **30** (1990) 81-96

Spectral database systems: a comparison

Wendy A. Warr

Wendy Warr & Associates, 6 Berwick Court, Holmes Chapel, Cheshire CW4 7HZ, England

Introduction

Very many research papers have appeared in the field of Computer Aided Structure Elucidation (CASE — not to be confused with a well-known CASE system that is actually named CASE [1]). Gray has written an excellent recent review of the discipline [2]. Small [3] earlier wrote a general overview.

Structure elucidation, or the identification of an unknown compound, involves three steps. Firstly, the known biological, chemical and spectral data are used to derive structural constraints, that is, to determine chemical substructures that are consistent with the known facts. All possible structures for the unknown are then generated by a combinatorial process: the various substructures consistent with the spectral and other data are combined in appropriate ways that are still consistent with the data. Finally the candidate structures are assessed and further work is carried out to determine which one structure is the most feasible.

Computers can be used to assist in various stages of the structure elucidation process. One technique is file search, or library search, in which the spectrum of an unknown is compared with a library or database of the structures and spectra of known compounds. In practice it is unlikely that an exact match would be found because of the problems of exact replication of spectra and because the database is likely to contain too few compounds. File search is therefore based on similarity techniques. The output of a search comprises those spectra which are most similar to that of the unknown compound, together with an indication of the degree of similarity in each case. If an exceptionally similar spectrum is found, an exact match may be assumed; if similarity is much less, possible chemical substructures can be determined.

Substructure detection may involve file search, pattern recognition (statistical) techniques or rule matching methods (also referrred to as decision systems or artificial intelligence techniques). A large database of high quality spectra for a broad range of compounds is needed if the system is to be useful and reliable. The reader is referred to other sources [4,5] for listings of available databases. Heller has recently written a useful summary [6].

Work on file search systems has fallen into four areas: development of similarity metrics, establishment of quality indices for the spectral data, devising techniques for speeding up searches and the establishment of databanks of high quality data. Many problems have been encountered in building the necessary databases [4,6-10]. Pretsch *et al.* [7] have summarised the requirements of spectral databases.

Mass spectral databases were amongst the earliest to be built, since the data is relatively easy to computerise [10]. Carbon-13 NMR spectra can also be stored relatively easily as position/intensity pairs and database building began many years ago. Assignments (the relating of each spectral peak to the precise carbon atom that gives rise to that peak) can be entered more easily in modern systems. C-13 NMR spectra have the advantage that a signal is guaranteed in the spectrum for every carbon atom in the molecule. This one-to-one relationship makes C-13 NMR databases particularly useful for deducing the structure of an unknown compound from its spectrum. Less reliably, the C-13 NMR spectrum of a given structure can be predicted using a C-13 NMR spectral database. Systems for prediction of mass spectra and infrared spectra are less well developed.

Gray [2] discusses spectral prediction as a CASE tool. After reviewing algorithms that have been used in combinatorial searches and structure generators he discusses the evaluation of the generated candidate structures. He does not recommend the use of spectral prediction alone for exclusion of candidates but it *is* useful for the ranking of structures.

The C-13 NMR spectrum of a candidate can be predicted by identifying the structural environment (preferably the stereochemical environment) of each carbon atom and looking up that environment in a highly detailed table relating substructures to chemical shift ranges. It is best if the method of substructure representation used allows the environment of a carbon atom to be described with details of all atoms out to a radius of at least four bonds, since shifts can be very sensitive to the relative positions of substituents on rings, for example. A five-bond substructural model means that a substituent at a *para* position can be taken into account. A good system will report a measure of the quality of the prediction and indicate the substructures used.

The three most commonly used approaches in predicting C-13 NMR spectra are database retrieval methods [9,11-15], linear additivity relationships [16-18] and empirical modelling techniques [19,20], latterly using neural networks [19].

A previous paper [4] summarised some of the challenges facing the builders of spectral databases (handling chemical structural, numeric, and textual data; quality control; funding and motivation; and human factors) and described several specific systems. The current paper covers only those spectral database systems which permit graphics chemical structure display or chemical substructure searching. Two systems, SpecInfo and CSEARCH, are selected for more detailed comparison.

Chemical structure handling on computers

A recent book has covered this subject in detail [21]. Graphics systems for handling chemical structures have become commonplace, with 2D chemical structure entry systems using a mouse. Foremost amongst these are MACCS, ChemBase and ISIS from Molecular Design Ltd (MDL) for handling in-house databases; and systems for accessing the CAS ONLINE databases on STN International. One 'front-end' to assist in the latter is STN Express. The structure drawing module for STN Express is PsiBase, which is also used in the Sadtler spectral system PC SEARCH.

Recently systems (such as MACCS-3D) for three-dimensional substructure searching have started to appear.

Links between spectral database systems and the commonest substructure searching systems

Analytical chemists have a spectra-oriented approach rather than a structure-centred one and some of them see no need for substructure or exact structure search. There are several other reasons, both human and technical, why systems such as MACCS have not been adequately linked to spectral database systems such as Sadtler's CSEARCH or Chemical Concepts' SpecInfo. A number of organisations, however have built proprietary systems for storing their spectral data within ChemBase or MACCS databases. These systems do not, however, allow sophisticated spectral library searching.

Spectral database systems have not yet caught up with developments in the field of three-dimensional substructure searching.

Some PC-based software

Sprouse Scientific's Quick-Search software allows users to create and search libraries of IR spectra and display, but not search, chemical structures. The PC version of the NIST/EPA/NIH mass spectral database, in its fourth version, contains 63,000 mass spectra, almost all of them with associated chemical structures, and numerous search facilities (but not substructure search) are possible. Preston Scientific's NMR database contains over 10,000 F-19 spectra and N-15 and P-31 modules are available too. There is a substructure search system related to this database but plans for supporting it are in state of flux. Sadtler's PC spectral search libraries are the most sophisticated of the range of PC products. Search of 160,000 IR spectra (about 80,000 with chemical structures), and also 34,000 C-NMR spectra (with structures) is possible in a system under Microsoft Windows that links with PsiBase under Windows for substructure searching.

Online, workstation or VAX-based systems

Databases on the Chemical Information System (CIS)

There are three relevant databases available online on the CIS [10,22]. CNMR contains C-13 NMR spectra for over 11,000 compounds. It can be searched by chemical shift. IRSS is an infrared search system, with data from EPA and from the Boris Kidric Institute in Yugoslavia, allowing retrieval of known spectra and analysis of unknowns. WMSSS, the Wiley Mass Spectral Search System, also incorporates the NIST/EPA/NIH database. The structures of all compounds on the CIS are substructure searchable using the Structure and Nomenclature Search System (SANSS) [23].

SpecInfo in-house version

This is a structure-related spectroscopic database system for VAX/VMS computers in-house or for use online. It was written by Bremser and co-workers [11-14] at BASF starting in the 1970s. Development is now a co-operative effort of BASF and Chemical Concepts in Weinheim (who market the system) [24,25]. Not only is SpecInfo multidimensional (that is, it handles more than one type of spectral technique) but it also incorporates chemical structure handling software.

Together with chemical structures the following data are included:

82,400	C-NMR
14,700	11-B, 15-N, 17-O 19-F and 31-P NMR
19,900	IR (14,300 full spectra; 5,600 additional peak lists)
150,000	MS

Quality control is carried out in close co-operation with BASF, the Max- Planck-Institut für Kohlenforschung in Mülheim, the Institute for Spectrochemistry and Applied Spectroscopy in Dortmund and various industrial experts. Further program development is done with TU Munich, TU Vienna, Toyohashi University, Berg-akadamie Freiburg and others. The project is supported by the German Federal Ministry of Research and Technology [BMFT].

The following searches are possible: spectral identity, spectral similarity, presence of spectral lines (variable number of missing lines), IR bands, mass peaks and neutral loss reactions, full and partial structure, similar structure, systematic CA name or partial name, CA Registry number, and ranges of molecular formula and molecular weight.

Hit lists resulting from spectral similarity searches are analysed statistically for structural information. Relevant partial structures can be extracted automatically, yielding important clues to the structure of an unknown compound. NMR spectra of arbitrary structures can be simulated. Coupling constants of arbitrary structures are estimated in a manner analogous to the chemical shift prediction.

The in-house system uses colour graphic displays. At ICI there are many VT340 users and Tektronix 4105 is a graphics standard. There are about 105 users in ICI, in an essentially client-server environment [26].

Full connection tables are stored for each compound. Similarity searching is done by functional group encoding. The system is based on Bremser's HOSE (Hierarchically Ordered Spherical Description of Environment) codes or HORD (Hierarchically Ordered Ring Description) ring codes. Until recently these codes were stored in inverted files but in the recent Version 2 of in-house SpecInfo, data storage and retrieval is based on the relational database management sytem SYBASE, and various indexes support fast data retrieval. In the in-house system, structure queries are entered by naming various fragments and giving numbers for ring positions. The input is textual but the structure is displayed graphically. (The online version, discussed later, has the advantage that structures can built offline graphically, using a mouse, with STN Express or built online using the STN structure menu.) In the in-house version of SpecInfo, structure handling is therefore not as user-friendly or sophisticated as in, say, MACCS, but MOLfile input and output is supported. (MOLfile is MDL's connection table standard which was recently made public [27].) An ISIS link is planned.

Spectral data can be displayed both graphically and as a listing.

Chemical Concepts plan to incorporate MassLib (see below) into SpecInfo in the near future and then carry out the following enhancements:

- integrate proton NMR storage, classification and search algorithms developed at BASF.
- add stereochemical structure display
- incorporate prediction of MS peaks (developed at TU Munich)
- produce an X-Windows interface.

BASF, Chemical Concepts, Toyohashi and Sumitomo have an R&D contract attempting to automate structure elucidation completely, making use of the structure generator CHEMICS [28,29] which is linked to SpecInfo as an independent module.

Varmuza's EDAS software has recently been linked to SpecInfo as a separate module to allow the statistical analysis of SpecInfo hitlists using pattern recognition techniques [30].

In the autumn of 1992 Chemical Concepts and collaborators are beginning several new development projects. The HOSE code will be extended. Stereochemistry descriptors and 3D coordinates will eventually be added. (3D structures will be generated from 2D connection tables using a rule-based program ALCOGEN.) The SpecTree project at BASF aims to automate data flow, store raw data, apply barcode systems for information input and define standards concerning exchange formats.

SpecInfo online

SpecInfo online on STN International in December 1991 incorporated 88,300 NMR spectra and 16,700 IR spectra but no mass spectra. Mass spectra will hopefully be available in December 1992. Three data generation packages are available:

- SPECAL for estimating NMR spectroscopic information for a query structure
- COUPCAL for estimating coupling constants for a query structure
- CHESS to search for chemical structures identical to a query structure or to search for similar structures using HOSE and HORD codes.

Structures are input using a slightly modified version of the Messenger Structure command, or in STN Express (Version 3.1 is required). Structure searching is done using the RUN CHESS command to execute a structure code similarity search. Various search options are available, e.g., structure identity (SI) and structure functionality (SF). The familiar Messenger substructure search system is not used.

The online version is in some ways less user-friendly than the in-house one and unfortunately mass spectra will not be available until December 1992. An online system is also not ideal for organisations which handle a lot of novel compounds and wish to integrate in-house and publicly available spectra.

SpecInfo Online replaces the C13NMR/IR database on STN which will be withdrawn at the end of October 1992.

MassLib

This set of programs was built by Henneberg [31,32] at the Max-Planck-Institut für Kohlenforschung, for the evaluation of low resolution mass spectra. The program can be used on VAX/VMS, ULTRIX or HP UX computers. It is now being marketed by Chemical Concepts but Henneberg *et al.* are still writing software for MassLib. The original system implemented both Wiley and NIST mass spectral databases and allowed a user to maintain his own library in addition. Structures could be displayed but not searched. Search for identical and similar structures is now possible. The mass spectrum similarity search in MassLib is better than that in SpecInfo [26], so ICI makes both MassLib and SpecInfo open to ICI users. As mentioned above, Chemical Concepts plans to incorporate the MassLib SISCOM (Search for Identical and Similar Components) algorithm in SpecInfo.

CSEARCH

This system for handling C-13 NMR spectra was developed by Robien and coworkers at the University of Vienna [33,34] and Robien is still incorporating new algorithms [35,36] while Sadtler commercialises the system. The database comprises 27,000 of Robien's literature abstracted spectra and 33,500 fully assigned spectra from Sadtler: 60,500 in all. Sadtler are negotiating to add new data (i.e., multinuclear spectra) to the system.

CSEARCH runs under X-Windows on the Silicon Graphics family of workstations, Sun SPARCstation, Sun 4, VAX and VAXstation. The user-friendly colour graphics interface is a particular advantage of this system. Stereochemical display is due to be added in December 1992.

The system is used at KaliChemie, Shell, Schering, Hoechst, Bayer, Neste and AGFA and at least four other companies have it on trial.

With the tools provided a user can:

- read spectrometer generated peak lists directly
- search spectra for identical chemical shift patterns
- request all functional groups for a chemical shift
- request chemical shift range for a functional group
- search by either full structure or substructures
- combine chemical shift with structure searches
- retrieve data based on author, source or properties
- estimate spectra from a structure proposal.

CSEARCH allows a spectroscopist to create his own database. Password protection at 9 levels, automatic data checking, and assignment comparisons are featured.

Before a search, the user chooses a database or databases. He searches multiple databases as if they were a contiguous set. Structure input is by the Robien line notation (Line Note), SMD, JCAMP-CS or MOLfile. The user can restrict his search by ringsize, molecular formula, homologous series, or molecular weight, or choose no restriction at all. The initial screen search uses three-atom fragments, molecular formula, and number of rings. Structures and hitlists can be stored. After a search the user has four choices of display: structures, spectra, structures and spectra, or datasheet.

Moving the cursor to point to any carbon atom in the structure will result in an unambiguous identification. A box is drawn around the corresponding carbon atom(s) in the assignment list. A prominent triangle points to the line(s) in the simulated C-13 spectrum which correspond(s) to the chemical shift position(s) of the selected carbon atom(s).

If a peak list is entered, the results of an identity/similarity search are a simulated spectrum of the exact compound if it is in the database and other spectra that are the nearest matches. Structures that correspond to these spectra can be displayed four to a screen.

Like SpecInfo, CSEARCH uses Bremser's HOSE and HORD codes but this is not obvious to the user. When a structure is displayed, CSEARCH explains what substructures it used in finding/predicting that structure by highlighting them within the display of the complete structure.

Data Entry

It is important for companies and analytical chemists to be able to incorporate their own data into a commercially available database system. This both increases the size of the database, which has general benefits, and introduces different structures that might be particularly relevant locally.

Unfortunately both CSEARCH and SpecInfo have defects in their data entry software. Originally, both had user-hostile PC data entry modules. This has now changed. ICI has discontinued use of PC data entry and VAX entry is the only option at present. Shell report that there is now a 'part-graphical' data input procedure for the X-Windows version of CSEARCH. One important advantage of CSEARCH is the point and click interface which provides an easy assignment method for customers to record unique interpretations into CSEARCH databases.

Neither system has a proper molecule editor, although there are interfaces to MDL and other software, as already mentioned, and Sadtler have progressed well with the development of a molecule editor. Both systems can upload ChemBase databases by means of an SD-file interface.

The SpecTree project described above will provide major enhancements to SpecInfo in the future.

ICI have developed a prototype interface which takes data from the JEOL NMR spectrometer and associates it with atoms in MOLfiles for input to the SpecInfo-based DAMIT system. Chemical Concepts say that NMR peaklists can be imported from Bruker, Varian or JEOL spectrometers, IR spectra in JCAMP-DX format and mass spectra as EPA or JCAMP-DX files. Special routines are then used to check the format of the input spectra and compare NMR spectra with calculated ones to detect errors and wrong peak assignments. CSEARCH also has data validation routines, as described above. Data can be imported in Varian (old and new), Bruker (old and new) and FELIX compatible form. Robien's development version has NMRI, WinNMR and JEOL interfaces also. Data capture from instruments is obviously very important but problems with standards hinder fast progress. There are problems with IR file standards and there is a lack of an international standard for NMR and mass spectra data files.

Comparison of CSEARCH and SpecInfo

It might be claimed that some of the software in both systems was written to 'academic' standards rather than to professional software engineering standards. If so, this situation is gradually changing as increasing commercialisation takes place. Shell have said that the CSEARCH software is well-structured.

Neither system has a 'proper' molecule editor. (Sadtler are known to be working on one.) Shell generate SMD files using a separate molecule editor and enter structures using Robien's line notation. SpecInfo now supports MOLfile input and output. Sadtler are also now offering a MOLfile interface for CSEARCH. For some companies even closer links with MACCS would be desirable and it remains to be seen what Chemical Concepts and Sadtler will negotiate with MDL.

Once ChemDraw Plus is available for use under X-Windows, ChemDraw users will be able to use this to generate MOLfiles for input to CSEARCH.

SpecInfo has the advantage of close links with Chemical Abstracts because the database is on STN International.

SpecInfo in-house offers multinuclear NMR, IR and mass spectra, CSEARCH only C-13 NMR. Sadtler are planning to introduce a product for use with CSEARCH, for searching IR spectra and structures in January 1993. Chemical Concepts are planning eventually to introduce a tool for the prediction of IR spectra (based on earlier work by Bremser *et al.* [37]) but there is a long way to go yet. A new BMFT-funded project in this area, being worked on at Bergakademie Freiburg and TU Munich, is underway.

CSEARCH has a much more attractive user interface than SpecInfo: user- friendly and intuitive with much more attractive structure displays. However, the laying out of menu items along three layers of a wall of 'bricks' at the top of the screen, and the appearance of other options at the bottom/right is not a computer-industry standard.

SpecInfo is suitable for use on a Wide Area Network and with a wide variety of cheap terminals (albeit with consequent restrictions on the appeal of the user interface). With CSEARCH you could link some local Silicon Graphics work-stations but offering access to one database worldwide could be more difficult. Some people claim that you would need an expensive client workstation at every site requiring access to a networked host database on a VAX, because X-Windows is very slow on cheaper hardware. Sadtler claim that it is really the bandwith of the network that is most important to the speed of the system. Tests are being carried out in the UK on the efficiency of using a VAX version on a Wide Area Network.

CSEARCH allows two groups of databases, A-M and N-Z (in practice N-Y), and access can be restricted to some of these. Shell likes this approach. ICI prefers the earliest SpecInfo approach because they want all data accessible to all users. Now that the SYBASE version of SpecInfo has been developed, individual users can be given different access privileges and spectral collections in any desired combination can be made accessible to various users or kept private.

The 'homologous series' display or increment function is unique to CSEARCH: one spectrum is displayed above another and a set of link lines indicates how a peak has moved in one spectrum compared with the other. The effect of a substituent on a chemical shift can thus be clearly illustrated.

CSEARCH can use HOSE codes to 5 bond levels, SpecInfo to only 4. In practice SpecInfo is often only able to use a 3-bond substructural model because of a 96-bit limit to the HOSE codes. The advantage that CSEARCH has here was explained in the introduction. However, as stated above, SpecInfo HOSE codes will eventually be extended.

In CSEARCH, during spectral prediction, there is an option for detailed analysis of the database entries contributing to each estimated chemical shift. Histograms of chemical shifts allow selection of substructures by chemical shift range, enabling analysis of stereochemical, conformational or solvent effects.

In CSEARCH Boolean logic can be used to link structure and peak searches. In SpecInfo you intersect the hitlists from separate peak and structure searches, but the user could easily define a macro to simplify a query to just one command.

Literature searching is possible with CSEARCH. It is available indirectly *via* STN for SpecInfo. Perhaps this literature data is even better albeit indirect, but it is not clear that most spectroscopists care much about literature search anyway.

The initial SpecInfo documentation in ICI was written within ICI in the days when inadequate public documentation was available (and what was available was not in English). Chemical Concepts say that a new manual has been written for Version 2 of SpecInfo. The initial CSEARCH manual is very clear but it lacks an index and there is a large appendix with no page numbers. Sadtler have said that this will be corrected.

Acknowledgements

The author is immensely grateful to the following people who promptly read an early draft of this paper and provided so many useful comments that the final version is vastly improved:

- Wolfgang Bremser of Rheinische Olefinwerke
- Christine Faber of STN Columbus
- Steve Hammond of Shell Research Ltd
- Angus Hearmon of ICI Wilton Materials Research Centre

- Reinhard Neudert of BASF
- Michael Penk of Chemical Concepts
- Wolfgang Robien of the University of Vienna
- Paul Stanley of ICI Agrochemicals
- Bruce Woods and Rod Farlee of BIO-RAD Sadtler Division

References

[1] Munk, M.E., Farkas, M., Lipkis, A.H., Christie B. 'Computer-Assisted Chemical Structure Analysis'. *Mikrochim. Acta* **1986,** II, 199-215.

[2] Gray, N.A.B. 'Computer-Aided Structure Elucidation'. In *Chemical Structure Systems;* Ash, J.E., Warr, W.A., Willett, P., Eds; Ellis Horwood: Chichester, 1991, pp. 263-298.

[3] Small, G.W. 'Automated Spectral Interpretation'. *Anal. Chem.* **1987,** *59,* 535A-546A.

[4] Warr, W.A. 'Spectral Databases'. *Chemom. Intell. Lab. Syst.* **1991,** *10,* 279-292.

[5] Lias, S.G. 'Numeric Databases for Chemical Analysis'. *J. Res. Natl. Inst. Stand. Technol.* ***1989,*** *94(1),* 25-35.

[6] Heller, S.R. 'Computerised Spectroscopy Databases'. *Chem Int.* **1991,** *13(6),* 235-238.

[7] Pretsch, E., Farkas, M., Fürst, A. 'Requirements of Spectral Databases'. In *Scientific and Technical Data in a New Era; Proceedings of the 11th CODATA Conference*; Glaeser P.S., Ed.; Hemisphere Publishing Company: New York, 1990, pp. 176-179.

[8] Rumble, J.R., Lide, D.R. 'Chemical and Spectral Databases: A Look into the Future'. *J. Chem. Inf. Comput. Sci.* **1985,** *25,* 231-235.

[9] *Computer-supported Spèctroscopic Databases* ; Zupan, J. Ed.; Ellis Horwood: Chichester, 1986.

[10] Heller, S.R. 'The Chemical Information System and Spectral Databases'. *J. Chem. Inf. Comput. Sci.* **1985,** *25,* 224-231.

[11] Bremser, W., Fachinger, W. 'Multidimensional Spectroscopy'. *Magn. Reson. Chem.* **1985,** *23,* 1056-1071.

[12] Bremser, W. 'Structure Elucidation and Artificial Intelligence' *Angew. Chem. Int. Ed. Engl.* **1988,** *27,* 247-260.

[13] Bremser, W., Neudert, R. 'Automation in the Spectroscopic Laboratory — Solutions and Perspectives'. *Eur. Spectrosc. News* **1987,** *N75,* 10-27.

[14] Neudert, R., Bremser, W.,Wagner, H. 'Multidimensional Computer Evaluation of Mass Spectra'. *Org. Mass Spectrom.* **1987,** *22,* 321-329.

[15] Small, G.W. 'Database Retrieval Techniques for Carbon-13 Nuclear Magnetic Resonance Spectral Simulation'. *J. Chem. Inf. Comput. Sci.* **1992,** *32*, 279-285.

[16] Lah, L., Tusar, M.; Zupan, J. 'Simulation of ^{13}C NMR Spectra'. *Tetrahedron Comput. Methodol.* **1989,** *2(1)*, 5-15.

[17] Pretsch, E., Fürst, A., Robien, W. 'Parameter Set for the Prediction of the Carbon-13 NMR Chemical Shifts of sp2- and sp-hybridized Carbon Atoms in Organic Compounds'. *Anal Chim. Acta* **1991,** *248(2)*, 415-428.

[18] Pretsch, E., Fürst, A., Badertscher, M., Bürgin, R., Munk, M. 'C13Shift: A Computer Program for the Prediction of ^{13}C NMR Spectra Based on an Open Set of Additivity Rules'. *J. Chem. Inf. Comput. Sci.* **1992,** *32*, 291-295.

[19] Anker, L.S., Jurs, P.C. 'Prediction of Carbon-13 Nuclear Magnetic Resonance Chemical Shifts by Artificial Neural Networks'. *Anal. Chem.* **1992,** *64*, 1157-1164.

[20] Jurs, P.C., Ball, J.W., Anker, L.S., Friedman, T.L. 'Carbon-13 Nuclear Magnetic Resonance Spectrum Simulation'. *J. Chem. Inf. Comput. Sci.* **1992,** *32*, 272-278.

[21] *Chemical Structure Systems;* Ash, J.E., Warr, W.A., Willett, P., Eds; Ellis Horwood: Chichester, 1991.

[22] Ash, J.E., Warr, W.A. 'Databanks'. In *Chemical Structure Systems;* Ash, J.E., Warr, W.A., Willett, P., Eds; Ellis Horwood: Chichester, 1991, pp. 154-191.

[23] Warr, W.A. 'Systems for Chemical Structure Handling'. In *Chemical Structure Systems;* Ash, J.E.; Warr, W.A.; Willett, P., Eds; Ellis Horwood: Chichester, 1991, pp. 88-125.

[24] Weller, M.G. 'Spectral Databases and SpecInfo Online'. In *Proceedings of the 15th International Online Information Meeting London 10-12 December 1991*; Raitt, D.I., Ed.; Learned Information: Oxford, 1991, pp. 61-71.

[25] Bremser, W., Grzonka, M. 'SpecInfo — A Multidimensional Spectroscopic Interpretation System'. *Mikrochim. Acta* **1991,** II, 483-491.

[26] Hearmon, R. A. 'Spectroscopy Data Handling'. *Spectrosc. Int.* **1991,** *3(7)*, 14-18.

[27] Dalby, A., Nourse, J.G., Hounshell, W.D., Gushurst, A.K.I., Grier, D., Leland, B.A., Laufer, J. 'Description of Several Chemical Structure File Formats Used by Computer Programs Developed at Molecular Design Limited'. *J. Chem. Inf. Comput. Sci.* **1992,** *32*, 244-255.

[28] Sasaki, S-I., Kudo, Y. 'Structure Elucidation System Using Structural Information from Multisources: CHEMICS'. *J. Chem. Inf. Comput. Sci.* **1985,** *25*, 252-257.

[29] Funatsu, K., Miyabayashi, N., Sasaki, S-I. 'Further Development of Structure Generation in the Automated Structure Elucidation System CHEMICS'. *J. Chem. Inf. Comput. Sci.* **1988**, *28*, 18-28.

[30] *Pattern Recognition in Chemistry*; Varmuza, K., Ed.; Springer-Verlag: Berlin, 1980.

[31] Damen, H., Henneberg, D., Weimann, B. 'SISCOM — A New Library Search System for Mass Spectra'. *Anal. Chim. Acta* **1978**, *103*, 289-302.

[32] Domokos, L., Henneberg, D., Weimann, B. 'Optimization of Search Algorithms for a Mass Spectra Library'. *Anal. Chim. Acta* **1983**, *150*, 37-44.

[33] Kalchhauser, H., Robien, W. 'CSEARCH: a Computer Program for Identification of Organic Compounds and Fully Automated Assignment of Carbon-13 Nuclear Magnetic Resonance Spectra'. *J. Chem. Inf. Comput. Sci.* **1985**, *25*, 103-108.

[34] Robien, W. 'Computer-assisted Structure Elucidation of Organic Compounds. III. Automatic Fragment Generation from Carbon-13 NMR Spectra'. *Mikrochim. Acta* **1986**, *2(1-6)*, 271-279.

[35] Chen, L., Robien, W. 'MCSS: A New Algorithm for Perception of Maximal Common Substructures and its Application to NMR Spectral Studies. 1. The Algorithm'. *J. Chem. Inf. Comput. Sci.* **1992**. In press.

[36] Chen, L., Robien, W. 'MCSS: A New Algorithm for Perception of Maximal Common Substructures and its Application to NMR Spectral Studies. 2. Applications'. *J. Chem. Inf. Comput. Sci.* **1992**. In press.

[37] Passlack, M., Bremser, W. 'IDIOTS — Structure-oriented Databank System for the Identification and Interpretation of Infrared Spectra'. In *Computer-supported Spectroscopic Databases*; Zupan, J. Ed.; Ellis Horwood: Chichester, 1986, pp. 92-117.

Chemical information processing in structure elucidation

M. E. Munk, V. K. Velu, M. S. Madison, E. W. Robb, M. Badertscher, B. D. Christie and M. Razinger

Department of Chemistry and Biochemistry Arizona State University Tempe, AZ 85287-1604 USA

Introduction

Today, research in the chemical sciences is characterised by enormous diversity. Yet there are some conspicuous common elements: the acquisition, processing and reduction of chemical information, and the manipulation of chemical structure. It is in the execution of such tasks that the computer has made a significant difference. Those scientists taking full advantage of the power of the computer have been rewarded with substantially increased levels of productivity and, in some cases, with workable solutions to otherwise unwieldy problems.

The characterisation of the structure of an organic compound on the basis of its chemical and spectral properties, which is of widespread importance in the chemical, biological and medical sciences, has long attracted the attention of investigators interested in developing computer-based solutions to non-numerical problems in chemistry. It is a well-defined problem domain with a large body of pertinent knowledge. It requires the acquisition and interpretation of large amounts of data — mostly spectral data — and draws heavily from the corpus of software tools for the manipulation of chemical structure. Numerous reports of efforts in the development of software to augment the productivity of chemists engaged in structure elucidation have appeared in recent years. Systems such as CHEMICS [1], EPIOS [2], ACCESS [3] and SESAMI [4] integrate some spectrum interpretation capability with constrained structure generation. CONGEN [5] and GENOA [6], although highly interactive and powerful, are structure generators only and rely on the chemist for the interpretation of spectral data and the input of the resulting structural information.

SESAMI is being developed as a comprehensive system for computer-*enhanced* structure elucidation. Its purpose is to complement, not replace the efforts of the chemist. In describing the nature of SESAMI's role, it is helpful to briefly review *conventional*, i.e., computer-unassisted, structure elucidation. If cast in broad strokes, conventional structure elucidation can be viewed as a process consisting of two separate and distinct steps. The first stage is time-consuming and multi-step in nature. It involves collecting and interpreting the results of chemical and spectroscopic experiments, most likely running and interpreting additional experiments and finally expressing the outcome as a manageable number of complete molecular structures compatible with all of the data collected. What is a 'manageable' number? It varies with the problem and the chemist doing the work, but less than 50 is what most chemists would hope for. In the second stage, the final assignment comes relatively quickly because experienced chemists are proficient at readily identifying the correct structure of an unknown from among a limited number of alternatives. Thus, it is the first of the two steps that is the bottleneck. If the chemist's involvement could begin with the second step, i.e., with a manageable number of alternatives, much time would be saved and substantially increased productivity achieved.

The goal of the SESAMI project follows readily from such a description: the development of computer software capable of directly reducing the collective spectral properties of an unknown to a *manageable* number of compatible molecular structures. The chemist is still a participant in the process, but is spared the most time-intensive step. Thus, automated structure elucidation is not the goal, computer-*enhanced* structure elucidation is.

In software development chemical data has been eliminated as a source of computer-derived structural information. Quite often chemical data are time-consuming to acquire, but, more important, it is believed that the collective spectral properties alone, if wisely chosen, can in many cases be sufficiently information-rich to narrow the plausible structures to a small number.

Program overview

Structure elucidation requires both spectrum interpretation and structure generation. SESAMI already possesses some capabilities in each of these areas. SESAMI is a second generation program and a conceptual departure from its predecessor, CASE. The system is currently at an early stage of development with limited structure elucidation capability. An example of problem solving is described later.

Figure 1 shows the information flow in SESAMI. INTERPRET is a two-track spectrum interpretation procedure. On one track (PRUNE), the molecular formula and collective spectral properties of the compound of unknown structure are processed to give rise to a set of uniformly-sized, explicitly-defined 'basic units of structure' that are compatible with the input. These basic units of structure can be thought of as the *structural building units* for the structure generator, COCOA. On the other track (INFER), the same input leads to substructural inferences. These are substructures which, in contrast to the basic units of structure, can be of any size, and any degree of inexplicitness and overlap. These inferences serve as the *constraints* on the structure generating process, thereby reducing the number of plausible structures produced. COCOA is the structure generator that exhaustively constructs all plausible molecular structures compatible with the dual input.

The structural diversity of compounds found in nature and even in the synthetic chemist's laboratory is enormous. In recognition of this, SESAMI must be able to produce the entire range of structures compatible with the observed spectral data, *without exception.* If the output is to exclude *no* plausible alternative, the shortlist of basic units of structure (the ACF shortlist in Figure 1) must exclude *no* fragment

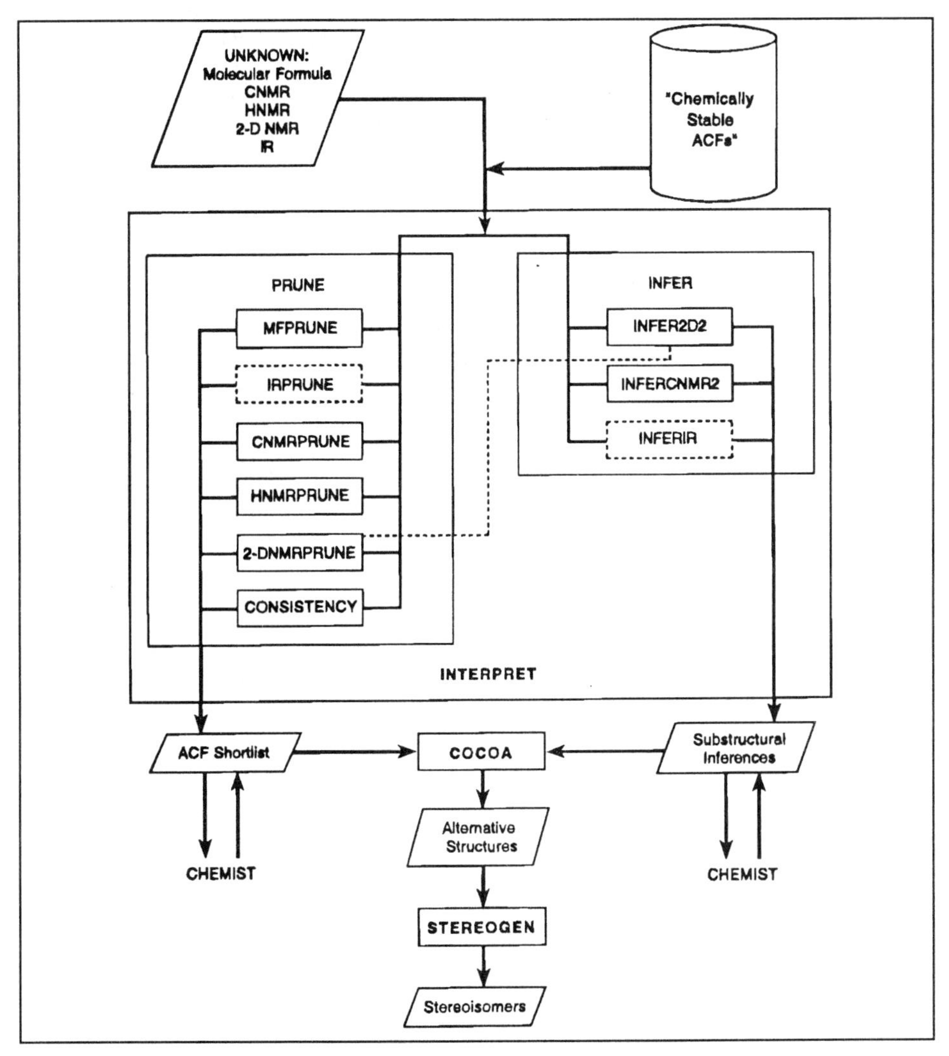

Figure 1: Information flow in the SESAMI system

compatible with the spectral data. This requirement is met by taking the set of *all* possible basic units of structure and deleting from it those that are incompatible with the observed spectral data. That is the role of PRUNE.

This approach to the generation of the 'shortlist' imposes two requirements on the nature of the basic unit of structure. First, this fragment must be small enough to ensure that the number of all possible basic units of structure is not too large to be efficiently handled by the computer. Second, the fragment must be large enough to possess distinctive spectral properties. Regrettably, these are conflicting requirements and cannot be met simultaneously.

A workable, but not ideal compromise under these circumstances is an atom-centered fragment that has one concentric layer of nearest neighbors. For convenience, this structural unit is called an ACF. As a unit of structure it is not sufficiently large to possess distinctive spectroscopic properties, but early results suggest it may still serve the intended purpose.

Element groups are the building blocks of ACFs. An element group defines a particular non-hydrogen atom, its attached hydrogen atoms, if any, and the nature of each partial bond by which it can join to other element groups. For example, the three possible element groups of oxygen are: hydroxyl oxygen (OH), ether oxygen (O), and carboxyl oxygen (=O). The central element group of the ACF must be valence-satisfied, but at least one first-layer element group must not be, e.g., $=CHCH_2O$, a methylene-centered ACF.

The ACF is a precisely defined substructural unit which can be conveniently and canonically represented for computer manipulation. It can also be correlated to its spectral properties. For initial study, those elements are included which are commonly encountered in organic compounds: carbon, trivalent nitrogen, oxygen, divalent sulfur and each of the monovalent halogens. Important structural features excluded by these elemental limitations can be conveniently treated as 'super' element groups. The nitro group was added in this way.

The computer-generated exhaustive list of ACFs contained a total of 13,703 fragments. Deleting those that would clearly lead to chemical instability in molecules containing them [e.g., $CH_2C(OH)_3$] gave rise to an 'exhaustive' list of 5088 *chemically stable* ACFs (Figure 1) from which PRUNE removes those ACFs incompatible with the observed spectral data.

Spectrum interpretation-INFER

INFER is modular in nature and consists of separate routines, the output of each of which is one or more substructures predicted to be present or absent in the unknown. This is the usual way to express the interpretation of spectral data. No restrictions are placed on the number of substructures inferred, the size of the substructures, the degree of ambiguity in defining the substructure, or the extent to which the substructures derived from the same or different routines may overlap. Alternative substructural interpretations of the spectral data, which are not uncommon, may also be produced.

At the present time, only two substructural inference makers are fully operational; INFER2D2 and INFERCNMR2. Considerable work has been done on INFERIR and it should be incorporated into SESAMI soon. SESAMI is highly interactive. Structural information known to the chemist from whatever the source can be conveniently entered. Thus, the chemist can be viewed as just another substructural inference maker.

Two-dimensional NMR spectroscopy is a powerful structural probe and therefore ideally suited as a substructural inference maker for complex compounds. Since in many cases, the usually acquired 2D NMR data alone are not sufficiently information-rich to narrow the possible structural assignments to a small number, this method complements, rather than replaces other spectral sources of structural information.

More than any other method, two-dimensional NMR spectroscopy influenced the development of INFER and the new structure generator COCOA. The goal was to utilise all of the various structural correlations produced by 2D NMR experiments. Since COCOA currently recognises only connectivity, only through-bond correlations can be utilised.

Some of the most widely used 2D NMR experiments in structure elucidation are COSY, giving mainly three-bond hydrogen-hydrogen correlations, and LONG-RANGE HETCOR and HMBC, giving long-range hydrogen-carbon correlations. If the data from these experiments are to be fully and efficiently utilised, one-bond hydrogen-carbon correlations (HETCOR, HMQC) are required by SESAMI. One-bond carbon-carbon correlations (2-D INADEQUATE), although very informative, are less widely used because of the difficulty in conducting the experiment.

INFER2D2 is a second generation program. It is at the same time simpler and more powerful than its predecessor INFER2D. To understand the problems that need to be addressed in interpreting 2D NMR data, it is instructive to look first at INFER2D.[7] The function of INFER2D,is to reveal carbon-carbon atom connectivity, i.e., to produce units of connected carbon atoms present in the unknown. It operates via a two-step process. In the first step INFER2D derives carbon-carbon *signal* connectivity, that is, units of 'connected' carbon signals. In the absence of molecular symmetry in the unknown, interpretation is complete since then carbon-carbon signal connectivity is the same as carbon-carbon atom connectivity. In the presence of the molecular symmetry, a second and rather complex step [7] is needed to convert carbon-carbon signal connectivity to carbon-carbon atom connectivity. Since the number of carbon signals in the spectrum of a compound with symmetry is less than the number of carbon atoms, a given carbon signal may represent more than one carbon atom, but the exact number of carbon atoms for each signal cannot be determined from the spectrum. The result is that two or more different substructural interpretations are common in cases of molecular symmetry.

With this approach, a difficult problem is encountered in the reduction of long-range hydrogen-carbon correlations to carbon substructure, even in the absence of molecular symmetry. The long-range HETCOR and HMBC experiments often produce many separate hydrogen-carbon correlations. In the usual case, the number of intervening bonds between the correlated hydrogen and carbon atoms cannot be precisely identified. It is usually two or three bonds, but can be four bonds. Thus, each correlation is ambiguous in the sense that it is consistent with two and possibly three different substructural interpretations, i.e., HCC, HCAC or HCAAC, where atom A can be any non-hydrogen element. If the goal is to deduce explicit substructures, there is a problem because the set of observed *ambiguous* correlations will give rise to many different sets of explicit correlations. Each such set corresponds to a different set of substructural inferences, i.e., a different interpretation. For example, 10 such long-range correlations limited to either two or three intervening bonds (a rather small number of correlations compared to what can be commonly encountered in a real-world structure elucidation problem) would give rise to 2^{10}, or 1024 different sets of 10 unambiguous correlations. Thus, in a procedure requiring unambiguous structural information, each of these sets would have to be generated and each would form the basis of a separate structure generation step, an inefficient process at best. The need for a more effective interpreter was clearly evident.

The key to overcoming these obstacles was the realisation that symmetry considerations and multiple interpretations of the data need not be the concern of this interpretation program. Rather, these tasks can be addressed during structure generation. Thus, the role of the second generation program, INFER2D2, is merely to provide carbon-carbon signal connectivity. Structure generation does the rest of the 'interpretation.' COCOA [8], in contrast to other structure generators, can process such information directly and prospectively without any preprocessing. The revised program, INFER2D2, performs the following simple tasks:

1. It unambiguously determines carbon-carbon signal connectivity from the combination of COSY and HETCOR/HMQC data by identifying carbon signals that correspond to carbon atoms bearing coupled vicinal hydrogens. In the absence of molecular symmetry, the program builds substructures from the COSY data, but in the presence of symmetry only signal connectivity is passed to COCOA.
2. One-bond carbon-carbon correlations (2D-INADEQUATE) are also reduced to carbon-carbon signal connectivity. Again, substructures are built only in the absence of molecular symmetry.
3. Each ambiguous long-range hydrogen-carbon correlation (LONG-RANGE HETCOR, HMBC)for example, those limited to two or three intervening bonds yield an inference requiring a pair of carbon atoms to be either directly connected to one another or separated by one other atom.

Each of the inferences produced, unambiguous or not, is passed *directly* to the structure generator COCOA. Bond type — single, double or triple — is not assigned by INFER2D2; only connectivity. Additionally, the program does not reveal information about attached hetero atoms.

INFERCNMR2, also a second generation program, is an interpretive library search routine. Its function is the same as that of its predecessor [9]; to retrieve substructures from the reference compounds of a library of *assigned* carbon-13 NMR spectra that are predicted to be present in an unknown compound. Both spectral and structural information (connection tables) are required.

Program input consists of the chemical shift of each signal in the spectrum of the unknown and its multiplicity due to one-bond carbon-hydrogen coupling. The molecular formula of the unknown is also needed. The output of the program is one or more explicitly defined substructures predicted to be present in the unknown, and the confidence level of each prediction. The program suffers the same limitations of all library search systems; substructures not present in the data base cannot be retrieved.

There are various approaches to retrieving substructural information about an unknown from a library of assigned carbon-13 NMR spectra. INFERCNMR2 is basically a subspectrum matching routine. The premise implicit in such a procedure is that if a subspectrum of a reference entry matches a subspectrum of an unknown, the substructure assigned to the reference subspectrum is also present in the unknown. Overall structural similarity between the reference compound and the unknown is not necessary in order to retrieve a substructure common to both.

In the initial step of the search strategy, the program retrieves all reference spectra in the library, at least four signals of which match signals in the spectrum of the unknown. That number is arbitrary and can be increased by the user. Matched signals have chemical shifts within 3.0ppm of one another and the same multiplicity. That window can be changed by the user as well. In a second step, the connection table of each retrieved reference entry is searched to identify those reference subspectra that correspond to a set of at least four *connected* carbon atoms, that is, to a substructure. In the final step, the program estimates the accuracy of each substructure prediction. If a substructure predicted to be present by INFERCNMR2 is handed to COCOA, every structure generated will contain that substructure. If that substructure is invalid, every structure produced by SESAMI will be invalid. Thus, it is important to know the reliability of a substructure predicted by INFERCNMR2 before handing it to COCOA.

A statistical model known as logistic analysis [10] was used to develop a measure of prediction reliability. A random set of 5000 assigned carbon-13 NMR spectra was taken from the library of 13,000 spectra. This is the *training set*. Using INFERCNMR2, each spectrum in the training set was compared to each of the 13,000 spectra in the database. A maximum tolerance of ±3.0 ppm was set. Approximately 1.72M predictions, i.e., substructures of four carbons or more, were obtained and the validity of each was determined using a substructure searching routine. For ease of handling, the output was divided into two sets of about 800,000 predictions. They each produced similar results. Those derived from one of the two sets are described here.

Logistic analysis leads to a probability function:

$$p_i = \frac{e^y}{1+e^y}$$

where p_i is the probability that a given substructure prediction is correct and y takes the form:

$$y = \beta_o + \sum x_i \beta_i + \sum x_i x_j \beta_{ij}$$

where β_o is the intercept, x_i, x_j are independent variables and β_i, β_{ij} are corresponding weighting coefficients. Twelve independent variables that influence prediction reliability have been identified:

1. root mean square error
2. width of maximum matching tolerance (ppm)
3. number of signals matched

4. number of singlets in unknown spectrum
5. number of doublets in unknown spectrum
6. number of triplets in unknown spectrum
7. number of quartets in unknown spectrum
8. number of times a given substance is retrieved from the reference library
9. number of residual bonding sites in predicted substructure
10. relative molecular size of the reference compound
11. number of signals in the unknown
12. solvent used — same or different

In solving for y, the summation is taken over all twelve variables. For this so-called prequadratic model, there are a total of 78 terms. Since it is known if each of the 800,000 predictions is valid or not, and since the values of the 12 variables for each prediction can be readily determined, statistical analysis was used to calculate optimum values of the weighting coefficients for all 78 terms. A value of y was then calculated for each of the 800,000 predictions and from there, a value of p_i, which ranges between 0 and 1. For the 800,000 observations, a close correspondence between a calculated p_i value and actual prediction accuracy was demonstrated by tabulating the number of valid and invalid predictions for the set of predictions with calculated p_i values equal to and greater than some selected pi value. For example, a p_i value of 0.95 should suggest a 95% probability of a prediction being correct, and comes close to doing so.

Program performance was evaluated using five test sets of 200 carbon-13 NMR spectra of known structures that were *not* part of the original database of 13,000 spectra. The original database of 13,000 spectra served as the reference library. The results for each of the five test sets are shown in Table 1. This presentation is designed to reveal the capacity of the probability function to discriminate between valid and invalid predictions.

The maximum matching tolerance used in this study was ±3.0 ppm. At this tolerance, the accuracy of the subspectrum matching routine is about 65% (Table 1, Matching Routine Performance). Thus, 2 out of 3 predictions made by the routine itself are correct. The probability function improves on that discrimination, but at a cost: the loss of some valid information. For example, the function can discriminate between valid and invalid predictions 90% of the time, but only a little more than 20% of the valid information is captured (Program Discrimination). At 95% accuracy, less than 10% of the valid information is retrieved; at 98%, only about 3%. The low information capture is not as serious as it seems, however. At an accuracy of 98%, on average somewhere between 10 and 25 substructure predictions will be retrieved for each unknown spectrum. There can be redundancies among these predictions, and of course with any given unknown, there can be few, if any predictions made.

INFERIR is a neural network-based interpreter. Although many different pattern recognition techniques have been used in automated spectrum interpretation with varying degrees of success, recent attention has focused on the application of simulated neural networks. This approach has shown promise in the interpretation of infrared spectra, enabling the presence or absence of a variety of structural features to be recognised with acceptable accuracy in a compound of unknown structure. [11,12] The neural network has an advantage over other methods of pattern recognition worth noting. The rules relating the output pattern (the list of structural features) to the input pattern (spectral information) need not be specified, or even known. It is the network itself that deduces the 'rules' during the process of training. This is important, since often these rules relating spectral and structural properties are either so complicated or poorly understood that construction of a rule-based system is impractical.

Test Set	Matching Routine Performance			Program Discrimination			Average Predictions
	Predictions			% Valid Predictions Captured at			Unknown
	Total	Valid	% Correct	90% Accuracy	95% Accuracy	98% Accuracy	98% Accuracy
1	192,827	126,337	65.5	23.7	9.5	3.9	25.4
2	186,658	120,194	64.5	22.3	8.8	3.4	21.5
3	162,089	107,497	66.3	24.0	9.8	3.7	20.8
4	180,561	119,381	66.1	19.6	6.8	2.3	14.5
5	173,442	114,332	65.9	20.3	6.7	1.9	11.1

Table 1: INFERCNMR2 Prediction accuracy

INFERIR utilises a two-layer, feed-forward network with one layer of hidden units. The input vector is a coded representation of the infrared spectrum. The spectral range is divided into 256 intervals, each of which serves as one input unit. If the spectrum has a peak in an interval, the input value for that unit is a number between 0 and 1, in proportion to the strength of the absorption. The output is also a vector, each unit of which is assigned to a specific structural feature. Its value usually varies between 0 (absent) and 1 (present) as well. Some units correspond to the traditional functional groups and substituents of organic chemistry, while others are a larger assemblage of atoms.

A training set is needed to train the network; in this case, a library of infrared spectra of compounds of known structure, the latter stored as connection tables. Using substructure searching methods, the correct value — 0 if the substructure is absent, 1 if it is present — can be assigned to each output unit for each spectrum in the training set. Using the iterative learning technique of back-propagation [11], the network is trained to approach those target values for these known compounds as closely as possible. The trained network can then be applied to the prediction of the presence or absence of these structural features in compounds of unknown structure. The output value for each unit can be related to the probability of its assigned structural feature being present or absent in the unknown. It is expected that a substructural inference maker (INFERIR) based on this approach will be an integral part of INFER very soon.

Spectrum interpretation-PRUNE

PRUNE is the second track of INTERPRET and is responsible for producing the ACF shortlist, the uniformly-sized fragments that are the structural building blocks in structure generation. PRUNE, like INFER, is modular in nature (Figure 1). The starting point for PRUNE is the exhaustive list of chemically stable ACFs. MF-PRUNE deletes all ACFs not compatible with the molecular formula of the unknown.

IRPRUNE is not fully operational yet. A neural network-based program is being developed to reliably predict the absence of functionalities easily detected by infrared spectroscopy, for example, the nitro group and the carbonyl group. That information will be used to delete ACFs surviving MFPRUNE.

CNMRPRUNE now takes the list of ACFs surviving MFPRUNE and deletes those carbon-centered ACFs not compatible with the carbon-13 NMR spectrum of the unknown. The surviving carbon-centered ACFs are organised in a specific manner. The program compiles a list of plausible carbon-centered ACFs for each signal in the carbon-13 NMR spectrum of the unknown such that the assigned chemical shift range and signal multiplicity of the central carbon atom of each carbon-centered ACF in a particular carbon signal group matches the observed chemical shift and multiplicity of that signal. The routine uses a database containing allowed carbon-13 NMR chemical shift ranges and signal multiplicities for the central carbon atom of all carbon-centered ACFs.

Next, the list of surviving ACFs is pruned on the basis of proton NMR data. HNMRPRUNE is similar to CNMRPRUNE in its function and operation. It takes the group of carbon-centered ACFs compiled by CNMRPRUNE for each carbon signal and deletes any ACF with a proton-bearing central carbon atom whose assigned proton chemical shift range and allowed multiplicity pattern are not matched in the observed proton NMR spectrum of the unknown. A database is required which contains allowed proton NMR chemical shift ranges and multiplicity patterns for hydrogens attached to the central carbon atom of carbon-centered ACFs. The signal multiplicity pattern is only used to a limited extent at the present time.

If 2-D NMR data have been entered for the unknown, the correlations made by INFER2D2 are passed to 2-DNMRPRUNE. The routine deletes ACFs within each carbon signal group of ACFs whose central carbon connectivity conflicts with the signal connectivity inferred by INFER2D2.

PRUNE also includes a routine to maintain the internal consistency among surviving ACFs. It does so by deleting wherever possible those ACFs causing the inconsistencies. The surviving list of ACFs is subjected repeatedly to this routine until no further deletions are obtained.

The ACFs that survive this final step constitute the ACF shortlist. However, the chemist can edit the ACF shortlist, thereby further reducing its size, by using supplementary structural information derived from any source.

If PRUNE were perfect, the ACF shortlist would contain only those ACFs actually a part of the unknown structure. But the ACF is too small a fragment to permit a distinction to be made between each and every ACF solely based on spectral properties. Thus, in practice, the ACF shortlist will usually contain many more invalid ACFs those not part ofthe unknown than valid ones. Although it is desirable to have an ACF shortlist as short as possible, the presence of invalid ACFs does not pose a problem. Most will be rejected in structure generation. However, there *is* a major problem if PRUNE deletes a valid ACF; in that case the correct structure cannot be generated. PRUNE was therefore designed on the principle that it is better to retain an invalid ACF on the ACF shortlist than to delete a valid one. For this reason, the assigned chemical shift ranges in the databases of CNMRPRUNE and HNMRPRUNE are set more broad than narrow.

Structure generation

The dual output of INTERPRET is the surviving ACFs that are the structural building units in structure generation and the collected substructural inferences that serve as the constraints on structure generationis handed directly to COCOA. It is important to note that the exhaustiveness of the list of chemically stable ACFs insures that no plausible structure will be excluded.

COCOA is a recently developed structure generator that represents a conceptual departure from currently available structure generating procedures. Its advantages are:

1. It uses all structural information produced by INTERPRET *prospectively.*
2. Its efficiency improves as the number of substructural inferences increases.
3. It can utilise potentially overlapping substructures produced by INTERPRET without any preprocessing.
4. It can utilise alternative interpretations of the spectroscopic data without any preprocessing.

Stereoisomer generation

A newly developed program, STEREOGEN, allows the user access to all possible stereoisomers of any or all of the topological structures generated by COCOA. The only input required is a *canonical* connection table, the format in which COCOA's output is stored. Once generated, each stereoisomer is stored internally as an augmented connection table in which the configuration of each stereocenter in the molecule is assigned as either +1 or -1. The output for each stereoisomer is a 2.5-D *representation,* a vector consisting of a number of elements, each +1 or -1, equal to the number of stereocenters in the structure. This vector can be transformed into a conventional 3.0-D representation by the chemist using a simple procedure.

In its operation, the program first identifies all stereocenters as either 'true' or 'pseudo' and then exhaustively and irredundantly generates all possible stereoisomers. It is generally applicable to structures with and without symmetry and has been tested on a wide range of structure types. At the present time, the program does not possess the intelligence to exclude all stereoisomers which are too structurally strained to exist under normal conditions, but such refinements are planned.

^{13}C NMR		^{1}H NMR			
Shift (ppm)	Mult	Shift (ppm)	Integral	Mult	Exch[a]
197.73	S	7.70	1		
191.26	S	6.80	1	S	
166.98	D	6.60	1	S	
150.59	S	5.95	2	S	
147.03	S	5.55	1		
132.04	S	5.20	1		
131.42	D	5.05	1		
130.69	S	4.95	1		
119.50	T	4.05	1		
108.82	D	3.70	1		
108.67	D	3.40	1		
101.76	T	3.15	1		
96.27	D	2.85	1		
82.29	S	2.75	1		
49.01	T				
37.46	T				
32.08	T				

[a] *An 'E' is entered if a signal disappears upon addition of D_2O.*

Table 2: One-D ^{1}H and ^{13}C NMR data for the Wasserman Compound

Problem solving

The following example of problem solving illustrates some of the current capabilities of SESAMI. A compound of molecular formula $C_{17}H_{15}NO_4$ was kindly provided by Professor Harry Wasserman (Yale University). The SESAMI system is largely an NMR-based structure elucidation system at this time in which two-

Signal[a]	Signal2[a]	Min[b]	Max[c]
C166.98	H7.70	1	1
C108.82	H6.80	1	1
C108.67	H6.60	1	1
C101.76	H5.95	1	1
C131.42	H5.55	1	1
C119.50	H5.20	1	1
C119.50	H5.05	1	1
C96.27	H4.95	1	1
C49.01	H4.05	1	1
C49.01	H3.70	1	1
C37.46	H3.40	1	1
C32.08	H3.15	1	1
C37.46	H2.85	1	1
C32.08	H2.75	1	1
H7.70	H4.95	3	3
H5.55	H5.20	3	3
H5.55	H5.05	3	3
H5.55	H3.40	3	3
H5.55	H2.85	3	3
H4.05	H3.15	3	3
H3.70	H3.15	3	3
H3.70	H2.75	3	3
C130.69	H2.75	2	3
C130.69	H3.15	2	3
C132.04	H2.75	2	3
C82.29	H2.85	2	3
C82.29	H3.40	2	3
C119.50	H2.85	2	3
C119.50	H3.40	2	3
C131.42	H3.40	2	3

Signal[a]	Signal2[a]	Min[b]	Max[c]
C191.26	H2.85	2	3
C130.69	H4.05	2	3
C82.29	H4.95	2	3
C166.98	H4.95	2	3
C191.26	H4.95	2	3
C37.46	H5.20	2	3
C147.03	H5.95	2	3
C150.59	H5.95	2	3
C32.08	H6.60	2	3
C132.04	H6.60	2	3
C147.03	H6.60	2	3
C150.59	H6.60	2	3
C130.69	H6.80	2	3
C150.59	H6.80	2	3
C197.73	H6.80	2	3
C82.29	H7.70	2	3
C191.26	H7.70	2	3
C32.08	H3.70	2	3
C32.08	H4.05	2	3
C49.01	H2.75	2	3
C49.01	H3.15	2	3
C82.29	H3.70	2	3
C96.27	H7.70	2	3
C131.42	H2.85	2	3
C132.04	H3.15	2	3
C147.03	H6.80	2	3
C166.98	H3.70	2	3
C166.98	H4.05	2	3
C197.73	H2.85	2	3

[a] Element, chemical shift (ppm).

[b] Minimum number of intervening bonds.

[c] Maximum number of intervening bonds

Table 3. 2-D NMR correlations for the Wasserman Compound

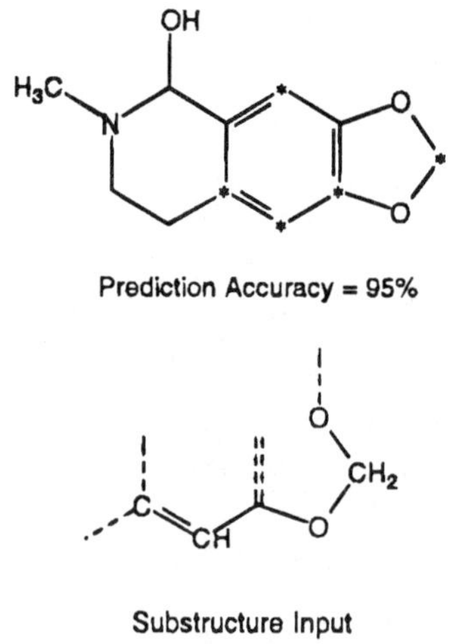

Prediction Accuracy = 95%

Substructure Input

Figure 2: Output of INFERCNMR2

dimensional NMR data and INFER2D2 play a central role. One-dimensional carbon-13 and proton NMR data are required as input (Table 2). Two-dimensional NMR experiments (HETCOR/HMQC, COSY and LONG-RANGE HETCOR/HMBC) gave rise to the signal correlations shown in Table 3. Keyboard-entered input consists simply of the chemical shifts of coupled signals, element type and the minimum and maximum number of intervening bonds. One-bond hydrogen-carbon correlations have been entered first in the Table; three-bond hydrogen-hydrogen correlations next; and last, all long-range hydrogen-carbon correlations that do not distinguish between two and three intervening bonds. Each of the 37 long-range hydrogen-carbon correlations serves as a separate constraint on COCOA and each has two different structural interpretations. Thirty-seven *either–or* statements lead to 2^{37} or about 137 billion different sets of 37 unambiguous inferences, a large number that dramatically conveys the degree of ambiguity in this information. If initial preprocessing into all possible sets of 37 unambiguous inferences was required, followed by treatment of each such set as a separate structure generation problem, solution of this problem would be impractical. But COCOA can contend with a large number of alternative inferences and use the structural information inherent in them prospectively to eliminate invalid structures before they are generated.

Using a matching tolerance of ±1.5ppm, a single substructure was inferred by INFERCNMR2 at the 95% accuracy level. The output of INFERCNMR2 displays a predicted substructure embedded in the reference library entry from which it was retrieved (Figure 2). The predicted substructure has four carbon atoms (shown as asterisks) but includes the oxygen atoms attached directly to matched carbon atoms. A fifth matched carbon is shown, but it is *not* part of a substructure, that is, it is unconnected. This substructure is handed to COCOA as a constraint.

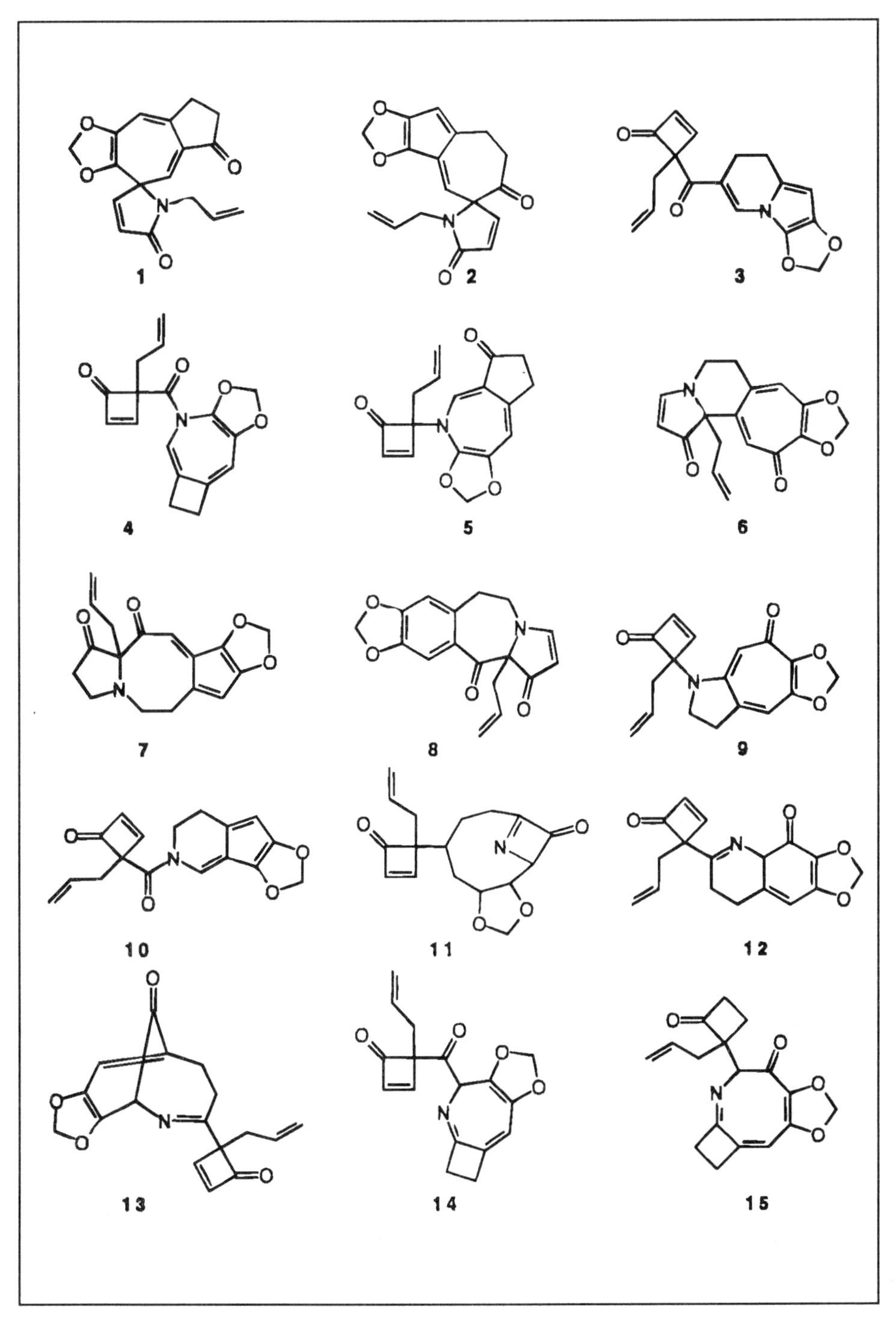

Figure 3: SESAMI output for the Wasserman Compound

User interaction was limited to editing of the ACF shortlist. Recall that PRUNE, which is responsible for creating the ACF shortlist, operates on the principle that it is better to retain an invalid ACF than to discard a valid one. Put differently: It is better for SESAMI to produce a broader range of alternative structures than to risk eliminating the correct structure. However, users, drawing on their experience/intuition, may choose to exclude certain ACFs believed to be unlikely, thereby possibly further narrowing the number of structures generated.

In this problem, the groups of ACFs assigned by PRUNE to the signals at δ197, 191 and 166 ppm included only ACFs with sp^2 central carbon atoms, but both carbon-carbon and carbon-oxygen double bonds were permitted. In the editing, the signals at δ197 and 191 ppm were required to correspond only to carbonyl carbon. PRUNE allowed both sp^2 and sp^3 central carbon atoms for the ACFs assigned to chemical shifts δ147, 132, 131 and 130 ppm, but as a user decision, ACFs with sp^3 central carbon atoms were excluded.

SESAMI produced a total of 15 structures (Figure 3). It is left to the user to distinguish between them. Some candidates may possess unlikely structural features in the view of the user and can be eliminated, for example, the cyclobutenones, of which there are nine. It is important to note that in distinguishing among the 15 structures produced by SESAMI, the user has the assurance that no other structure equally compatible with the data has been overlooked. This is not always the case in assignments made in conventional structure elucidation.

Returning to the output of INFERCNMR2 (Figure 2), note that two signals for benzene ring carbons failed to match within the tolerance set. The chemist could decide to take an added risk and pass the entire disubstituted dioxymethylene benzene fragment to COCOA. If that is done, a single structure (8), the correct structure, is generated by SESAMI.

Summary

The strategy on which SESAMI is based appears to offer considerable promise as a framework for a comprehensive system of computer-enhanced structure elucidation. In particular, the seamless link between spectrum interpretation and structure generation, the ability to utilise all 2-D NMR-derived, through-bond atom correlations, even those that are ambiguous, the effective treatment of symmetry, and the interactive nature of the program enhance its power and versatility.

Acknowledgements

The financial support of this research by the National Institutes of Health (GM37963), The Upjohn Company, Sterling Drug, Inc. and Merck Research Laboratories is gratefully acknowledged. We are also indebted to Professor Michael Driscoll, Department of Mathematics, Arizona State University, for invaluable guidance in the study using logistic analysis.

References

1. Funatsu, K., Miyabayashi, N., Sasaki, S. *J. Chem. Inf. Comput. Sci.* **1988**, *28*, 18.

2. Dubois, J. E., Carabedian, M., Ancian, B. C. R. *Seances Acad. Sci. Ser. C* **1980**, *290*, 383.

3. Bremser, W., Fachinger, W. *Magn. Reson. Chem.* **1985**, *23*, 1056.

4. Christie, B. D., Munk, M. E. *J. Am. Chem. Soc.* **1991**, *113*, 3750.

5. Carhart, R. E., Smith, D. H., Brown, H., Djerassi, C. *J. Am. Chem. Soc.* **1975**, *97*, 5755.

6. Carhart, R. E., Smith, D. H., Gray, N. A. B., Nourse, J. G., Djerassi, C. *J.Org. Chem.* **1981**, *46*, 1708.

7. Christie, B. D., Munk, M. E. *Anal. Chim. Acta* **1987**, *200*, 347.

8. Christie, B. D., Munk, M. E. *J. Chem. Inf. Comput. Sci.* **1988**, *28*,, 87.

9. Shelley, C. A., Munk, M. E. *Anal. Chem.* **1982**, *54*, 516.

10. Neter, J., Wasserman, W., Kutner, M. H. 'Applied Statistical Methods', 1985, Chap. 10, Irwin, Homewood, Illinois.

11. Robb, E. W., Munk, M. E. *Mikrochim. Acta* **1990**, *I*, 131.

12. Munk, M. E., Madison, M. S., Robb, E. W. *Mikrochim. Acta* **1991**, *II*, 505.

Aspects of protein modelling at the University of York: a review of the current research programmne

R. E. Hubbard, T. J. Oldfield, P. Herzyk and J. Zelinka

Department of Chemistry, University of York, Heslington, York YO1 5DD, England

1. Introduction

We are experiencing an explosion in the quantity, diversity and complexity of protein structural information that is being produced. Each week sees the publication of not only the detailed structural analysis of a major protein system determined by crystallography or nmr, but also an increasing range of low resolution indications of structure, and the more tentative results of molecular modelling. In this paper, we describe some of the research projects at York concerned with the analysis and management of these data, and how this information can be used to produce models of molecular structure. In addition, we describe current work in developing techniques for characterising the chemical and steric aspects of protein binding sites and how this information can be used to search databases or build new small molecules that may satisfy these restraints.

This paper is divided into four sections each of which outlines a particular area of research at York.

2. Abstracting structural information from stereo diagrams

With the rapid pace of modern methods of macromolecular structure determination, there is an increasing number of research articles describing the three dimensional structure of proteins. One of the frustrations in reading these articles is that often the coordinates on which diagrams are based are unavailable for local study, comparison and modelling.The Brookhaven Data Bank [1] acts as central repository for protein structural data, and coordinates for most protein structures eventually appear in the database. However, this can take some time. In most cases, the coordinates are caught up in the process of being deposited, validated and published in the database. Occasionally, the coordinates are unavailable for commercial reasons or because the research group wishes to exploit the structure further. Whatever the reason, the result is that structural information reported in a research paper is not available for further analysis.

These problems were recognised many years ago by Michael Rossman, who produced a computer program for generating a three dimensional structure from the information contained in CA stereo diagrams [2]. This involved measuring or

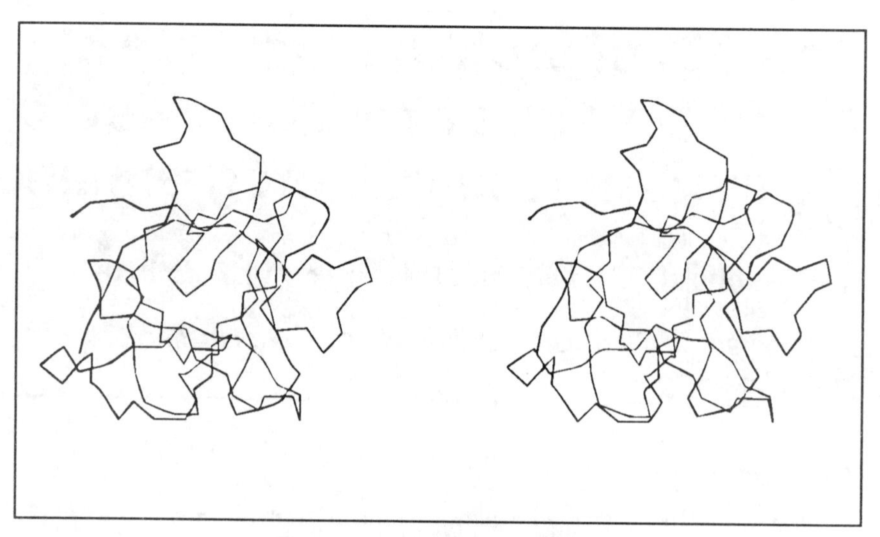

Figure 1. 1il-b stereo diagram

digitising the two dimensional coordinates of a left and right eye view image (which are assumed to be related by a 6 degree rotation), and from these coordinates refining a consistent three dimensional model of the protein molecule.

We have developed a novel technique for generating three dimensional coordinates from stereo diagrams. The main advantage of the procedure is that it is more reliable, robust and rapid than others we have tried. The following is a summary of the essential features of the approach.

The left and right eye view parts of a published stereo diagram (such as that shown in Figure 1) are digitised and transferred to a workstation. The images are scaled and displayed on the appropriate parts of the screen. When the workstation is switched to stereo mode, the published diagram then appears on the screen in stereo. A three dimensional model is then built to coincide with the stereo image using a CA backbone. After initial scaling and positioning of one end of the CA backbone, additional CAs are added, assuming that the distance between successive CAs in a sequence is 3.8Å with the CA-CA-CA angle and CA-CA-CA-CA dihedral as variables. An analysis of high resolution crystal structures reveals that there are clearly defined regions for these pseudo angles and dihedrals within proteins [3]. This empirical observation can be used to guide building, being particularly effective in resolving ambiguities in which direction (+/- z) the chain is proceeding.

Once the complete chain is built, a manual refinement phase is entered, in which each CA position and the appropriate pieces of the diagram are shown on a larger scale. The user can then adjust the position of the CA atom, this time with the length of the CA-CA bond being allowed to vary as well. A polyalanine chain is then generated from the CA backbone from a database of protein fragments, and side chains (if required) added using a standard molecular modelling program.

The critical test of the procedure is the precision with which the program can reproduce the coordinates of a known structure. Two test case proteins were used, recombinant porcine myoglobin [4] and human interleukin 1 beta [5]. The first is an alpha helical protein, the second a beta sheet protein and both had been solved to 2Å at York.

A stereo alpha carbon trace was generated for each protein and printed on a postscript laser printer as a stereo figure. In both cases a stereo angle of 6 degrees was used. The procedure to generate the three dimensional coordinates of a poly alanine structure was followed for both molecules. Table 1 shows the results of the RMS residuals between the known coordinates and the built proteins. The structures took about an hour each to build.

	Number of Residues	CA RMSD	N, CA, C, O,CB RMSD
Myoglobin	153	1.38	1.58
Interleukin	153	0.62	1.21

Table 1. Quality of structures generated with abstract

A potential application of structures generated by this approach is to provide coordinates to solve a similar or homologous protein by molecular replacement. This requires the abstracted structure to be good enough to find a solution to a rotation function search. Tests with both myoglobin and interleukin showed that the correct solution to a rotation search could always be found.

We have extended the basic ideas of this approach to tackle the problem of building models from 'lower resolution' mono images such as that shown in Figure 2. Again, the image is scanned and displayed on the workstation, but in this case, regular pieces of secondary structure such as alpha helices and beta sheets are available for fitting to the image. At the same time, the program allows for 'bump checking' between these pieces of structure. Using this approach, we have successfully produced models of quite large molecular systems from very little information. A particular example is the generation of a model for the reverse transcriptase / duplex complex from the cartoon representations in the *Science* article by Steitz *et al.*[6]

3. Low resolution modelling and restraint satisfaction

Molecular dynamics techniques have proved very powerful at generating structures from distance constraints [7]. In these calculations, a molecular mechanics description in terms of the energies of the bonds, angles, improper torsions and simplified repulsive van der Waals of the molecule, is combined with an energetic term measuring how well a structure satisfies particular distance constraints. This model can compute the energy for any particular combination. In simulated annealing, a molecular dynamics simulation is initially performed at high temperature (1000K) to allow large movements of the atoms, followed by molecular dynamics with slow cooling of the temperature. The principal is that the atoms will arrange themselves in the low energy ground state, provided the initial temperature is sufficiently high

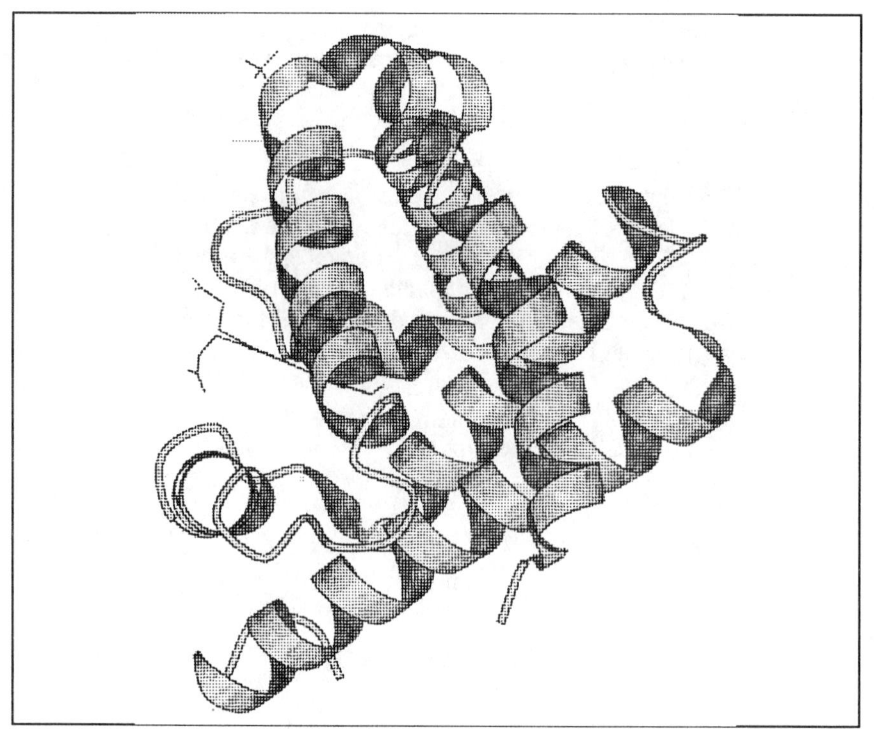

Figure 2. Molscript figure of say myoglobin

and the cooling is carried out slowly enough. In practice, the technique does not necessarily find the global energy minimum, but is a powerful way of moving over potential energy barriers to satisfy restraints.

These calculations require considerable computer time. A single simulation of a peptide of 50 amino acids requires some two hours computer time on an SGI 4D/310 computer. We found that it was necessary to perform at least 50 of these calculations to generate a reasonable ensemble of molecules for analysis, thus requiring a day or so of computer time for each set of conditions. We have been investigating methods whereby the calculations can be speeded up, by reducing the detail of the atomic representation of the molecule under study.

There have been a number of previous studies on the use of reduced representations in molecular mechanics calculations. These have either reduced the number of degrees of freedom in the model (working in torsion angle space for example [8]) or carefully reduced the representation of the molecular model to preserve the shape and chemical properties as much as possible [9]. These representations have proved only partially successful for true molecular simulations, that is simulating the dynamics of molecules to determine the thermodynamic and structural properties of an ensemble. This is probably because such simulations are sensitive to the detailed and sometimes subtle interactions between individual atoms.

We felt that the reduced representations required for restraint satisfaction could however be very simplistic. In restraint satisfaction, the aim is to pull the molecule into an approximation of a shape that will satisfy the restraints and then use more sophisticated molecular representations to refine the structure. In many ways, this is analogous to the approach adopted by some in using hybrid distance geometry — simulated annealing protocols. We felt, however, that reduced representation molecule mechanics models could also be applied in more general constraint satisfaction techniques, in particular when linked to interactive graphics as in for example the virtual reality project.

We have produced as reduced a representation as possible for the amino acids. A single atom is taken to represent the main chain (at the CA position) and side chains are reduced so as to preserve their overall shape as much as possible. This gives side chain descriptions of 0, 1, 2 or 3 atoms. We have taken little regard for the chemistry of the side chain atoms. In restraint satisfaction, the overwhelming forces involve the satisfaction of the restraints, with the van der Waals packing next in importance. Differentiating the subtleties of chemical interactions is a lower priority, and best handled by an all or extended atom representation at a refinement stage.

The positions of these virtual atoms were derived by simple geometric consideration of the position of the real atoms that they represent. Similarly, the restraint list required significant modification (and reduction) to reflect the new structure. Finally, we conducted a survey of some 84 distinct protein structures, solved to high resolution, to derive parameters to describe the virtual bonds, angles and dihedrals.

This representation of the molecule was compiled for the XPLOR program [10], and all calculations used a shortened version of the standard YASAP protocol [11]. In general, 50 structures were calculated and these structures were analysed for restraint satisfaction (root mean square deviation from the restraint upper bound), low potential energy (the final energy in the XPLOR procedure) and deviation from the family of structures (rms deviation from all structures). The final value of the potential energy target function (E_{final}) can be used only to compare the quality of structures obtained with different protocols within a representation, as the individual energy terms are so different. The results of all these calculations are presented in Table 2.

The time taken to produce a single structure falls dramatically from 5100 secs for the all atom representation to 162 seconds for the Herzyk representation. A more useful measure is the amount of time taken to produce an acceptable structure shown in the final column of Table 3. It can be seen that the best Herzyk protocol takes less than a quarter of the time of the all atom representation to produce an accepted structure. Interestingly, as the number of steps in the protocols is reduced for the Head-Gordon representation, the amount of time per accepted structure rises. But for the Herzyk representation, the time per accepted structure falls. This implies that there may still be efficiencies to be gained by further pruning of the simulated annealing protocol.

Residue	**Protein representation**		
	Herzyk	**Head-Gordon**	**All atoms**
GLY	1	3	7
ALA	2	4	10
VAL	2	4	16
CYS	2	5	11
PRO	2	5	15
SER	2	5	11
THR	2	5	14
MET	3	5	17
LEU	3	5	19
ILE	3	5	19
ASP	3	5	12
ASN	3	6	14
PHE	3	6	20
TYR	3	7	21
HIS	3	8	17
GLU	4	6	15
GLN	4	7	17
LYS	4	6	22
ARG	4	9	24
TRP	4	8	24

Table 2: Number of atoms in 20 aminoacids in different representations

	N[a]	RMSV[b] [Å]	E_{final}[c] [kcal mol^{-1}Å^{-1}]	RMSD[d] [Å]	<SA> vs SA[e] [Å]	CPU time[f] [s]	CPU time[h] [s]/good
ALL[g]	43	.003-.016	87.1-116.1	0.81-1.97	1.06 (.20)	5100	5950
HEAD[g]	22	.000-.003	40.6-46.1	1.41-2.20	1.13 (.24)	1141	2600
HERZ[g]	6	.008-.013	39.1-50.3	2.10-2.75	1.14 (.25)	162	1350

[a] Number of crambin structures (out of 50) accepted for the final analysis.

[b] Range of the root mean square violation of NOE restraint upper limit.

[c] Range of the potential energy target function at the end of the structure determination.

[d] Range of the root mean square deviation from crystall structure over the backbone atoms.

[e] Average root mean square deviation of the analysed structures from the mean structure, calculated using the backbone atoms.

[f] CPU time for determination of one structure.

[g] ALL, GORD, and HERZ stand for all atom, Head-Gordon, and Herzyk representations, respectively.

[h] Average number of seconds to produce an accepted structure.

Table 3 Comparison of Crambin structures determined using different protein representations and different timings of YASAP protocol

4. Suggesting novel molecules for binding sites — HOOK

The binding of small organic molecules to proteins and enzymes is central to the metabolism, function and control of most biological systems. With the expansion of structural molecular biology, detailed descriptions of an increasing number of proteins which bind ligands are becoming available. This has led to computational strategies for characterising the chemistry of ligand binding sites, and the development of a number of programs which suggest molecules which may bind to these sites.

A number of programs have been reported recently which attempt to direct the design of small molecules to bind to these sites. The DOCK program [12] describes a binding site in terms of a set of spheres, and uses this geometric description to search for molecules that may bind, essentially ignoring the chemistry of the interaction. The CAVEAT program [13] requires a definition of the chemically important functionality in terms of a binding vector. The program then scans a database for an equivalent set of vectors. More recently, a report has appeared of the CLIX program [14] which considers both the geometry and chemistry of binding and the LUDI [15] program which constructs molecules from fragments that reflect chemical functionality within the site.

The most widely used technique to characterise the chemical nature of the binding sites of proteins and enzymes is the approach of Goodford [16]. The energy of interaction between a functional probe, represented as a point charge with defined van der Waals and hydrogen bonding properties, is calculated on a grid in the binding site. Subsequent contouring of the interaction energy has allowed successful identification of potential binding sites for particular functional groups. The CLIX program uses the output from the Goodford approach to define regions of interaction.

More recently, Miranker and Karplus have developed an alternative approach using molecular simulation techniques on multiple copies of functional groups to produce a functionality map of binding sites [17]. There are three distinctive features of the MCSS (multiple copy simultaneous search) approach. Firstly, more realistic probes of chemical functionality are possible. Instead of a point probe, functional groups such as ethanoic acid, methyl ammonium and N-methyl acetamide can be used. Secondly, the resulting functionality map accurately identifies the minimum energy interaction sites and provides a geometry for the interaction with a resulting orientation for the functional group. Finally, the technique produces an extensive description of the chemistry of the binding site in terms of discrete, localised binding sites.

These techniques define positions within the binding site of a protein at which particular functional groups could interact. The aim of the HOOK program [18] we have been developing is to search for rigid molecules that will link these functional sites together at the same time as satisfying the chemical and steric requirements of a known protein binding site.

We start with a functionality map of the binding site of a protein. This can come from a variety of sources. In the examples presented here, functionality maps were produced using the Multiple Copy Simultaneous Search (MCSS) method of Miranker and Karplus [17], or by inspection of known protein-ligand complexes. This map defines a number of functional sites each representing a position and orientation at which a particular functional group interacts favourably with the binding site in the protein.

In the HOOK program, this set of functional sites is screened against a series of molecules. These molecules can either be taken from existing three dimensional databases, or constructed from carbon skeletons. The aim is to overlap bonds in the functional sites with appropriate bonds on the skeletons such that the skeleton will bind to the binding site. This search occurs in a number of stages, each designed to reduce the computational demands of the search. These stages include an incremental assessment of the number of functional sites that can be linked by a skeleton, and consideration of the bonded geometry of the final molecule produced by fusing the skeleton and functional sites (distance, angle and dihedrals). An important component of the search is the use of a simplified scoring regime, which discards molecules which make unfavourable van der Waals interactions with the binding site.

The development of this approach is still at an early stage. However, it is already apparent that the program will find molecules that by all the established criteria do fit into the binding site. In particular, the program may have some value in identifying novel substrate architectures. Our main concern at present is gaining access to the appropriate databases and synthetic organic chemistry to properly validate the approach.

5 Data and project management

During a project, a wide variety of different information is generated. This may be crystallographic data and derived structures, the sequence of operations in determining a structure, NOEs from nmr data, modelled structures, biological data, pictures generated by the project or from other publications. Ideally, it should be possible to keep all these data together under one database, for subsequent analysis and retrieval. These types of data explosions are particularly acute in large structural groups where the number of projects under way at any one time can cause serious problems for data, in particular for being able to share it effectively. For example, at York there are at any one time some 20 crystallographic projects under way, with the plethora of experimental datasets, refinement protocols, intermediate models, etc. spread over some 20 Gbytes of disk space. Unless you can find the right person who remembers whereabouts on which disk a particular piece of information is, then it is very difficult to find, and more importantly to validate, the origin of any piece of data.

The aim of this project is to develop a flexible data management system, capable of tracking a wide variety of computer stored information. As with any system, it will only be effective given certain disciplines and working practices, but does provide an attractive working environment for retrieving data. A key feature of the system is its flexibility. The data schema, and the types of operations possible on the data, are all user definable, making use of a central core set of routines. Because of this, the project has evolved in a number of different ways.

Our initial demonstrator was to develop a robust system in which all structural data for a project could be stored for later retrieval. In addition, we wanted to introduce a limited set of data manipulation routines. An example of the use of this level of the database is for somebody in the laboratory to be able to sit down at a workstation and compute and display any type of electron density map, for any structure solved in the lab. Without having to ask somebody else.

Figures 3 and 4 show two different ways in which the database can be viewed. In Figure 3 is presented a typical database schema (it must be stressed that *all* these data structures can be modified or customised by the user very simply within the overall system). The Universe is the top level of the database. Under this there can be any number of *projects*, which can additionally have any number of levels of projects underneath them. In York, for example, we expect to maintain just one

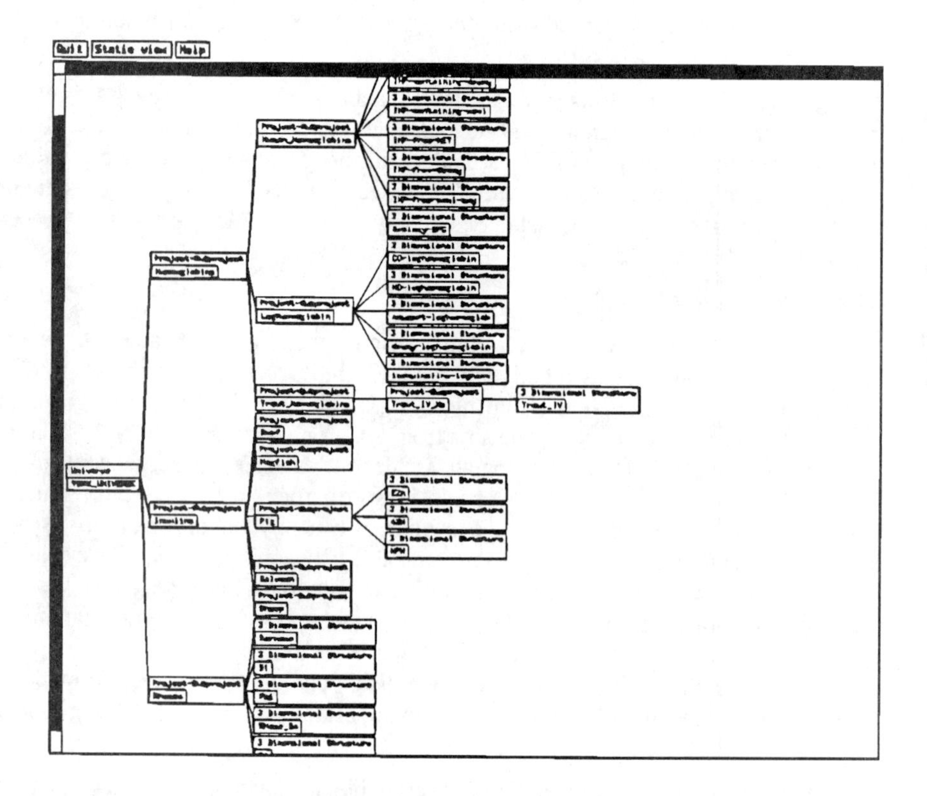

Figure 3. View of schema

Universe. The first level of projects will then be insulin, haemoglobin, barnase, gyrase, cellulase, lipase, etc. Taking the insulin example, there will then be a second level of projects, say native and mutant insulins. Under the native insulins, we could divide the projects in a number of different ways, but let us take for example a division on species. There will be beef, pig, sheep, hagfish, salmon, etc. Then under the pig insulin project, there will be a number of *structures*, for example, 2Zn insulin, 4Zn insulin, NPH insulin, etc. The definition of what is a structure and what is a project is not rigid within a system, but we are currently experimenting with the definition (for our own database) that a structure is defined by the chemical composition of the molecule. So for example, if we had done the 2Zn insulin crystal structure at different temperatures, or if there had been different crystal forms, then these would all be part of the same structure.

For each structure, there are then 'experimental' classes such as sequence, x-ray crystallography, molecular modelling, NMR spectroscopy, etc. The various data types associated with each of these is reasonably clear from the diagram.

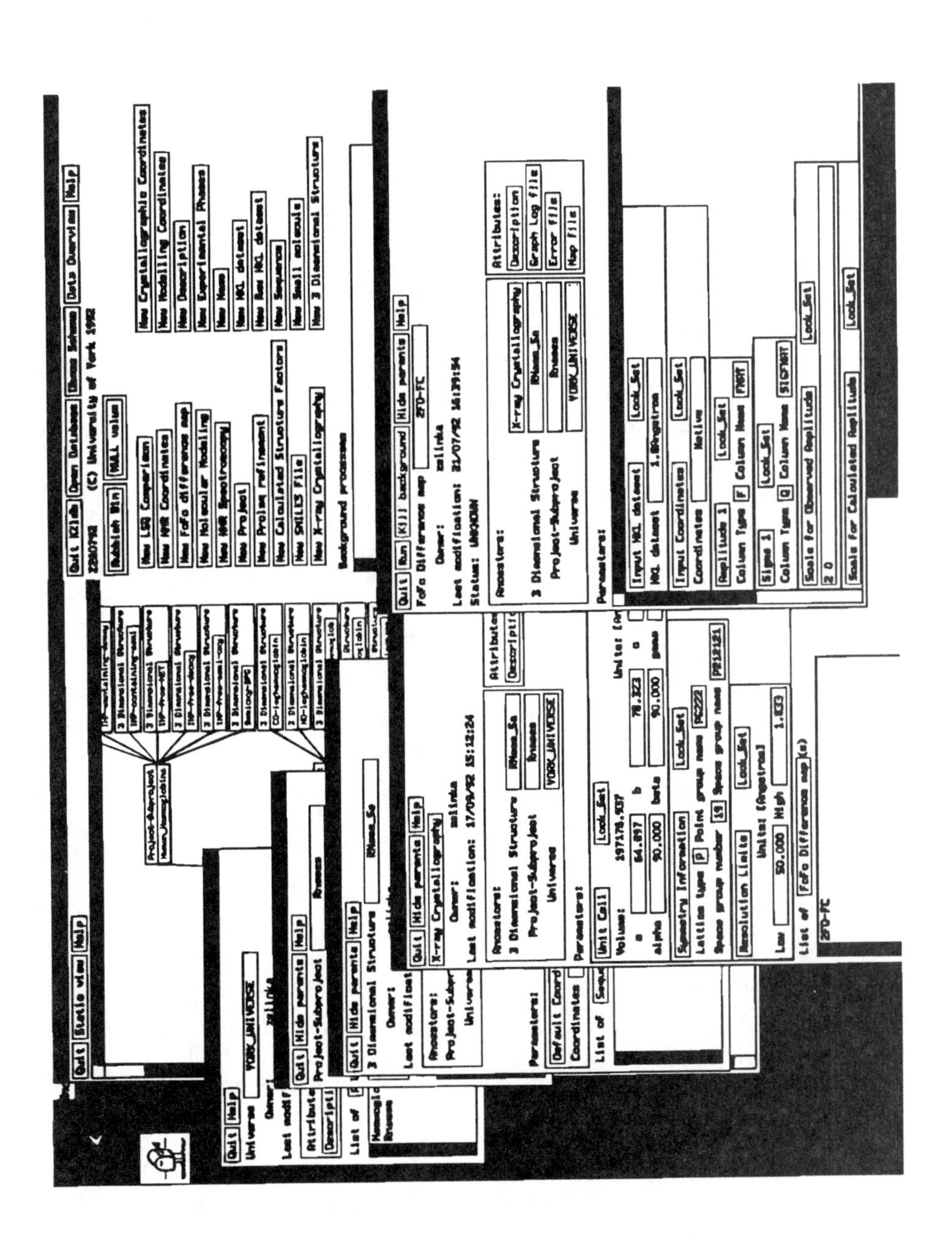

Figure 4. View of database

The alternative view of the data in the top left of Figure 4 shows the actual database information stored at each of these levels — what is labelled the 'Data Overview'. This shows the names of the different data entries at each point in the hierarchy.

To understand how this database is used, is best done by sitting at a workstation and trying to access information. Figure 4 shows a session in progress on a very simple database. I will try to describe some of the features of the database with reference to this figure.

The top of the main window (labelled XZldb) contains commands for opening the database and producing different views of the data. The user can get to a particular piece of data, either by moving down the hierarchy, or by picking an item on the data overview window.

An important interaction mechanism in the database is picking up data types and moving them graphically from place to place within the database. So for example, if a project window is open, to open a new sub-project, the user would pick up the New Project object from the XZldb window, and move it into the project window.

The other window displayed in Figure 4 is the 2Fo-Fc map for a ribonuclease structure. The window contains all the relevant information about the map, and the information necessary to recalculate the map from new coordinates, etc.

This has been only a cursory description of the database customised for one particular application. An extension of the data management project is to use it to manage and access the wide range of available computer-readable information. For example, systems have been developed in which the database is central to operating and recording the use of a wide variety of sequence database search programs, molecular modelling programs, storing the results of the calculations, screen images or session files from modelling, scanned images from published papers, etc.

This system provides the mechanism whereby a structural chemist can organise his use of computational tools, but as with any system, relies on a certain amount of organisation on the part of the user. But given that discipline, it could provide a very attractive way of both working with the wide range of structural information becoming available, and allowing the user to store away in a logical way the results of such studies.

Acknowledgements

The work of the Protein Structural Research Group at York is supported by the SERC, MRC, Wellcome Trust, Celltech, Molecular Simulations, Juvenile Diabetes Fund, Glaxo Group Research, IBM (UK) and NOVO-Nordisk

References

[1] Bernstein F.C., Koetzel T.F., Williams G.J.B., Meyer E.F., Brice M.D., Rogers J.R., Kennard O., Shimanouchi T. and Tasumi M.J (1977) *J. Mol. Biol.* **112** pp 535-542

[2] M. Rossmann, program STEREO distributed as part of the Protein Data Bank from Brookhaven.

[3] Oldfield and Hubbard, manuscript in preparation

[4] Smerdon S.J., Oldfield T.J., Dodson E.J., Dodson G.G., Hubbard R.E. and Wilkinson A.J. (1990) *Acta Cryst* **B46** pp 370-377

[5] Brookhaven coordinate sets 1ilb, 2ilb, 4ilb and 5ilb. Also Oldfield *et al.*, manuscript in preparation.

[6] Kohlstadt L.A., Wang J., Friedman J.M., Rice P.A. and Steitz T.A. (1992) *Science* **256** pp 1783-1790

[7] Clore G.M and Gronenborn A.M. (1989), *Critical Reviews in Biochem and Mol. Biol.* **24** pp 479-564

[8] Guntert P., Braun W. and Wuthrich K. (1991), *J. Mol. Biol.* **217** pp 517-530

[9] Head Gordon T. and Brooks C. L. (1991) *Biopolymers* **31** pp 77-100

[10] Brunger A.T. XPLOR reference manual version 2.1

[11] Nilges M., Clore G.M. and Gronenborn A.M (1988) *FEBS Lett* **239** pp 129-136

[12] Desjarlais R.L., Sheridan R.P., Seibel G.L., Dixon J.S and Kuntz, I.D. (1988) *J. Med Chem* **31** pp 722-729

[13] The Caveat Program, P. Bartlett, Personal Communication

[14] Lawrence M. C. and Davis P.C. (1992) *Prot Struct. Func Gen* **13** pp 31-41

[15] Bohm H-J., (1991) *J. Comp. Aided. Mol. Design* **6** pp 61-78

[16] Goodford P. J. (1985) *J. Med. Chem.* **28** pp 849-857

[17] Miranker A. and Karplus M. (1991) *Prot Struct Func Gen* **11** pp 29-34

[18] Eisen M., Hubbard R.E., Karplus M. and Wiley D.C. manuscript in preparation

Integrated chemical information systems

Dennis H. Smith, James Barstow, Raymond E. Carhart, David Grier, John Laufer, and Anthony J. Schaller

Molecular Design, Limited, 2132 Farallon Drive, San Leandro, CA 94577, USA

Introduction

The challenges facing information systems in the 1990s involve a matrix of complex requirements. The migration from centralised mainframe computing to desktop personal computing has created opportunities and challenges in moving access to and control of information closer to the end-user. During this evolution, we have seen data repositories migrate as well; from centralised hierarchical databases, to relational databases (centralised and distributed), and (much more gradually) object-oriented databases. A common aspect of all of these transitions is that the cost of conversion has been too high in many cases to permit old software and applications to be replaced in favour of new. The effect is the emergence of duplicate or redundant information residing among multiple sources. In some instances, this situation has been accelerated through mergers and acquisitions, which doubled or tripled the number of information sources and hardware platforms in a given organisation within a relatively short timeframe.

Within this new wave of computing, there has been a drive to move the access control for information to the desktop of the user. Graphical user interface standards were introduced in order to increase ease of use and provide a consistent look and feel to desktop applications. This provided some assistance in shielding the complexity of the various systems. However it did not address the difficulties in accessing the different sources of information which existed through heterogeneous database systems (database systems of different vendors and/or different architectures) on heterogeneous platforms. The user was left to manually merge the information and perform the cross-checks required to insure the correct information was reported (especially when redundant information existed).

The new computing environment

These industry trends are driving specific changes in the scientific computing environment. Terminals that were connected by asynchronous communication links to remote, centralised mainframes are giving way to dedicated desktop workstations or personal computers (PC). As users grow weary of inconsistent response times due to intermittent CPU-intensive compute jobs, or the increasing load placed on hosts by additional users, the notion of dedicated compute resources provided to each user is extremely attractive. Personal computers empowered the end-user with the ability to fulfil his/her own computing need. The services for printing, file sharing, data access, etc. were still needed so these systems were linked to remote hosts or dedicated *servers* via high speed *networks.*

A *server* refers to a processor (or processors in the case of a multi-processor architecture) which is dedicated to fulfilling the requests of attached workstations (Figure 1). Servers can operate for remote disk storage (disk server); file sharing and storage (file server); computing services (compute server); or acting as a centralised print manager of attached printer devices (print server). We predict that the trend from centralised mainframe computing to network based workstations will continue at an accelerated pace.

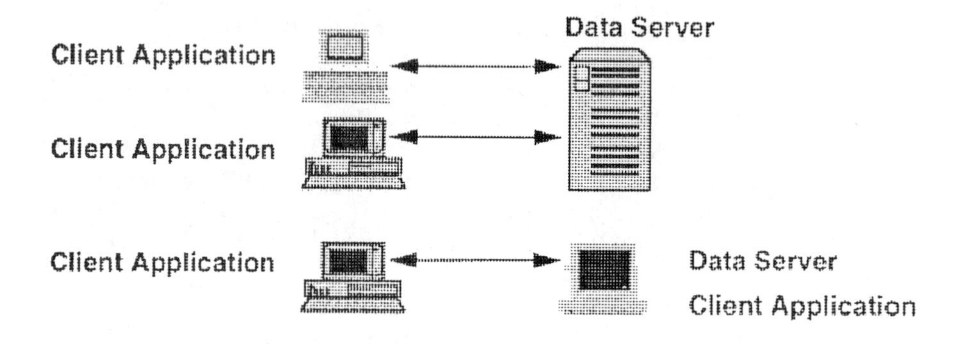

Figure 1 Example of a Client–Server application environment

A *network* refers to a communications medium which provides connectivity among workstations, servers and hosts. These networks transmit 1) requests for computations from a workstation to a server; 2) results of the searches or calculations back to the workstations, 3) file access requests to service read/write operations, 4) print jobs to network attached printers. Transmission speeds of networks are orders of magnitude higher than asynchronous telecommunication. Even these speeds are constantly being extended as is evident with 100 megabit/second Ethernet becoming a reality over common telephone wiring. (unshielded twisted pair)!

In scientific R & D, it was expected that workstations would be available as an integral part of the desktop or laptop of every scientist, including all associated technical and administrative support staff. In addition to the hardware, powerful software would be purchased, installed and used effectively by well educated (but not necessarily 'computer literate') persons as opposed to computer programmers. This software would conform to user interface guidelines and standards thereby limiting the amount of training and 'interface shock' that a user would need to endure when switching from one application package to another.

It was also observed that evolutionary changes were taking place with respect to database management systems (DBMS). More and more data were being collected in relational and other DBMS regarding laboratory data or experimental results. In addition, it was felt that other data such as free text, chemical and biological structures, and even images would make up the domain of information within which the systems would require access. Effective integration of these diverse data sources was seen as necessary for the comprehension of all of these data, especially with regards to presentation to others (e.g. regulatory agencies).

The database types, from which this information integration would be initially supported, included the chemical database systems (CDBMS) produced by Molecular Design Limited (MDL) that were accessible via the REACCS and MACCS-II systems. Support would also be provided for non-chemical data sources such as relational database management systems (RDBMS).

ISIS project objectives

To respond to these dramatic changes in computing environments, Molecular Design Limited launched a project in 1987 to build a distributed information system to access chemical and non-chemical information. ISIS (Integrated Scientific Information System) was designed to provide access to information in a manner that would shield the complexity of the underlying database systems from the end user. A graphical user interface would be provided to enable access to the data in a consistent manner with the ease-of-use that is characteristic of a 'point and click' interface. Datatypes and data relations would be provided in a way that would be consistent across the spectrum of heterogeneous database sources. The system would ensure a view of information that provided *location transparency* by shielding the end-user from the actual data components residing within the underlying databases. This integrated perspective would give the user a 'merged' view of data across the organisation. Access to information would also not be limited to a single particular host system. Data would reside on multiple hosts with integration requirements stretching across networks. Information links would be required to support periodic query and update of corporate databases and/or other project databases elsewhere within an organisation.

In addition to fulfilling the access requirements for coupling in the various database types, ISIS would also provide *data transparency* in that the system would shield the user from the underlying organisational data model relative to each database The end-user would not need to know if the data were stored in an indexed sequential access (ISAM) format, a relational format or a format using a relative index. Instead, the software would use the concept of an external schema local to each user or application that would permit the data between different sources to be linked in various organisational models.

An 'open architecture' was an additional requirement. Therefore, we designed in ISIS/Host (see below) a set of methods or entry points by which a customer (or third party vendor) could define links into their respective data repositories. These links could be used to build specific *gateways* to the underlying databases that would be integrated into the overall heterogeneous view. This would provide extensibility to allow integration of data that may be critical to a vendor or customer. Further, an *application programming interface* (API) was provided to allow the integrated databases to be accessible to other application systems. These systems could utilise a C binding interface to access data transparently through the ISIS/Host API. This could define a migration pathway from the old to the new systems.

Finally, the software architecture to be utilised would be a client-server model. The server would reside on a host computer and provide access to the data that resided there. In this case, the host would be some type of mini or mainframe system (although the paradigm could be extended to workstation servers). The server component would have responsibility for receiving requests from the client for data and coordinating the search of data from the various database sources. Once the data were retrieved, the results would be passed back to the client for presentation to the user. The client would reside on the desktop workstation servicing the needs of the user for query validation, local database access/storage, and data presentation. In addition, it was expected that the client software would also provide the capability for the user to construct the type of query that was required as well as generate useful forms to customise the look and feel of the presentation.

ISIS project implementation

The ISIS project proceeded from design to product during the period 1988–91. Close contact was maintained with customers to understand the needs and desires for this distributed solution. Functionality was prototyped, revised, demonstrated and further revised. During this period, the determination of hardware was made with respect to customer requirements. These requirements dictated support of the Macintosh and other Intel based personal computers supporting Microsoft Windows as clients. These desktop systems provided the basis for a graphical user interface which operated in a consistent manner across the platforms, while obeying the user interface style guide for each system.

As the majority of the MACCS and REACCS sites were operating on the Digital Equipment VAX hardware, the first choice for the server software was VMS with CDBMS and RDBMS access provided (Figure 2). With respect to the relational systems, customer requirements dictated support of RDB/VMS, ORACLE and INGRES. From the beginning, it was our intention that it would be straightforward to add other access to data sources via the Open Gateway (see below).

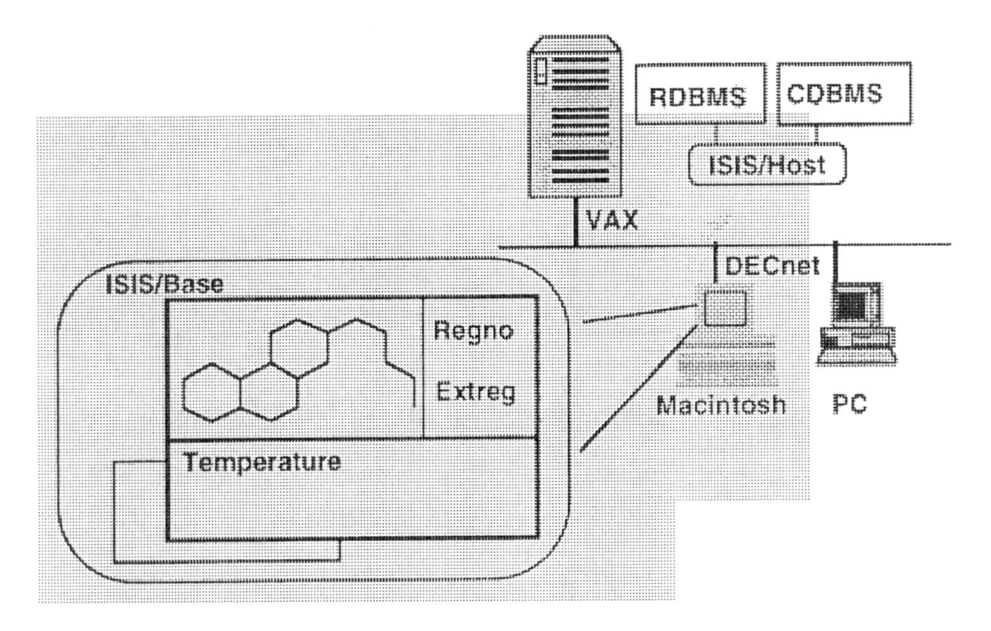

Figure 2. The ISIS Client-Server application model

ISIS matured into several software components (see Figure 3). ISIS/Base is a client application that resides on the desktop. It manages a forms-based graphical user interface (GUI) through which information is presented and search predicates are constructed. Forms are constructed based upon the schema dictionary defined via an *HView* (this view is a local external schema that specifies the relationships between multi-vendor data sources. These relationships define the integration). Customisation of the forms is possible to meet the needs of the user or organisation. A procedural language (PL) is available to permit customisation of the desktop interface. Local desktop database facilities are provided for storing chemical information (structures and reactions). ISIS/Draw is a second application that provides the drawing capability which one would expect in a GUI environment, with the benefit of ensuring chemical significance. ISIS/Host is the server application that resides on the host. It contains several components: a communication layer; a programmable API to allow other applications to interface to the ISIS/Host; and the integration core. The integration core has the responsibility to take the schema definition of the HView and create the logical links between gateways as well as splitting the transactions (query and update) for dispatch. Furthermore, ISIS/Host provides the individual database gateways which map the transaction fragments for interfacing to the native database mechanism.

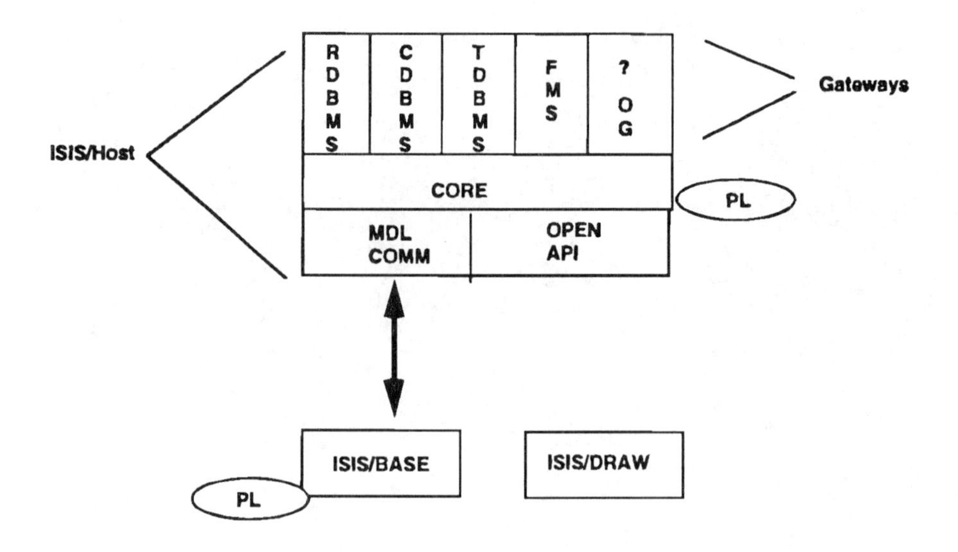

Figure 3 ISIS Product Architecture

Relational Algebra Base

ISIS is consistent with relational concepts; however these concepts were extended to be more aligned with the NF2 (non-first normal form) model. ISIS treats data as tables (relations) or collections of joined tables, albeit with some restrictions on how they can be manipulated. As a portion of the NF2 support, one of our datatypes is a RELATION, allowing us to represent hierarchies of data in a natural way. Native datatypes in the MACCS, REACCS, RDBMS, are translated to ISIS datatypes such as INTEGER, TEXT, REAL, REAL RANGE and DATE. Other native datatypes such as those used to hold molecules or reactions are treated as abstract or composite datatypes, each with their own operators and functions.

SQL concepts are used at certain levels of the operation. For example, the SELECT command within our command-line interface to Host follows syntax similar to an SQL SELECT statement (Figure 4) except for extensions that are required to map to the linkage of the schema that constructs the heterogeneous view. Our goal was to implement a complete and extensible query language. SQL names are used, when possible, for operators. ISIS/Base communicates with ISIS/Host in a composite language that includes the operators of the various databases.

SELECT column | function | expression, . . .

FROM table | view, . . .

WHERE query-conditions

GROUP BY column, . . . *HAVING* query-conditions

ORDER BY column [ASC|DESC]

Figure 4. Syntax of the ISIS/Host extended SQL SELECT statement.

Although ISIS offers the power of a query language to advanced users, most do not require this knowledge. The system uses the power of the graphical user interface to search for and present information via a form.

Challenges for ISIS/Host integration

One of the difficulties was to build a mechanism by which heterogeneous data could be related to construct a composite view by which the end-user or developer could access the information in a transparent fashion. Several approaches are possible to resolve this. One method is a 'top down' approach in which a site-wide schema is constructed that contains all information related to an overall federated view. The drawback to this solution is that it requires that tables, fields, etc. of all databases involved in heterogeneous access be brought together in one schema before they can be accessed. As part of the merging process, decisions must be made as to field naming (where redundant data are merged but may be named differently in the underlying separate databases), data type conversion (i.e. merging of a field stored as integer in one system; character in the other), and removing the security barriers so that users have permission to access fields or tables that are not normally used in their work.

The chosen approach was to use a 'bottom-up' strategy. Rather than organising all possible fields into one massive view of information, the concept of an HView was employed. The HView is a local, external schema that allows the application developer or end user to define the domain of information relative to the gateways, tables/entities and fields that they wish to view in an integrated manner. This provides a quick mechanism to define multi-databases that may be specific to an end-user, developer, or project group.

In terms of organisation methods for how data would be linked together within the HView, a model is provided that closely represents how end-users think of the data. Generally this is in terms of a hierarchy. Traditional database theory dictates that one speaks of master versus detail records. Each master record refers to a number of detail records. An example of this is evident through systems such as IMS (Information Management System) or REACCS. Within REACCS there can be several levels of the hierarchy; each REACCS reaction may consist of many variations, each of which may contain multiple references, each of which could contain many authors. Even in MACCS, which is generally thought of as a non-hierarchical, flat database system, end-users often treat a flexible text field as though it were a table of detail records, each delimited by a carriage return. Column searching is then used to search for specific fields within the detail records.

SQL is certainly compatible with a hierarchical view of data although it requires modifications to adequately refer to the positional relationships that may exist with respect to a parent-child. The relational concepts of SQL permit representing any data architecture, including a hierarchical or a network architecture. Figure 5 depicts an example that was obtained from Oracle. It is the 'Order By Regno' clause and the 'Break On Regno No Dups' statement that define the nature of the hierarchy. The 'Order By' clause insures that rows with the same Regno are grouped together. (The 'Group By' clause could theoretically be used instead to insure Regnos are grouped together, without specifying their order as well. Oracle does not allow a 'Group By' clause in this context but ISIS does provide the function.) The 'Break No Dups' statement tells Oracle to print each Regno only once, which results in a visually layered or hierarchical table. Note that the hierarchy can easily be inverted as in Figure 6, just by changing the 'Order By' clause and the 'Break' statement. (The 'Break' statement is not an SQL statement *per se*, but rather a formatting command. It is available in both Oracle and Ingres.)

```
Break on Regno No Dups
Select          Regno, Author
From            Demo
Order By        Regno
```

REGNO	AUTHOR
1	A Van Zon
	N G Kundu
	S Akabori
2	H Schilling
	N G Kundu
3	A Costa
	S Akabori
4	H Schilling

Figure 5

```
Break on Author No Dups
Select          Regno, Author
From            Demo
Order By        Author
```

REGNO	AUTHOR
3	A Costa
1	A Van Zon
2	H Schilling
4	
1	N G Kundu
2	
1	S Akabori
3	

Figure 6

Although a hierarchical display can be obtained within SQL, it becomes quite clumsy and nearly impossible to handle multiple levels and multiple branches of the hierarchy. The HView concept is provided to orchestrate such integration.

As you can see from Figure 7, within the HView, rules are established to identify the various data sources that will be part of the linked view.

```
HView MX_RDBMS
            TREE    ISISMXDB
                    DEVICE MACCSDB
                    DATABASE ISISMX:
                    TINFO KEY MOLREGNO
                    TNAME MOL
                    PASSWORD
            TREE    INVENTORY
                    DEVICE ORACLE
                    USERNAME <prompt> ORACLE Username?:
                    PASSWORD <prompt_noecho> ORACLE Password?:
                    TNAME DEMO_INVENTORY
            TREE    CHECKOUT BASEDON INVENTORY
                    TNAME DEMO_INVENTORY
            LINK    ISISMXDB MOL>(CORP_ID) OVER INVENTORY (CORP_ID)
            LINK    INVENTORY DEMO_INVENTORY>(BOTTLE_ID) OVER -
                    CHECKOUT (BOTTLE_ID)
```

Figure 7 A sample HView

For each data source or device, the participating table, tree or *segment* is identified. Link statements are provided to establish the nature of the hierarchy as far as what fields are used to pass values as criteria for locating the child records. This 'linking' concept provides the logical join to the various relations or records that exist either above or below in the hierarchy. Another way to display and think of an HView is the assembled view shown in Figure 8. Here we see that each relation is treated as a field with the type PARENT. Each PARENT field can store multiple sub-records, each consisting of the primary fields (non- PARENT fields) beneath the PARENT in the hierarchy.

Rxns				PARENT
	RxnID			NUMBER
	RxnSymbol			CHAR
	Keywords			CHAR
	*Reaction			REACTION
	Variations			PARENT
		LineNo		NUMBER
		References		PARENT
			Author	CHAR
			Journal	CHAR
			Volume	NUMBER
			No	NUMBER
			Year	NUMBER
			Page	NUMBER
	Solvents			PARENT
			Regno	NUMBER
			Grade	CHAR

Figure 8 Sample resulting view

Each reaction can store multiple variations, each of which can store multiple references and solvents. This display demonstrates how an HView can represent a REACCS database. Each PARENT data type of the HView will map to a PARENT data type in the REACCS database. Alternatively, an HView can represent data actually stored in an RDBMS or in a combination of MACCS and an RDBMS. In this case, the joins between the relations in the HView are in the definition in the HView. To the user, it will be irrelevant as to where the data are actually stored. The HView defines completely how ISIS/Host will treat the data regardless of location. In particular, search queries can refer to any field or combination of fields in an HView. The same search query can refer to fields at different levels of the HView hierarchy.

Datatypes for each gateway are defined where they make sense relative to the underlying database gateway. For example, a MOLECULE data type may not exist for data in an RDBMS. Similar data types are expected for fields that are used to join or link the parents and children.

Ideally, functions and operators should treat a MOLECULE data type the same way no matter where the molecule is used. If a molecule in a package such as ISIS/Base behaves differently and allows different operators and functions from a molecule in MACCS, the two molecules should by strict standards be different data types. However ISIS follows a slightly relaxed guideline that, for a particular data type, every operator and function works in a similar manner (when possible). This latter exception permits the treatment of MACCS and ISIS/Base molecule data types as the same data type. There are some cases where idiosyncratic behaviour exists, e.g.. some RDBMS text searches are case-dependent, while the CDBMS searches are not; REACCS and MACCS do not handle % in the middle of a text query while the RDBMS gateways do.

Fields within the HView are referred to by field *path names*. A path name is a concatenation of the name of all PARENTS above the field in the hierarchy, followed by the field name itself, each separated by a '>'. A valid field name for the HView fragment in Figure 8 is Rxns>Variations>References>Author. A field is often referred to with a *partial pathname*, truncated on the left, as long as the partial pathname is unambiguous. The preceding field name could be abbreviated to Author. The name of a PARENT data type can be used in a SELECT command, or on a form box, to refer to *all* the fields below it in the hierarchy.

The PARENT data type in an HView allows a hierarchical field name to refer to a tree of data. In some cases, it is useful to use a hierarchical field name to refer to data that are logically grouped together, even if the data are not actually hierarchical. For example, in the HView depicted in Figure 9, there may be fields that contain data related to the chemist that created the compound. These data are not hierarchical; there is only one chemist per reaction in this example. However, it would still be useful to be able to refer to the fields as Chemist.Name, Chemist.Dept, Chemist. Hours, or to the entire group of fields as Chemist. Certainly an alternative method in the relational context would be to perform a join between two relations. In the event that there were multiple chemists per compound, however the hierarchical field name provides the added benefit of avoiding the duplicity of values that can occur with a join (i.e. duplicate value for Compound and Chemist for every occurrence of Chemist in the secondary table.

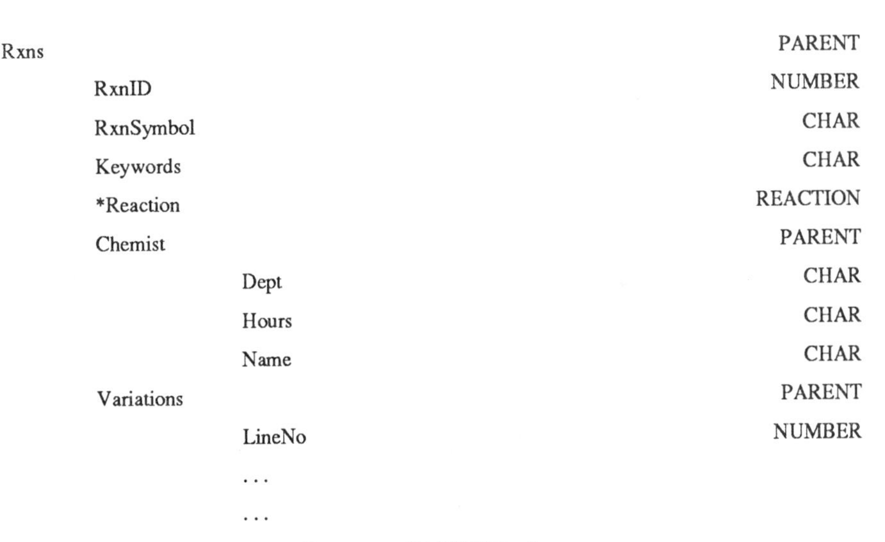

Figure 9 Example of PARENT reference

HViews do not merely specify relationships between individual segments in homogeneous sources or heterogeneous sources. The goal of the system is to integrate data from distributed heterogeneous sources. As a result, it is possible to define HViews that refer to other HViews; a custom HView can link together any combination of other HViews. These HViews can refer to databases on the same or different hosts provided that they are linked together via a supported protocol. Currently, the protocols which are supported include DECNET and TCP/IP.

When the HView is first accessed, ISIS/Host will access the referenced devices or databases and create an *integrated data dictionary.* The layout of the fields as specified in the HView is validated, then the field information (including names and types) is transmitted to ISIS/Base. Thus to the workstation, data links between host sources are transparent. The view of data that appears available for access by the user can consist of actual data components from MACCS, REACCS, ORACLE, Ingres, flat ASCII file, etc.

The original model used for Host was to enable process creation of the server at the time the user requested a remote connection. Under the circumstance in which the workstation supports DECNET protocol, it is possible to open a connection via transparent DECNET. As part of initialising the connection, an ISIS/ Host server process is started on the host system. The transport layer is implemented in an 'object oriented' manner so that other communications transports can be easily plugged into in place of DECNET (i.e. VMS Mailbox, VMS subtask, and Appletalk). Situations where other protocols are used (e.g. TCP/IP) dictate that another model for process startup must be used. One consideration is a *broker* scheme.

In the broker scenario, a process awaits connection requests via a well known service name. Once the request for Host service is received back from a workstation user, the broker starts the server process and communicates the address information to the requesting program. Dialogue continues directly between the workstation program (ISIS/Base) and the host program (ISIS/Host). Although this is functional, the early design was not robust enough to deal with large groups of users attempting simultaneous connection.

As an alternative, the concept of a *conduit* is used. Under this mechanism, a known process awaits connections via TCP/IP, and upon receipt, starts a Host process using DECNET. The ISIS/Host program continues to communicate via DECNET using the conduit which translates the data to/from TCP/IP for communication with ISIS/Base.

In the course of developing the host server, several tools were developed to debug the specialised nature of the developed code. One of these tools in use by customers and field sales personnel is HSearch. HSearch is a host-based tool which accesses the heterogeneous data through the HViews and supports a command line user interface. This tool is used for debugging as well as data retrieval. It is so widely used that it is now being developed into a fully supported product.

Desktop integration challenges

The operating environment on the desktop workstation is a windowed, graphical use interface for access to both local workstation and remote host application software.

Although the environment of Base provides a 'Query By Form' capability in which default forms are available to quickly generate queries and review results based upon the HView definition, it is not possible to provide only one user interface that is useful for every end-user. Therefore, it has a customisable component (PL which is described below), to permit the building of specific interfaces for viewing information according to the special needs or reporting requirements of individuals or their organisation.

Storage of local information is also provided with Base. There exists a need to allow users or project groups to maintain their own data for manipulation and avoid the communications overhead for all data operations.

Various methods are used for moving information between applications on the desktop. Cut and paste using the system clipboard is one standard method. The clipboard receives data from one application to then be moved to another application. Although this technique seems to offer a clean method for sharing information, MDL discovered various situations in which limitations or a lack of standards existed.

On the PC and Macintosh, the clipboard supports a variety of data formats including one for graphical images. Unfortunately, for most applications this graphical format is different from its native format. Thus when an image is pasted into a receiving application it cannot be moved back to the creating application and edited. In order to solve this problem, MDL hid native sketch data in every image that it moves to the clipboard. When an image is pasted into ISIS/Draw from the clipboard, it is first searched for hidden native data which is used if found. If the native data are not there, the image is treated as a separate picture and cannot be edited.

This scheme works well in almost all cases since most receiving applications display the image and ignore the sketch data. Unfortunately, there are some applications which either delete or corrupt these data. An example of this would be the use of a text editor or word processor in which structure drawings are pasted, for example within a printed document. This document may later be referenced by another reader who cuts the structure from the document and pastes it into ISIS/Base for use as a query. When this happens, moving the image back to the ISIS/Base or ISIS/Draw application for editing is not possible.

On OSF/Motif the problems are much greater since there is no standard graphical image format for the clipboard. Under this system, ISIS is forced to export Encapsulated PostScript files which are then imported into receiving applications.

Each of these boxes or regions can have relevant information or text that are attached to the nearby symbol. In the event that the symbol is moved, the information moves with it. Another chemistry-related concern was with the registration of chemical formulae. MDL needed to construct a new Font Type that intelligently subscripted numbers in chemical formulas.

Early promoters of the ISIS project indicated that the desktop applications, Base and Draw, required operation on a number of platforms (i.e. Macintosh, MS-Windows, and OSF/Motif). In order to quickly port the Base and Draw applications to these systems but yet retain the style characteristics that provide individual identity to these GUI, a library was created called MDL Windows.

MDL Windows provides a layer between the native resources of each GUI and the functions required by Base and Draw. As a result, a single set of machine-independent code can be moved from one platform to another. As a result, the application operates in a consistent manner across all desktops but provides the 'look and feel' appropriate to each specific desktop environment.

To facilitate the development of extensions not included in the 'out of the box' functionality, PL, or Procedural Language is provided to allow further customisation. Implemented as a subset of PASCAL, it allows users or application developers to construct easy point and click interfaces to tasks that may be required.

With the first release of the product, the PL intrinsics operated in a manner which was exclusive of the included functionality for Base operations (i.e. Search, Update, etc.). The programmer assumed responsibility for all of these operations once the decision was made to implement a customised solution. Customer feed-back indicated that extensions were more often desired to augment the delivered functions and not entirely replace them. As a result, the new release of PL allows the programmer to trap the events that are received by the Base application, as well as menu selections, in order to give him/her the option for executing a separate section of PL code and bypassing/returning to the normally packaged functionality. This ability to inter-mingle customisation with standard features provides a powerful mechanism to incrementally build more complex systems while retaining the best aspects of the bundled features.

Chemistry Challenges

Access to the data in existing MACCS and REACCS databases by the Host software is provided using software gateways that utilise the same searching methods existing in the mainframe-based chemistry applications. The idea of a 'molecule structure' and 'reaction structure' field is formalised and operators exist to represent structural searches. Thus an exact-match molecule search is posed to ISIS as the query shown in Figure 10.

MOLSTRUCTURE = <molecule-query>

Figure 10 Query construct for exact-match molecule search

All other structural operators are represented by special names; '=' is the only operator which is overloaded to deal with the structure fields. Some of the other structural operators that are supported are shown in Figure 11.

SSS	molecule substructure	*RSS*	reaction substructure
TAUTOMER	molecule tautomer	*SUBSET*	reaction subset
ISOMER	molecular stereoisomer	*FLEXMATCH*	flexible exact match
SIMILAR	molecule or reaction similarity		

Figure 11 Structural operators

The MACCS and REACCS gateways serve to collect the query, which is presented to them as a field number, operator, and molecule or reaction connection-table. There may be other operands that are optional to modify the search. The query is then re-packaged into a form at which the native database engine can accept, and passed over to the engine for searching.

MACCS and REACCS data are represented using a set of data types based upon the ANSI SQL standard, with extensions as required to represent specific MACCS and REACCS datatypes. New datatypes are added (for example) to represent molecule/reaction structures, and the REACCS 'real-range' data type.

Access to REACCS data is presented much as it is in REACCS. The field names and types are essentially the same, and the hierarchical structure is maintained in the view presented by the gateway. The gateway makes certain 'special' fields act like normal data fields; this is required by the ISIS model of search and retrieval. These include molecule and reaction structures, molecular formula, and reactant and product molecule internal registry numbers. Data search queries are translated into the appropriate REACCS search, and submitted to the query engine.

Each REACCS database is logically presented as two hierarchical 'trees', one rooted in reactions, the other in molecules. Using the standard Host integration capability, these can be combined into one hierarchy, so that users can view or search fields from both trees at once. Internally, the gateway recognises that the two trees correspond to the same database, and it can use efficient REACCS list conversion routines to bridge the two trees.

MACCS data present more difficulties. Whereas REACCS has an essentially SQL-like set of field types, text, and numeric search operators, MACCS has idiosyncratic datatypes, and supports only a BETWEEN operator for numeric and text searching. In order to present a uniform view to the Host 'Integration Core', the MACCS gateway must therefore perform more work in translating queries and/or taking over some of the searching burden in order to provide a more SQL-like set of operators. In addition, the MACCS datatypes must be presented in a more regular fashion.

For example, a MACCS 'numeric' field can store two numbers and a text descriptor. The numbers can be the same, in which case they are presented at the MACCS interface as if they are just one number, they can be the lower and upper bounds of a range, or they can be two independent values. The way in which the two numbers are treated depends entirely upon the commands the user uses to search, update, or retrieve those numbers.

The MACCS gateway regularises the numeric field by presenting only one view of the field. However, in order to support the three different interpretations of the field, that view is customisable so that the user can see either a single numeric field, a pair of numeric fields, or a single numeric field whose type is 'range of values.'

The MACCS database is essentially a single table, with a set of fields such as DENSITY, MELTING.POINT, etc. As in the REACCS gateway, certain 'special' MACCS fields are presented as normal data fields, for example molecule structure. MACCS also has some data types which are capable of storing multiple lines of information. For example, DENSITY might be a multi-line numeric field, storing densities measured at various temperatures. The MACCS gateway represents these multi-line datatypes as sub-tables, allowing one to search and manipulate each line separately.

A MACCS-3D database is presented by the gateway as 'two trees', one rooted in 2D molecules, the other in 3D models. Like the REACCS reaction and molecule trees, the 2D and 3D trees can be combined using Host integration so that queries and data retrieval can pertain to fields in both trees at once. Internally, MACCS 'transform list' technology is used to convert results from one tree to the other.

Deployment challenges

During the delivery of the first release of ISIS, one of the difficult challenges for MDL was managing customer expectations of the first several releases of the software in accordance with the overall vision for distributed computing. The desire of customers to quickly decentralise and distribute the tasks of registration and searching led various groups to tackle difficult problems such as heterogeneous database update that no vendor to date has fully resolved.

In addition, the desire to have computing systems on every bench or desk with systems deployed to the end-user is often slowed by the need to build the support infrastructure required to manage networks of hundreds or thousands of personal computers. In many cases, success is achieved with deployment of applications focused on fulfilling needs of the end-user scientists in assisting their day to day information requirements. These can be seen by the following examples:

Hoechst AG developed an early application and rolled it out to users as the single user-interface to the chemical and relational databases. They delivered functional systems into production within two months after final product release. An example of the systems that they have deployed is provided by their PRISMA Substance Information system.

The product information system (PRISMA) progressed through various stages of development. A data analysis was carried out using an entity relationship model in which data residing in a 20 year old IMS database was transferred while taking into account additional new requirements. A list of the data fields was created, and necessary relations were formed from the entities. The prototype of PRISMA was realised using a simple MACCS-II database. At the same time, an Oracle database with the 20 relations and a corresponding MACCS-II database were prepared. The principle, that the data should be structured independently of the tool used to access it, has been largely held to. HViews have been written and forms within ISIS were developed so that the database could be used with over 120,000 structures. Work continues with additional functionality being added via PL.

ICI Americas concentrated on production applications that provided time savings. Within the first few months of ISIS production release, they built a system to combine the use of electronic mail with the product. Results can be quickly posted from a screening laboratory to a medicinal chemist, or the compound registration group can mail information electronically back to the chemist so screening may begin. Within the mail message, information is provided from the relational systems (Ingres) as well as the structure drawing. The ISIS/Base front-end application allows the scientist to enter information, with error checking occurring locally on the workstation, during creation of the submittal form. This reduces the need to recycle the form back from a central data entry group when mistakes are noted. As a result, an overall decrease in processing time is achieved while reducing the opportunity for errors. Once data are entered, they are applied to the chemistry database using the automated mechanisms that existed previously prior to ISIS. The functions of the underlying production application are preserved while eliminating the weaker aspects of the process.

The following examples offer a sampling of other general applications that have been constructed:

- For analysis of biological results, the screen depicted in Figure 12 represents a forms layout to support query of biological and structure information.

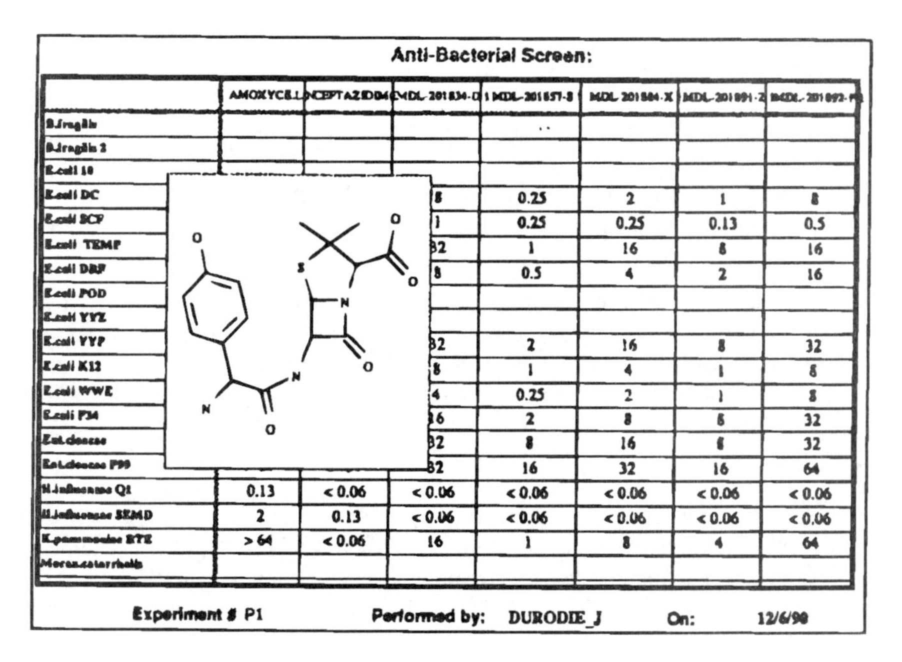

Anti-Bacterial Screen:

	AMOXYCILL	CEPTAZIDIM	MDL-201834-0	MDL-201857-8	MDL-201804-X	MDL-201891-2	MDL-201892-?
B.fragilis				..			
B.fragilis 2							
E.coli 10							
E.coli DC			8	0.25	2	1	8
E.coli SCF			1	0.25	0.25	0.13	0.5
E.coli TEMF			32	1	16	8	16
E.coli DRF			8	0.5	4	2	16
E.coli POD							
E.coli YYZ							
E.coli YYP			32	2	16	8	32
E.coli K12			8	1	4	1	8
E.coli WWE			4	0.25	2	1	8
E.coli F34			16	2	8	8	32
Ent.cloacae			32	8	16	8	32
Ent.cloacae P99			32	16	32	16	64
H.influenzae Q1	0.13	< 0.06	< 0.06	< 0.06	< 0.06	< 0.06	< 0.06
H.influenzae SEMD	2	0.13	< 0.06	< 0.06	< 0.06	< 0.06	< 0.06
K.pneumoniae STE	> 64	< 0.06	16	1	8	4	64
Morax.catarrhalis							

Experiment # P1 Performed by: DURODIE_J On: 12/6/90

Figure 12: Biological data review

- In order to support inventory queries regarding availability of compounds within a store room, the structure is used to provide an unambiguous way to allow the chemist to locate chemicals. In the example shown by Figure 13, the chemist has used the structure to locate information for ordering purposes.

ISISDEMO Structure and Inventory Display Form

Name	Felodipine,H-154/82,Hydac(TM),Plendil(TM)
Formula	$C_{18}H_{19}Cl_2NO_4$
Corp ID	MUSE33300004
Molecular Weight	384.26
Other Names	

Bottle #	Stockroom	Supplier	Amount	Purchase Date
4805.0000	E-1-20	RBI	88.34	1/3/82
6352.0000	A-1-16	EMKACHEM	45.48	6/5/80
30559.0000	D-5-17	B & J	40.48	12/3/77
7334.0000	B-2-18	FLUKA	66.36	11/12/83
19878.0000	E-5-22	PROLABO	65.35	10/1/85

Chemist	Laboratory	Check-Out Date	Check-In Date
WORDEN	B-229	6/26/90	10/12/90

Figure 13 Inventory management application

If it is in stock and sufficient quantity exists, then it may be electronically ordered. ISIS may suggest other compounds that could be substituted if it has the same structural features or physical properties.

This example demonstrates an application supporting automatic search of literature references via a search and browse technique among reactions which are pertinent to their work. A simple point and click interface enables the chemist to electronically mail the Library with requests for specific literature references (Figure 14).

Future considerations

Although the ISIS project overcame many of the challenges that were presented, there is still much work to be done. The tasks are not only limited to chemistry but exist in the domain of general computer science. Many potential solutions exist in a ill-defined state with virtuous intentions (or features) but complex and unresolved logistical difficulties.

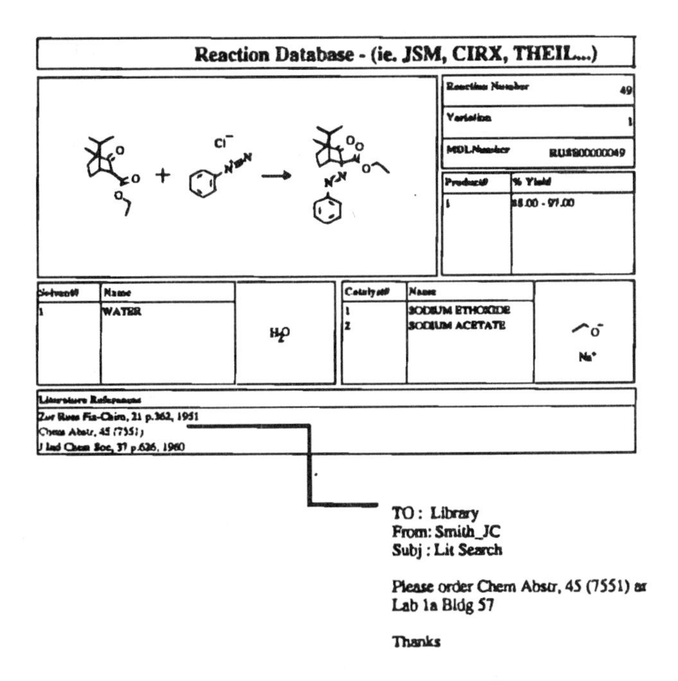

Figure 14: Literature search application

An example of the latter is the extension to desktop inter-application communication (IAC) known as OLE (*Object Linking Embedding*). OLE is a facility which permits objects to carry with them the knowledge of what application was used for their creation as well as the data representing the object. If the object is linked, the document provides only minimal storage for the data to which the object is linked (generally a filename), and the object can be updated automatically whenever the data in the original application change. For example, if a range of spreadsheet cells is inserted as a linked OLE object into a word processor document, the data are stored in a file created by the spreadsheet application, and only a link to the file is saved with the document.

For an embedded object, all the data associated with it is saved as part of the file in which they are embedded. If a range of spreadsheet cells is inserted as an embedded OLE object into a word processor document, the data in the cells would be saved with the document, along with the name of the spreadsheet application. The user can select this embedded object while working with the word processor document, and the spreadsheet application will be started automatically for editing those cells. The presentation and the behaviour of the data is the same for a linked and an embedded object.

The OLE concept has not fully matured with respect to remote systems. Links to a remote database could have unpredictable results in the event that the remote data are held in a locked state, renamed, or the remote site is unavailable. This aspect has been studied through similar issues in database research regarding referential integrity constraints across distributed databases.

In the area of access to distributed heterogeneous database systems, MDL is one of the first commercial vendors to provide a true integrated solution for searching. The work however does not end there. Areas that are under examination involve issues of transparent registration or update. The challenge exists to provide transparent support of update transactions to remote data that may exist on one or more machines without requiring elaborate programming or interaction from the end user. For example, it is not acceptable for a transparent solution to suddenly alert the user that the data on server BIO1 (Internet address 128.2.53.4) are not available for access, and request confirmation if they wish to continue or abort the update. In fact, research done in this area would indicate that even a two-phase commit model is inadequate to service the needs of heterogeneous systems. The MDL model is to co-exist with systems that are already operational with applications deployed in the native database applications. A more likely solution involves a type of voting protocol that operates against a quorum of database gateways. In this regard, a 'presumed commit' protocol is supported for updates to be pushed through or staged for the minority of database sources to come available. Added complexity occurs as this requires the intermediate result of transactional updates to be visible to other users and applications, with compensating transactions to back out the changes in the event that the update is eventually aborted. In an heterogeneous integrated model, these updates could operate over a period of seconds to several days, depending upon the availability of data, nature of the data and accessibility to information residing on remote systems.

Summary

This paper has reviewed the effort and challenges faced in implementing a product that meets the demands of the changing chemical information environment. With new technology providing enhanced capabilities on the desktop of the end-user, as well as information being distributed across multiple databases of varying types, the complexity of accessing these sources of information poses new requirements for systems in this decade. The ISIS product line is designed to provide transparent integration of data to allow the end-user and application developer to easily access heterogeneous databases on distributed hardware. Although many of the problems in this area have been addressed, there are future opportunities to resolve the more difficult challenges such as heterogeneous update, query optimisation, and recovery.

References

1) Smith. D.H., 'Consumer-Driven Access to Scientific and Technical Information, 1990s and Beyond', paper presented at The Division of Chemical Information Herman Skolnik Award Symposium, 197th Meeting of the American Chemical Society, Dallas, Texas, April 11, 1989.

Acknowledgements

ISIS is a trademark of Molecular Design Limited.

RDB, DECNET, VAX, VMS are trademarks of Digital Equipment Corporation.

ORACLE is a trademark of Oracle Systems Corporation.

INGRES is a trademark of ASK Computer Systems.

Macintosh is a trademark of Apple Computer.

MS-Windows and OLE are trademarks of Microsoft Corporation.

IBM is a trademark of International Business Machines.

Prediction of physical properties from chemical structures

Alexandru T. Balaban

Department of Organic Chemistry, Technical University Bucharest, Splaiul Independentei 313, 77207 Bucharest, Roumania

Abstract.

Chemical structures as reflected in their constitutional formulae are modelled by molecular graphs. For quantitative structure-property relationships, several models may be used; group contribution methods are the simplest, and quantum-chemical calculations are the most elaborate; as a trade- off between accuracy and complexity, an intermediate level is represented by correlations using topological index approaches. Such indices and their bases (local invariants) are reviewed, and examples of correlations are presented for thermodynamic properties from the author's own contributions.

1 What is chemical structure ?

Chemists classify all substances either into organic, inorganic, and organometallic ones, or according to the types of chemical bonds they possess: covalent (the strongest bonds), ionic, and metallic bonds. All the elements in the universe are able to form chemical bonds, and most of them do so in condensed states. All around us, we do not see free atoms of the 80 stable elements known to mankind, but we see these atoms combined into substances via chemical bonds. Only the noble gases give rise to stable substances as single atoms in condensed forms. In a cosmos where atoms seldom collide with other atoms, stable nuclides may exist as single atoms of any element, yet surprisingly during the last decades, many multiatomic molecules have been detected via their radiofrequency radiations, and it is suspected that among such aggregates buckminsterfullerene (C_{60}) may predominate.

In the following, we shall dwell on substances with covalent bonds, which indeed constitute more than 95% of the known compounds; till now, more than ten million substances have been indexed by *Chemical Abstracts.* One should be aware that chemists use the term 'substance' in a restrictive sense, implying 'pure,' *i.e.* meaning that all molecules in such a substance must be identical, otherwise we have mixtures, with the exception of proteins and polynucleotides.

Since the middle of the last century when the Structure Theory emerged due to efforts of Kekulé, Leibig, Wöhler, Berzelius, Butlerov, Avogadro, Cannizzaro, Couper, Crum Brown and others, it became clear that in organic substances atoms are bonded forming molecules. The diverse modes of linking various numbers and types of atoms into a molecule give rise to what is called constitution, and molecules with the same molecular formula but with different constitutions are 'constitutional isomers.' Through the work of Van't Hoff and Le Bel in organic chemistry, and of Werner in the chemistry of inorganic complexes, stereoisomerism was added to constitutional isomerism when it was discovered that covalent bonds have preferential directions in the tridimensional space.

Thus, when by combined qualitative and quantitative analysis chemists have obtained the molecular formula of a substance, they know what atoms compose the molecule but still do not know the structure of the molecule, *i.e.* how these atoms are linked together by covalent bonds. Nowadays, the molecular formula is easily obtained by high-resolution mass spectrometry. The constitution, *i.e.* the connectivity of each atom, is obtained via two dimensional ^1H- and ^{13}C-nuclear magnetic resonance; nuclear Overhauser effects indicate close proximity in space, and this also helps in deciding about stereochemical problems.

The molecular formula constitutes an easy and sure indexing facility, and this is the basis of the *Chemical Abstracts* registry system. However, in order to discriminate among isomers (all of which have the same molecular formula), one must have recourse to other systems. Registry numbers coined by Chemical Abstracts Service provide some help, but have no systematic meaning. The nomenclature rules elaborated by IUPAC [1] and *Chemical Abstracts* separately and sometimes divergently constitute at present the basis on which a literature search is made, yet these rules are quite elaborate and cumbersome. Graph-theoretical notions may provide a better solution, as advocated by Lozac'h and Goodson with their nodal nomenclature [2], or by Read with his centric approach [3]. Fortunately, it is now possible via computer programs and satellite links with central databases provided by Chemical Abstracts Service online to acquire information about chemical structures obviating the need of a conventional name, simply by drawing a constitutional (or stereochemical) formula on the computer screen. Despite all these problems, owing to the fact that there is no ambiguity in a structural formula, *chemistry is nowadays the best documented science*: in a matter of minutes, it is possible to learn whether any substance one may imagine is known, and if so, where to find the relevant information. The situation is quite different in mathematics, physics, biology, medicine, and even chemistry when instead of structural formulas one has to use words, because words are conventions, and therefore they are not unique and unambiguous.

2. Correlations between properties and chemical constitution

Quantum chemical calculations have arrived at present to the degree of precision where they can account for many properties of individual molecules, and this applies both to small molecules for which one can perform *ab initio* calculations, and to larger molecules for which semiempirical methods are the only ones that are feasible with present-day computers. However, bulk properties in which weak intermolecular forces (Van der Waals interactions, charge transfer, electrostatic interactions and hydrogen bonds) play the major role cannot yet be calculated with reliable results. Even less can such calculations, or molecular mechanics, account for drug-receptor interactions which determine biological activities. Yet the driving force for the synthesis of more than half a million new substances in each year is the search for new medicinal drugs; this effort, plus the high cost of drug testing, makes mathematical modelling for drug design a highly appreciated field. Any method which can spare the expenses for synthesis, treatment of effluents, cell cultures, animal and clinical testing, lowers the cost of medical treatment.

Predicting properties on the basis of chemical structure can also be based on quantitative structure-property (or for biology structure-activity) correlations: QSPR and QSAR, respectively. A specialised journal exists in this area: *Quantitative Structure-Activity Relationships.* Many papers also appear in other journals such as: *Drug Design, Mathematical Chemistry, Journal of Mathematical Chemistry, etc.*

For such correlations, the critical problem is the following one: while properties are expressed on numerical scales which give rise to a metric enabling an ordering of the given property along the scale, chemical structures are discrete entities, for which no metric is apparent, especially if one considers a set of isomers. Such a set can contain a very high number of structures, as shown by the numbers of the simplest organic substances, alkanes (Table 1).

n	Constitutional isomers	Stereoisomers	Constitutional isomers	Stereoisomers
	of alkanes		of alkyl groups	
1	1	1	1	1
2	1	1	1	1
3	1	1	2	2
4	2	2	4	5
5	3	3	8	11
6	5	5	17	28
7	9	11	39	74
8	18	24	89	199
9	35	55	211	551
10	75	136	507	1553
11	159	345	1238	4436
12	355	900	3057	12832
13	802	2412	7639	37496
14	1858	6563	19241	110500
15	4347	18127	48865	328092
20	366319	3396844	5622109	82300275
25	36797588	749329719	712566567	22688455980

Table 1 Numbers of alkanes and alkyl groups with n carbon atoms (see Ref 4)

For QSPR, the solution is to associate to each chemical structure one or several molecular descriptors, *i.e.* numbers which can then be correlated with the corresponding properties. Till now, the molecular descriptors have encompassed electronic parameters such as Hammett sigma constants for aromatic systems, steric parameters such as Taft constants, hydrophobic parameters which model the through-membrane transport *via* the n-octanol/water partition coefficient as advocated by Hansch. The simplest descriptors are numbers of atoms of each type in the respective molecules. However, for all isomers such numbers are the same, therefore other structural characteristics have been developed. They are termed topological indices and are described in the next section.

3. Topological Indices (TIs)

Chemical constitutional formulae can be viewed as graphs, *i.e.* as collections of points or vertices (symbolising atoms, usually non-hydrogen atoms) and lines (symbolising covalent bonds). To each graph one can associate local and global invariants, *i.e.* numbers which do not depend on the way one draws the graph, or the way one labels its vertices. Two such local invariants (LOVI's) are integer numbers: the *vertex degree* is the number of lines meeting at a vertex, and results as the sum over rows or columns in the adjacency matrix of the graph. Entries in this matrix are 1 for adjacent atoms (*i.e.* atoms connected by a line) or 0 otherwise. [Figure 1] The topological distance between two vertices is defined as the number of lines on the shortest path between these vertices. Another matrix which characterises a graph is the distance matrix, whose entries are the topological distances; like the adjacency matrix, it is symmetrical with respect to its main diagonal, has zeros on this diagonal, and entries 1 for adjacent vertices; all other entries, however, are numbers larger than 1. The sums over rows or columns are called *distance sums* and constitute a second local graph invariant.

By summing all vertex degrees for a set of isomers, one obtains the same number (twice the number q of lines); however, if one sums the squares of the vertex degrees, one obtains a simple topological index (TI), proposed by the Zagreb group. [5] Other quadratic indices, linearly related to the previous TI, have been proposed later. [6]

Historically, the first TI was invented by H.Wiener [7] and is the half-sum of all entries in the distance matrix, or of all distance sums. Unlike LOVIs, the TIs are global invariants, and may serve as a measure of the degree of branching of the molecule. Most of them also vary with the number of vertices in the molecular graph.

Hosoya's TI denoted by Z is an integer [8] which for acyclic graphs is the sum of the absolute values of the coefficients of the characteristic polynomial obtained from the adjacency matrix. All TIs described above are first-generation TIs, namely integer TIs based on integer LOVIs.

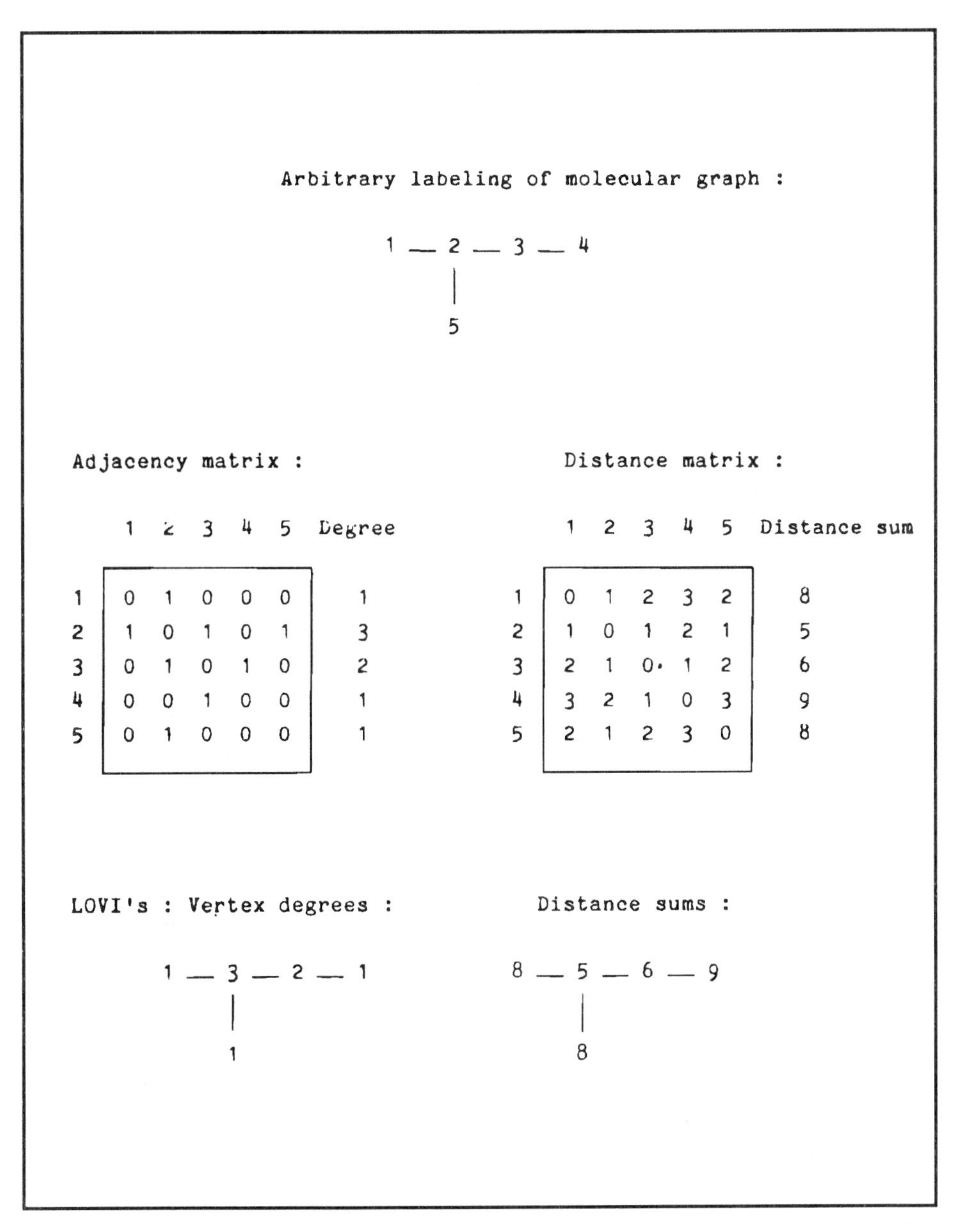

Figure 1. The adjacency and distance matrices of 2-methylbutane, and the derivation of integer local vertex invariants (LOVIs): vertex degrees as sums over rows or columns of the adjacency matrix, and distance sums by a similar operation with the distance matrix.

The TI which so far has been most used, especially for drug design, was invented by Randic and is a real number [9] (unlike previous TIs which were integers) obtained from vertex degrees v_i and v_j (endpoints of each line in the hydrogen-depleted molecular graph) according to formula (1) where the sum is over all lines:

$$\chi = \sum (v_i v_j)^{-\frac{1}{2}} \qquad (1)$$

Two monographs describe the use of this TI. [10]

All TIs are more or less degenerate because more than one structure can correspond to the same numerical value of the TI. The degeneracy is higher for the integer TIs than for the real-number TIs because the set of integers is smaller than that of real numbers. In order to reduce the degeneracy, Bonchev and Trinajstic applied information theory for the global TIs *via* Shannon's formula using binary logarithms ($\log_2 x$, denoted by 1b x) to the global TI, and obtained thereby real-number TIs. [11,12]

In 1982, I proposed two new TIs: the average distance connectivity J, based on topological distances d_i, d_j, according to formula (2) where n is the number of vertices and the number of lines in the hydrogen-depleted graph; the summation is over all edges, as in formula (1): [13]

$$J = \frac{n}{q-n+2} \sum (d_i d_j)^{-\frac{1}{2}} \qquad (2)$$

The degeneracy of this index is very low; j varies only with the 'shape' of the molecule, and less with the number of vertices. As an example, for an infinite linear alkane, J=3.14159 (*i.e.* the number pi). [14] An advantage of basing a TI on distances rather than vertex degrees is the wide range of distances; also, it is easier to code for the presence of multiple bondings (by considering the corresponding distance shorter) and of heteroatoms. [15]

The second TI is the mean square distance, also based on topological distances. [16] All these real-number TIs based on integer LOVIs have lower degeneracy than first-generation TIs and belong to the second-generation TIs.

During the last few years, we have started to develop third-generation TIs which are real numbers based on real-number LOVIs. [17] For this purpose,we have introduced several new real-number LOVIs. A few of them will be presented below.

The adjacency or the distance matrix (**A** or **D**, respectively) may be converted into a system of linear equations by introducing on their main diagonals a column vector, and by equating each row to a free term which is another column vector. These vectors may convey chemical information (atomic number **Z**, electronegativity or relative atomic radius with respect to carbon) topological data (*e.g.* vertex degree **V** or distance sum **S**), or even a constant. Thus a triplet is obtained, leading to a system of linear equations whose solutions are real-number LOVIs. an example is provided in Table 2 for LOVIs of 2-methylbutane based on the AZV triplet. [18]

Another approach was based on the following reasoning: vertex degrees provide information only on the closest (neighbour) atoms, whereas most of the weight in the distance sums is due to the most remote atoms. To arrive at intermediate categories, we constructed new matrices. One of these includes 'regressive vertex

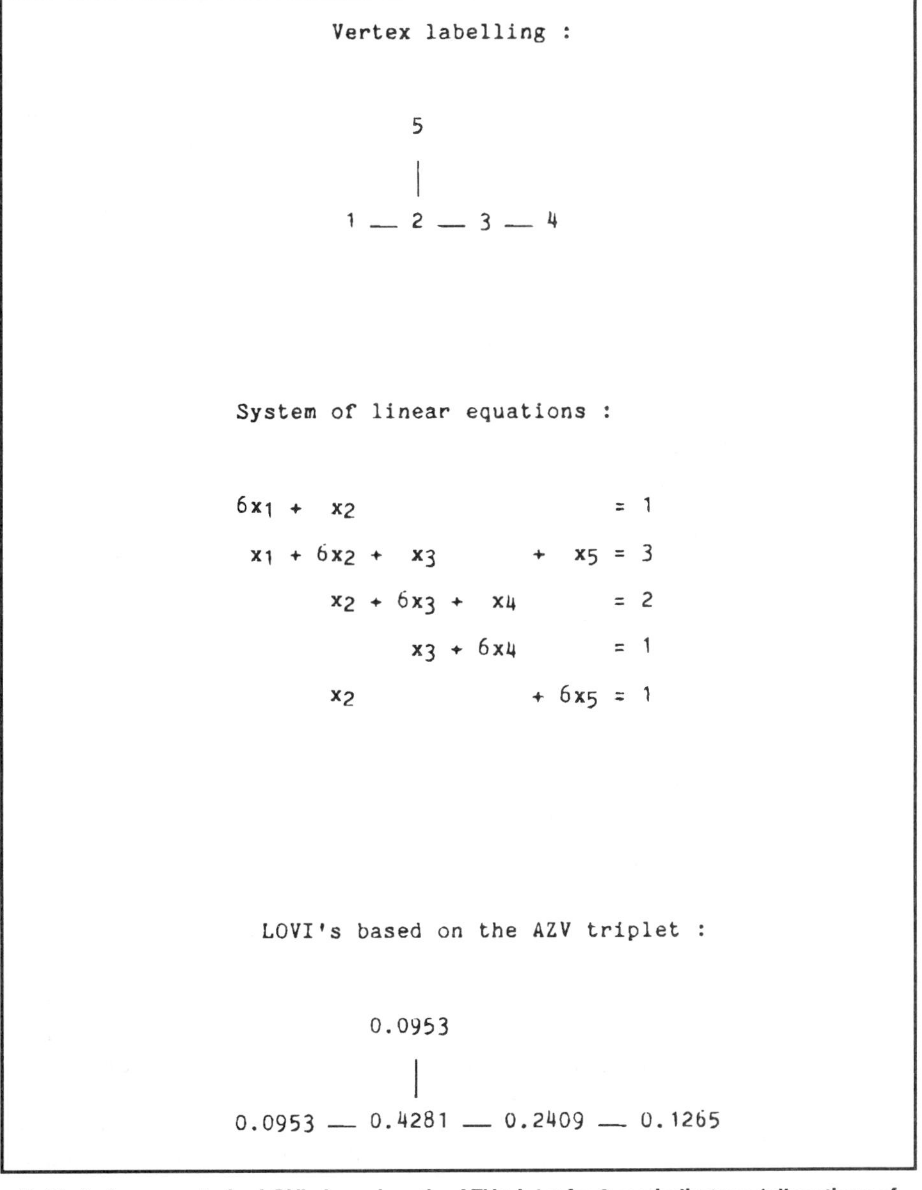

Table 2. An example for LOVIs based on the AZV triplet for 2-methylbutane (all vertices of the molecular graph have Z = 6)

degrees' and reduces gradually contributions due to vertices at larger distances. [19] A second approach modifies similarly distance sums. In the latter case, it is easier to introduce structural information on multiple bonding and/or heteroatoms into the TI. [20]

Eigenvectors corresponding to the smallest or largest eigenvalues of the adjacency or distance matrices can also be used as real- number LOVIs. [21]

Information theory on the distribution of distances for each vertex can be used locally, by means of Shannon-type formulas, for obtaining yet other types of real-number LOVIs. [22,23] Thus formulas (3)–(6) provide such LOVIs denoted by u, v,x, and y.

$$u_i = -\sum \left(\frac{j g_j}{S_i}\right) 1b \left(\frac{j}{S_i}\right) \qquad (3)$$

$$v_i = S_i \, 1b \, S_i - u_i \qquad (4)$$

$$y_i = -\sum j \, g_j \, 1b \, j \qquad (5)$$

$$x_i = s_i \, 1b \, s_i - y_i \qquad (6)$$

In all above formulas, summation extends over all j = 1,2,3, etc distances, and g_j denotes how many times distance j occurs from vertex i to any other vertex.

From such real-number LOVIs, one can obtain global TIs by means of various operations such as: summation of products for edges of LOVIs or of their squares of square roots; summation of products corresponding to the two endpoints; etc. For instance, changing equation (2) to apply to the four LOVIs mentioned above leads to TIs denoted by U, V, X, and Y respectively; analogously, summation of LOVIs resulting as solutions of linear equation systems yields TIs denoted by the three letters of the triplet giving rise to the linear equation system, *e.g.* AZV, or DSV.

All new LOVIs were validated by intramolecular comparison, and all new TIs by intermolecular comparison within isomeric sets of alkanes.

4 Predicting physical properties of organic compounds

Several reviews on estimation methods for physical properties are available. [24,25] In 1988, the Beilstein Institute (which publishes Beilstein's Handbook of Organic Chemistry, containing critically analysed data for every organic substance) together with the Bundesministerium für Forschung und Technologie (German Ministry for Research and Technology) sponsored a Workshop at Schloss Korb in Northern Italy whose proceedings were published under the title *Physical Property Prediction in Organic Chemistry.* [26] On more than 550 pages, an international group of 53 scientist presented lectures on the following areas: thermodynamic properties, environmental properties, statistical thermodynamics, interpolation methods, quantum mechanics, empirical methods, group contribution methods, structure-activity relationships, and solubility determination.

In the present lecture, I wish to summarise a few of the conclusions of that workshop, and to add some of our contributions based upon the use of topological indices.

Group contribution methods are by far the most simple and therefore the most used for estimating physical (especially thermodynamic) properties; they rely on the assumption that each group in the molecule has an individual contribution, independent of the presence of other groups, and that the total property results additively from these contributions. Molar volumes can be thus predicted fairly accurately. Also, ideal gas heats of formation for C_5 to C_9 alkanes may be thus determined, [27] but a more precise approach takes also in consideration the nearest neighbours to the atom or group. [28] However, in other cases, such additive methods such as UNIFAC fail, for instance in estimating activity coefficients of di-ethers the errors worsen as the distance between the oxygen heteroatoms decreases. [29] For alkanes, group contribution methods would yield identical calculated properties in the case of valence-isomers, corresponding to molecular graphs with the same partition of vertex degrees, *e.g.*2- and 3-methylpentanes which both have 3 CH_3 groups, 2 CH_2 groups, and one CH group. For higher alkanes, such degeneracies in calculated properties are more numerous.

5 Predicting physical properties by using Topological Indices

Since we have been interested mostly in developing new methodology rather than in adding to the literature on QSAR/QSPR, we needed properties which had been determined for large numbers of compounds with a certain degree of accuracy. This eliminated from the outset correlations with most biological activities, as the precision of these activities is fairly low.

Therefore our correlations, which were needed for calibrating our descriptors (both LOVIs and TIs), were concentrated mostly on thermodynamic properties: boiling points at normal pressure (normal boiling points), saturation pressures (allowing the determination of boiling points at any pressure), critical pressures and temperatures, heats of vaporisation, *etc.*

In principle, other bulk physical properties are also of interest, such as solubility data, melting points for crystalline compounds, densities, molar volumes, optical properties such as the refraction index. However, for these properties other methods than topological ones have better chances of success, because the phenomena involved are fairly complex. On the other hand, till now very complex phenomena such as odour (correlating with molecular shape and interactions with the olfactory receptors) have only recently started to be understood and modelled. Much better understood are properties of individual molecules such as colour (*i.e.* electronic absorption spectra), vibrational modes (*i.e.* infrared or Raman spectra), nuclear magnetic resonance spectra: for all these types of interaction between molecules and electromagnetic radiation, quantum chemical calculations yield excellent results.

The octane numbers of alkanes reflect chemical properties on the ease of self-ignition on compression; the anti-knock property of a gasoline is defined in terms of two conventional standards: a non-branched alkane (n-heptane) with the lowest octane number, O.N.=0, and a highly branched alkane (2,2,4-trimethylpentane or iso-octane) with O.N.=100. Significant correlations (r^2 =0.99) were found separately for heptanes and octanes based on a single TI, namely the centric index. [6] If the number of carbon atoms is added, biparametric correlations for the set of alkanes C_4 – C_8 were also satisfactory (r^2 =0.72). Slightly better (r^2 =0.90) correlations for the same set of 34 alkanes with four to eight carbon atoms were obtained in uniparametric correlations based on the mean square distance. [16] On using a wider range of chemical descriptors with the SAS statistical package [30] via principal component analysis, and on enlarging the data base to include 45 alkanes C_2 - C_8, 35 cycloalkanes devoid of steric strain C_5 - C_8, and 73 alkenes C_2 - C_{11}, gave for the alkanes a three-parameter correlation with r^2 =0.92, for the cycloalkanes a three-parameter correlation with r^2 =0.78, and for the alkenes a four-parameter correlation with r^2 =0.77. [31]

Normal boiling points of alkanes have been correlated with many topological descriptors. One of the best correlations for the set of all alkanes with 2 to 8 carbon atoms was based on the triad AZV described above: it gave r^2 =0.998 and a standard deviation of less than 3°C. [18]

For compounds containing other atoms than carbon and hydrogen, the situation is more complicated because heteroatoms may give rise to other kinds of molecular interactions such as hydrogen bonding. On excluding such interactions, we developed a correlation for mono- and poly-halogenated alkanes C_1 - C_4 (532 substances, r^2 =0.97, six-parameter equation [32] in terms of the Randic index and numbers of H, F, Cl, Br, and I atoms). A similar approach involving several molecular descriptors [33] (the most important of which proved to be the J index modified for the presence of heteroatoms) allowed a correlation with normal boiling points of acyclic ethers, peroxides, acetals, and their sulphur analogues. In this case, instead of univalent halogen atoms corresponding to heteroatoms represented by vertices of degree one in the molecular graph, we have vertices of degree two. An interesting observation concerns the effect of these heteroatoms on the boiling point, exemplified by Table 3 : replacement of a CH_2 group by an oxygen heteroatom has a small effect on the boiling point (b.p.), but each sulphur atom increases substantially the b.p. This is clearly seen in the six blocks into which Table 3 is divided.

C	H	O	S	Name	B.P.(°C)	Structure
6	14			n-Hexane	69.0	C-C-C-C-C-C
5	12	1		Butyl methyl ether	70.3	C-O-C-C-C-C
5	12	1		Ethyl Propyl ether	63.6	C-C-O-C-C-C
4	10	2		1,2-Dimethoxyethane	84.7	C-O-C-C-O-C
4	10	2		Ethoxy-methoxy-methane	67.0	C-O-C-O-C-C-
4	10	2		Diethyl-peroxide	63.0	C-C-O-O-C-C
5	12		1	Butyl methyl sulfide	123.2	C-S-C-C-C-C

5	12		1	Ethyl propyl sulfide	118.5	C-C-S-C-C-C
4	10		2	1,2-bis(Methylthio)ethane	183.0	C-S-C-C-S-C
4	10		2	Ethylthio-methylthio-methane	166.0	C-S-C-S-C-C
4	10		2	Diethyl-disulfide	154.0	C-C-S-S-C-C
7	16			n-Heptane	98.4	C-C-C-C-C-C-C
6	14	1		Methyl pentyl ether	99.5	C-O-C-C-C-C-C
6	14	1		Ethyl butyl ether	92.3	C-C-O-C-C-C-C
6	14	1		Dipropyl ether	90.1	C-C-C-O-C-C-C
5	12	2		1,3-Dimethoxypropane	104.5	C-O-C-C-C-O-C
5	12	2		1-Ethoxy-2-methoxy-ethane	102.0	C-O-C-C-O-C-C
5	12	2		Diethoxymethane	88.0	C-C-O-C-O-C-C
6	14		1	Methyl pentyl sulfide	145.0	C-S-C-C-C-C-C
6	14		1	Butyl ethyl sulfide	144.2	C-C-S-C-C-C-C
6	14		1	Dipropyl sulfide	142.8	C-C-C-S-C-C-C
5	12		2	bis(Ethylthio)-methane	181.0	C-C-S-C-S-C-C
5	12		2	Ethyl proply disulfide	173.7	C-C-S-S-C-C-C

Table 3. Normal Boiling points of n-alkanes with 6 or 7 atoms and of their analogues with one or two oxygen or sulphur heteroatoms

On closer inspection, however, finely tuned effects of the heteroatoms may be observed, irrespective if they are oxygen or sulphur atoms: displacement of the heteroatom from the margin of the chain towards its centre (or, what amounts to the same effect, lowering of the distance between two heteroatoms) lowers the b.p. All these effects are nicely reflected by the J index modified for the presence of heteroatoms. For 185 compounds, a three-parameter equation in terms of the Randic index, the modified J index, and the number of sulphur atoms, one obtains r^2 =0.97. If only 72 monoethers with 3-11 non-hydrogen atoms are taken into account, the two-parameter equation in terms of χ and J_{het} has r^2 =0.98; for 44 monosulfides, a similar equation has r^2 =0.99. [33] Such correlations are useful for validating databases, since they may detect experimental or indexing errors.

Critical data, namely critical temperatures (T_c), pressures (P_c) and volumes (V_c) for alkanes present a marked dependence on chemical structure, as seen in Table 4. Monoparametric quadratic correlations (7) with three TIs (U, X, and AZV) and biparametric linear correlations (8) with four TIs (J, X, Y, and V) were tested for each property M (the number of carbon atoms is N:

$$M = a + b.TI + c.N \quad (7)$$

$$M = a + b.TI + c.TI^2 \quad (8)$$

Results of correlations for 49 alkanes with 4 to 10 atoms for which data on T_c and P_c are available, and for 36 alkanes with 4 to 8 carbon atoms for which V_c data are available, gave good correlation coefficients and low standard deviations in terms of X, Y, and J for P_c and V_c, and in terms of U, Y, and AZV for T_c and V_c. In Table 4, for brevity, only 9 heptane isomers and 16 of the 17 octane isomers are presented, and the values of only three TIs are indicated. [34]

Alkane	J	Y	AZV	T_c (K)	P_c (atm)	V_c cm^3 per mol
C7	2.4475	1.7297	1.5303	540.2	27.0	432
2M-C6	2.6783	2.1070	1.4849	530.3	27.0	421
3M-C6	2.8218	2.4597	1.4924	535.2	27.8	404
24MM-C5	2.9532	2.6724	1.4375	519.7	27.0	418
3E-C5	2.9923	2.9330	1.5000	540.6	28.5	416
23MM-C5	3.1442	3.2531	1.4562	537.3	28.7	393
22MM-C5	3.1545	3.2097	1.3891	520.4	27.4	416
33MM-C5	3.3604	4.0959	1.4079	536.3	29.1	414
223MMM-C4	3.5412	4.6563	1.3646	531.1	29.2	398
C8	2.5301	1.5743	1.7803	568.8	24.5	492
2M-C7	2.7158	1.8121	1.7348	559.6	24.5	488
25MM-C6	2.9278	2.1249	1.6897	550.0	24.5	482
3M-C7	2.8621	2.0490	1.7426	563.6	25.1	464
4M-C7	2.9196	2.1650	1.7411	561.7	25.1	476
24MM-C6	3.0988	2.4512	1.6955	553.5	25.2	472
3E-C6	3.0744	2.4808	1.7489	565.4	25.7	455
23MM-C6	3.1708	2.6100	1.7052	563.4	25.9	468
34MM-C6	3.2925	2.8870	1.7127	568.8	26.6	466

3E-2M-C5	3.3549	3.0824	1.7111	567.0	26.7	443
234MMM-C5	3.4642	3.2648	1.6673	566.3	26.9	461
22MM-C6	3.1118	2.4356	1.6397	549.8	25.0	478
224MMM-C5	3.3889	3.0046	1.5905	543.9	25.3	468
33MM-C6	3.3734	3.0706	1.6552	562.0	26.2	443
3E-3M-C5	3.5832	3.7505	1.6723	576.5	27.7	455
223MMM-C5	3.6233	3.6700	1.6194	563.4	26.9	436
233MMM-C5	3.7083	4.0126	1.6290	573.5	27.8	455

Table 4. Topological indices and critical data of some alkanes (notation: Cn denotes a linear chain with n carbon atoms, M denotes methyl, and E ethyl)

Alkane	B	C
C7	2911.32	56.51
2M-C6	2845.06	53.60
3M-C6	2855.66	53.93
24MM-C5	2744.78	51.52
3E-C5	2882.44	53.26
23MM-C5	2850.64	51.33
22MM-C5	2740.15	49.85
33MM-C5	2829.10	47.83
223MMM-C4	2764.40	47.10
C8	3120.29	63.63
2M-C7	3079.63	59.46
25MM-C6	2964.06	58.74
3M-C7	3065.96	60.74
4M-C7	3057.05	60.59
24MM-C6	2965.44	58.36
3E-C6	3057.57	60.55
23MM-C6	3029.06	58.99
34MM-C6	3062.52	58.29
3E-2M-C5	3035.08	57.84
234MMM-C5	3028.09	55.62
22MM-C6	2932.56	58.08
224MMM-C5	2896.28	52.41
33MM-C6	3011.51	55.71

3E-3M-C5	3102.06	53.47
223MMM-C5	2981.56	54.73
233MMM-C5	3057.94	52.77

Table 5. Coefficients B and C of the Antoine equation for alkanes with 7 and 8 carbon atoms (notation as in Table 4)

For the same set of alkanes, Table 5 presents the coefficients B and C of the Antoine equation (9) which allows the calculation of saturation pressures (P, in Torr) as a function of the temperature (T, in degrees K)

$$l_n P = A - \frac{B}{(T-C)} \qquad (9)$$

The third empirical parameter in this equation, denoted by A, does not show any systematic variation among all alkanes with 4 to 10 carbon atoms, and is within the range 15.6 to 16.0. On the other hand, as seen from Table 5, the two coefficients B and C show a consistent decrease with increasing branching of the alkane. Using the same two types of equations (7) and (8) as above, one obtains good correlations with TIs such as Randic's X, AZV, and J. [34]

It should be noted that the ordering of alkanes in Tables 4 and 5 is according to increasing branching as indicated with mathematical support by Bertz; [35] this ordering coincides for heptanes with increasing J values, and also for octanes this ordering is well approximated by the J index; this is actually another validation of the J index by intermolecular comparison.

Till now, a handicap of these TIs was the fact that stereochemical features are not taken into account in constitutional formulas, hence in the TIs. However, recent efforts have been described for inputting three-dimensional information on dia-stereoisomerism (but not yet on enantiomerism) into topochemical-geometrical descriptors, similar to TIs. [36] It is hoped that these ideas will enlarge the field of applications, especially for biochemical correlations.

We conclude with a brief mention of biological applications of TIs. In addition to reviews on the use of the Randic index and of the refinements of this index by Kier and Hall, [10] four other reviews on biological uses of TIs have been published. [37-40]

Mathematical modelling of carcinogenic structures and of anticancer agents was described in a recent book. [40] Among the most recent and promising anti-cancer agents where correlations between structure and activity can be made, retinoids occupy a privileged place. In a study based on the J index modified for the presence of multiple bonds and of heteroatoms, 14 classical retinoid structures were examined by means of simple (both linear and non-linear) correlations. [41] A more powerful method, Optimised Approach based on the Structural Indices Set (OASIS) developed by Mekenyan and Bonchev [42] was applied in a joint investigation to the same group of compounds plus retinoic acid, and was able to yield good 3- and

4-variable correlations; in addition to the above-mentioned TI, significant descriptors in these correlations are the hydrophobicity, the steric requirements as described by molecular refraction, and the calculated dipole moment; the correlation factors are in the range r=0.96 – 0.97. [43]

A different approach, using pattern recognition together with molecular descriptors was employed by Jurs and coworkers [44] for correlations on the anti-tumour activity of 9-aniline-acridines; the same group performed, in addition to QSPR studies on boiling points of various compounds, [45] QSAR studies on chemical carcinogens. [46] Our correlations [47] between the chemical structure and the carcinogenic activity of polycyclic aromatic hydrocarbons used TIs,delocalization energies, and MTD (minimal steric differences). Milne reviewed {48] the computer-aided selection of chemicals for biological testing: an activity is associated with an atom-centred fragment, and the National Cancer Institute thus avoids testing compound that have low chances of being active.

A few other reviews on uses of TIs exist. [49-52]

6 Conclusions

It was shown that several approaches for predicting physical chemical, or biological properties of substances based on their chemical structure are feasible. The most rigorous and precise methods rely on quantum-chemical calculations which are very elaborate and expensive, both in terms of money and of time. An inexpensive and simple approach is empirical, and is based on group contribution methods, relying on the assumption that properties may be computed additively from contributions of atoms, or larger parts (such as substituents or functional groups) of molecules; this approach is the least precise and does not discriminate well among isomeric structures, but in many cases as for instance in rough approximations of some thermodynamic properties, can give rapidly satisfactory orientative results. An intermediate level of sophistication is represented by topolological correlations, exemplified by the use of topological indices (TIs), which are rapid and inexpensive, but empirical, and can serve well for correlations with imprecise data such as most biological activities for drug design.

References

1. IUPAC 'Nomenclature of Organic Chemistry' Pergamon Press, Oxford, 1979

2. N.Lozac'h, A.L.Goodson and W. H.Powell, *Angew Chem Internat Ed Engl*, 1979, **18**, 887; A L Goodson, *J Chem Inf Comput Sci*, 1980, **20**, 167, 172; *Croat Chem Acta*, 1983 **56**, 315.

3. R.C.Read, *J Chem Inf Comput Sci*, 1983, **23**, 135; 1985, **25**, 116.

4. R.W. Robinson, F.Harary, and A.T.Balaban, *Tetrahedron*, 1976, **32**,355; R C Read, in A.T.Balaban (editor), 'Chemical Applications of Graph Theory', Academic Press, London, 1976, p.25

5. I.Gutman, B.Ruscic, N.Trinajstic and C.F.Wilcox, Jr, *J Chem Phys*, 1975, **62**, 3399

6. A.T.Balaban, *Theor Chem Acta* (Berlin), 1979, **53**, 355

7. H.Wiener, *J Amer Chem Soc*, 1947, **69,** 17

8. H.Hosoya, *Bull Chem Soc Japan*, 1971, **44**, 2332

9. M.Randic, *J Amer Chem Soc*, 1975, **97**, 6609

10. L.B.Kier and L.H.Hall, 'Molecular Connectivity in Chemistry and Drug Research', Academic Press, New York, 1976; Molecular Connectivity in Structure-Activity Analysis, Research Studies Press, Wiley, New York, 1986

11. D.Bonchev and N.Trinajstic, *J Chem Phys*, 1977, 67, 4517

12. D.Bonchev, 'Information Thoretic Indices for Characterisation of Molecular Structures', Res Studies Press, Chichester, 1983

13. A.T.Balaban, *Chem Phys Lett*, 1982, **89**, 399; A.T.Balaban and L.V.Quitas, *Math Chem*, 1983, **14**, 213

14. A.T.Balaban, N.Ionescu-Pallas and T.S.Balaban, *Math Chem*, 1985, **17**, 121

15. A.T.Balaban, *Math Chem*, 1986, **21**, 115; A.T.Balaban and P.Filip, *ibid*, 1984, **16**, 163

16. A.T.Balaban, *Pure Appl Chem*, 1983, **55**, 199

17. A.T.Balaban, *J Chem Inf Comput Chem*, 1992, **32**, 23

18. P.Filip, T.S.Balaban and A.T.Balaban, *J Math Chem*, 1987, **1** 61

19. M.V.Diudea, O.Minailiuc and A.T.Balaban, *J Comput Chem*, 1991, **12**, 527

20. A.T.Balaban and M.V.Diudea, *J Chem Inf Comput Sci*, (submitted for publication)

21. A.T.Balaban, D.Ciubotariu and M.Medeleanu, *J Chem Inf Comp Sci*, 1991, **31**, 517

22. A.T.Balaban and T.S.Balaban, *J Math Chem*, 1991, **8**, 383

23. G.Klopman and C.Raychaudhury, *J Comput Chem*, 1988, **9**, 232; *J. Chem. Info. Comput.Chem.*, 1990, **30**, 12

24. R.C.Ried, J.M.Prausnitz and B.E.Poling, 'The Properties of Gases and Liquids', 4th edition, McGraw-Hill, New York, 1987

25. W.J.Lyman, W.F.Reehl and D.H.Rosenblatt, Handbook of Chemical Property Estimation Methods. 'Environmental Behaviour of Organic Compounds', McGraw-Hill, New York, 1982

26. C.Jochum, M.G.Hicks, and J.Sunkel (Eds), 'Physical Property Prediction in Organic Chemistry', Springer, Berlin, 1988

27. T.W.Copeman, P.M.Mathias and H.C.Klotz, in ref 26, p. 349

28. S.W.Benson, 'Thermochemical Kinetics', Wiley, New York, 1968; S.W. Benson, F.R.Cruickshank, D.M.Golden, G.R.Haugh, H.E.O'Neal, A.S. Rogers, R.Shaw and R.Walsh, *Chem Rev*, 1969, **69**, 269

29. H.S.Wu and S.I.Sandler, cited in ref 27

30. SAS Institute Inc, 'SAS Users Guide: Statistics, Version 5', SAS, Cary NC, 1985

31. A.T. Balaban, L.B. Kier and N.Joshi, *Math Chem* (in press)

32. A.T.Balaban, N.Joshi, L.B. Kier and L.H.Hall, *J Chem Inf Comput Sci*, 1992, **32**, 233

33. A.T.Balaban, L.B.Kier and N.Joshi, *J Chem Inf Comput Sci*, 1992, **32**, 237

34. A.T. Balaban and V. Feroiu, *Repts Mol Theory*, 1990, **1**, 133

35. S.H.Bertz, *Discrete Appl Maths*, 1988, **19**, 65

36. O.Mekenyan, D. Bonchev, N.Trinajstic and D.Peitchev, *Drug Design*, 1986, **36**, 421

37. A.Sabljic and N.Trinajstic, *Acta Pharma Jugosl*, 1981, **31**, 189

38. A.T. Balaban, A.Chiriac, I.Motoc and Z.Simon, 'Steric Fit in Quantitative Structure-Activity Relations', Lecture Notes in Chemistry No 15, Springer, Berlin, 1980

39. A.T. Balaban, I.Niculescu-Duvaz, Z.Simon, *Acta Pharm Jugosl*, 1987, **37**, 7

40. N.Voiculetz, A.T.Balaban, I.Niculescu-Duvaz and Z.Simon, 'Modeling of Cancer Genesis and Prevention', CRC Press, Boca Raton FL, 1990

41. A.T.Balaban, C.Catana, M.Dawson and I.Niculescu-Duvaz, *Rev Roum Chim*, 1990, **35** 997

42. O.Mekenyan, S. Karabunarliev and D. Bonchev, *Comput Chem*, 1990, **14**, 193

43. D.Bonchev, C.F.Mountain, W. A.Seitz and A.T.Balaban, *J Med Chem*, (submitted for publication)

44. P.C.Jurs, J.T.Chou and M.Yuan, *J Med Chem*, 1979, **22**, 476

45. D.T. Stanton and P. C.Jurs, *Analyt Chem*, 1992, **62**, 2323

46. D. Henry, P. C.Jurs and W. A.Denny, *J Med Chem*, 1982, **25**, 899

47. Z.Simon, A.T.Balaban, D.Ciubotariu and T.S.Balaban, *Rev Roum Chim*, 1985, **30**, 985

48. G.W. A.Milne and L.Hodes, in ref 26, p 79

49. N.Trinajstic, 'Chemical Graph Theory', CRC Press, Boca Raton FL, 2nd edition, 1992

50. A.T.Balaban, I.Motoc, D. Bonchev and O.Mekenyan, in Steric Effects in Drug Design, (Eds M Charton and I Motoc), Springer, Berlin, *Top Curr Chem*, 1983, **114**, 21

51. A.T.Balaban, *J Chem Inf Comput Sci*, 1985, **25**, 334

52. A.T. Balaban and T. S.Balaban, *J Chim Phys*, 1982, **89**, 1735

Index